Other Books by *Bernard Jaffe*

Outposts of Science
New World of Chemistry
Men of Science in America
Chemical Calculations
New World of Science (with R. W. Burnett and H. S. Zim)
Michelson and the Speed of Light
Chemistry, Science of Matter, Energy and Change
(with Gregory Choppin)
Chemistry Creates a New World
Moseley and the Numbering of the Elements

BERNARD JAFFE

CRUCIBLES
THE STORY OF
CHEMISTRY

From Ancient Alchemy to Nuclear Fission

NEW REVISED AND UPDATED
FOURTH EDITION

DOVER PUBLICATIONS, INC.
NEW YORK

Published in Canada by General Publishing Company, Ltd., 30 Lesmill Road, Don Mills, Toronto,
Ontario.
Published in the United Kingdom by Constable
and Company, Ltd., 10 Orange Street, London WC 2.

This Dover edition, first published in 1976, is a
revised and enlarged republication of the work
originally published by Simon and Schuster, New
York, in 1930.

International Standard Book Number: 0-486-23342-1
Library of Congress Catalog Card Number: 75-38070

Manufactured in the United States of America
Dover Publications, Inc.
180 Varick Street
New York, N.Y. 10014

To Rindy, Mandy and David

CONTENTS

"There is first the groping after causes, and then the struggle to frame laws. There are intellectual revolutions, bitter controversial conflicts, and the crash and wreck of fallen philosophies."

FRANCIS P. VENABLE

I

TREVISAN

HE LOOKS FOR GOLD IN A DUNGHILL

IN THE dark interior of an old laboratory cluttered with furnaces, crucibles, alembics, stills and bellows, bends an old man in the act of hardening two thousand hens' eggs in huge pots of boiling water. Carefully he removes the shells and gathers them into a great heap. These he heats in a gentle flame until they are white as snow, while his co-laborer separates the whites from the yolks and putrifies them all in the manure of white horses. For eight long years the strange products are distilled and redistilled for the extraction of a mysterious white liquid and a red oil.

With these potent universal solvents the two alchemists hope to fashion the "philosopher's stone." At last the day of final testing comes. Again the breath-taking suspense, again—Failure!—their stone will not turn a single one of the base metals into the elusive gold.

Secretly had the old man worked at first, for had not the Arabian master of alchemy, Geber himself, admonished his disciples: "For heaven's sake do not let the facility of making gold lead you to divulge this proceeding or to show it to any of those around you, to your wife, or your cherished child, and still less to any other person. If you do not heed this advice you will repent when repentance is too late. If you divulge this work, the world will be corrupted, for gold would then be made as easily as glass is made for the bazaars."

The quest of the Golden Grail obsessed him. As far back as he could remember, Bernard Trevisan had thought and dreamed of nothing else. Born in 1406 of a distinguished family of Padua, oldest of the northern Italian cities, he had been reared on his grandfather's stories of the great search of the alchemists. Stories of failures, all, but he would succeed where others had failed. Encouraged by his parents, Bernard began his great adventure at the age of fourteen. His family approved, for they hoped to multiply the

1

young heir's patrimony a thousandfold. But as the years of failure passed and his fortune slowly dwindled they lost faith as others had done. They pitied him and attributed his pursuit of alchemy to nothing short of madness.

But no failures or discouragement could dampen the hopes of the alchemist. Undeterred by the fiasco of the eggshell experiment, carried on with the aid of Gotfridus Leurier, a monk of Citeaux, he continued his labors with superhuman patience. "I shall find the seed," he whispered to himself, "which will grow into great harvests of gold. For does not a metal grow like a plant?"[1] "Lead and other metals would be gold if they had time. For 'twere absurd to think that nature in the earth bred gold perfect in the instant; something went before. There must be remoter matter. Nature doth first beget the imperfect, then proceeds she to the perfect. Besides, who doth not see in daily practice art can beget bees, hornets, beetles, wasps out of the carcasses and dung of creatures? And these are living creatures, far more perfect and excellent than metals."

For ten more long years, Bernard Trevisan followed the will-o'-the-wisp teachings of Rhazes and Geber. He dissolved and crystallized all kinds of minerals and natural salts. Once, twice, a dozen times, even hundreds of times, he dissolved, coagulated and calcined alum, copperas, and every conceivable animal and vegetable matter. Herbs, flowers, dung, flesh, excrement—all were treated with the same painstaking care. In alembics and pelicans, by decoction, reverberation, ascension, descension, fusion, ignition, elementation, rectification, evaporation, conjunction, elevation, sublimation, and endless other strange operations, he tried everything his tireless ingenuity could conjure.

"Gold is the most perfect of all metals," he murmured. "In gold God has completed His work with the stones and rocks of the earth. And since man is nature's noblest creature, out of man must come the secret of gold." Therefore he worked with the blood and the urine of man. These operations consumed twelve years and six thousand crowns. He was surrounded by a motley group of pretended seekers after the stone—by men who, knowing the Italian rich, offered him secrets which they neither understood nor possessed. His wealth dwindled slowly as he supported all manner of adepts, for he had not yet learned that where one honest adept of alchemy is found, ten thousand cheats abound.

Finally he became tired of the knaves who had reduced him almost to penury. He rid himself of these impostors and turned his attention to the obscure and mystic works of two other masters of alchemy, Johannes de Rupecissa and Sacrobosco. His faith in the philosopher's stone revived, this time he allied himself with a monk of the Order of St. Francis. This friar had told him how Pope John XXII, during the "Babylonian Captivity," maintained a famous laboratory at Avignon where he himself labored to make gold, and as he piled up a fortune of eighteen million florins, issued bulls against the competition of other alchemists.

Thrice ten times Bernard Trevisan rectified spirits of wine "till," as he said, "I could not find glasses strong enough to hold it." This liquor would not fail him, he thought. Again the test was made— the "stone" proved as unfruitful as ever. But the fire still burned hot within him. He buried himself in his dark dungeon of a laboratory, sweating and starving for fifteen more years in the search for the unattainable.

By now he had spent ten thousand crowns, and his health was very poor. But the fervor of the aging man was unabated. Almost maddened by failure, he betook himself to prayer, hoping that God in His goodness would select him as the deliverer of man from poverty. But the favor of the Lord was not visited upon him, and his friend, the Franciscan, died in the quest. Bernard Trevisan was alone once more.

He transported his laboratory to the shores of the Baltic Sea where he joined forces with a magistrate of the city of Trèves, who also belonged to that band of erring men impelled by an almost insane force to the strange search. "I am convinced," said this magistrate, "that the secret of the philosopher's stone lies in the salt of the sea. Let us rectify it day and night until it is as clear as crystal. This is the dark secret of the stone." So for more than a year they labored, but the *opus majus* still remained concealed.

Now Bernard, still fumbling in the dark, came upon another clue. Turning to silver and mercury he dissolved them in *aqua fortis,* a very strong acid. By concentrating the solutions over hot ashes obtained from foreign coals, he reduced their volumes to half. Then carefully he combined the two liquids, making sure not to lose a single drop, and poured the mixture into a clay crucible, which he placed in the open, exposed to the action of the sun's rays. "For does not the sun acting upon and within the earth form the metals?" he argued. "Is not gold merely its beams condensed to a yellow solid? Do not metals grow like vegetables? Have not diamonds been known to grow again in the same place where years before they had been mined?" He, too, had heard of mines being closed to give the metals an opportunity to grow larger. For another five years he worked with this sun-exposed mixture, filling phial after phial and waiting for the great change which never came.

Bernard Trevisan was now close to fifty years old, but the fire still burned within him with a full flame. Gathering his meager possessions, he set out in search of the true alchemists. His wanderings carried him to Germany, Spain, and France, where he sought out the famous gold searchers and conferred with them in the hope of finding the key that would put an end to his all-consuming desire.

He finally settled down in France, still working in his laboratory, when word reached him that Master Henry, Confessor to Emperor Frederick III, had finally discovered the secret formula of the stone. He started off to Vienna at once—and found a man after his own heart. Master Henry had been working all his life to solve the supreme riddle of transmutation. He was no deceiver, but a man of God, sincerely searching for the germ of gold. The two dreamers vowed eternal friendship, and that night Bernard, "the good," gave a banquet in honor of his new partner, to which he invited all the alchemists of the vicinity. At the banquet-table it was agreed that forty-two gold marks should be collected from the guests. Master Henry, contributing five marks, promised to multiply the coins fivefold in the crucible. Bernard added twenty marks, while his five last surviving comrades, who had kept him company on his travels, added their little share, borrowed from their patron.

In a glass vial of strange design Henry mixed yellow sulfur with a few drops of mercury. Holding the vial high over a fire, slowly he added a few grains of silver and some pure oil of olives. Before finally sealing the glass container with hot ashes and clay, he placed in it the forty-two gold marks and a minute quantity of molten lead. This strange mixture was placed in a crucible and buried in a red-hot fire. And while the alchemists ate and drank heartily, and chattered volubly of the great search of the centuries, the concoction in the vial boiled and bubbled unguarded in the kitchen furnace.

Patiently they waited until the vial was broken. The "experiment" was a failure. Master Henry could not understand. "Perhaps," he ventured, "some ingredient had been wanting." Others suggested that the phase of the moon and the position of the planets and stars were not propitious for such a momentous experiment. Yet was it not strange that when the crucible was emptied in the presence of the queer company that surrounded Bernard, only sixteen of the forty-two gold marks were salvaged? The other twenty-six had disappeared, perhaps to appease Hermes Trismegistus, the father of alchemy. This farce infuriated Trevisan, and he vowed to abandon the quest of the philosopher's stone.

For two weary months which seemed to go on and on forever, Bernard kept his pledge, but again that burning in his heart overcame cold reason, and his mind was set once more on retrieving his vanishing fortune through the stone. And now his thoughts turned to the cradle of alchemy—to Egypt, Palestine, Persia, Greece, Turkey, the Isle of Cyprus. For was not the father of alchemy identified with the grandson of Noah, who was intimately familiar with the philosopher's stone? Had not Sarah, the wife of Abraham, hidden an emerald tablet engraved with the cryptic directions for making gold? Had not Alexander the Great discovered it in a cave near Hebron? "Whatever is below is like that which is above, and that which is above is like that which is below, to accomplish the miracle of one thing." This he had read, and stranger things, too. "The father thereof is the sun, and the mother thereof is the moon, the wind carries it in its belly, and the nurse thereof is the earth. This thing has more fortitude than fortitude itself, because it will overcome every subtle thing and penetrate every solid thing. By it this world was formed." Here was the meaningful secret of the universal solvent which Hermes, the son of Osiris, King of Egypt, had discovered. Had not Jason and the Argonauts gone in search of the Golden Fleece, which was nothing else than a book of alchemy made of sheep-skin? And had not Gaius Diocletian, Roman Emperor in 290 A.D., ordered all books which treated of the admirable art of making gold committed to the flames, "apprehensive lest the opulence of the Egyptians should inspire them with confidence against the Empire"? Perhaps, thought Bernard, some of these books had escaped destruction. There, in the Greek colony of Alexandria, he would rummage through the scrolls of the ancients.

For four more years he made his pilgrimage. "In this affair," he wrote, "I spent upwards of eleven thousand crowns, and in fact, I was reduced to such poverty that I had but little money left, and yet I was more than sixty-two years of age." Soon he met another monk, who showed him a recipe for whitening pearls. The pearls were etched in the urine of an uncorrupted youth, coated with alum, and left to dry on what remained of the corrosive. Then they were heated in a mixture of mercury and fresh bitch's milk. Bernard watched the process, and behold—the whitest pearls he had ever seen! He was now ready to listen to this skilled adept. Upon security of the last remnant of his once-great estate, he persuaded a merchant to lend him eight thousand florins.

For three years he worked with this friar, treating a rare iron ore with vinegar in the hope of extracting the mystic fluid. He lived day and night in his dirty laboratory, losing his fortune to multiply it. So obsessed was he by this search that he had no time even to wash his hands or his beard. Finally, unable to eat or drink, he became so haggard and thin that he thought he had been poisoned by some of the deadly fumes in which he had been working. Failure again sapped his health, and the last of his estate was gone.

So alone, friendless, penniless, weary in mind and physically broken, Bernard Trevisan started for his home in Padua, only to find that his family would have nothing to do with him. Still he would not give up the search. Retiring to the Isle of Rhodes, he continued his work with yet another monk who professed to have a clue to the secret. The philosopher's stone remained as elusive as ever! Bernard had spent threescore years grappling with nature; he had lost thousands of crowns; he no longer had the strength even to stand before the furnace. Yet he continued the search.

Once more he returned to the study of the old philosophers. Perhaps he had missed some process in the writings of the ancient alchemists! For ten long years he read and reread every manuscript he could find, until one day he fell asleep and dreamed of a king and a magic fountain. He watched the heavenly bodies robe and disrobe. He could not understand, and in his dream he asked a priest, "What is all this?" and the priest answered: "God made one and ten, one hundred and one thousand, and two hundred thousand, and then multiplied the whole by ten." "But still I do not understand!" cried Bernard. "I will tell you no more," replied the priest, "for I am tired." Then Bernard awoke suddenly. He felt faint and knew the end was near.

> I did not think to die
> Till I had finished what I had to do
> I thought to pierce the eternal secret through
> With this my mortal eye.
> Grant me another year,
> God of my spirit, but a day, to win
> Something to satisfy this thirst within.
> I would *know* some thing here.
> Break for me but one seal that is unbroken.
> Speak for me but one word that is unspoken.

But the prayer of the dying alchemist was not answered.

> The fire beneath the crucible was out:
> The vessels of his mystic art lay round,
> Useless and cold as the ambitious hand
> That fashioned them, and the small rod,

Familiar to his touch for threescore years,
Lay on the alembic's rim, as if it still
Might vex the elements at its master's will.[2]

And thus, in 1490, died Bernard Trevisan.

As we peer down the vista of the past we find the delusion of transmutation holding the most prominent place in the minds of thinking men. Frenzied alchemy held the world in its grip for seventeen centuries and more of recorded history. This pseudoscience with its alluring goal and fascinating mysticism dominated the thoughts and actions of thousands. In the records of intellectual aberrations it holds a unique position. Even Roger Bacon of Oxford, easily the most learned man of his age, the monk who seven hundred years ago foresaw such modern scientific inventions as the steamship and the flying machine, believed in the possibility of solving this all-consuming problem.

Sir Isaac Newton, one of the clearest scientific thinkers of all time, bought and consulted books on alchemy as late as the eighteenth century. In his room in Trinity College, Cambridge, he built a little laboratory where he tried various experiments on transmutation. After leaving the university, he was still concerned with this problem, and wrote to Francis Aston, a friend who was planning a trip through Europe, to "observe the products of nature in several places, especially in mines, and if you meet with any transmutation those will be worth your noting. As for particulars these that follow are all that I can now think of. In Schemnitrium, Hungary, they change iron into copper by dissolving the iron in vitriolate water." He was intensely interested in a secret recipe with which a company in London was ready to multiply gold. Robert Boyle, President of the Royal Society, was also so impressed that he helped to procure the repeal of the Act of Parliament against multipliers of gold.

The power and influence of many of the alchemists can hardly be exaggerated. In nearly every court of Europe were men appointed by kings and emperors to transmute base metals, like lead and iron, into gold, and so advance the financial status of their kingdoms. Records exist which tell of the lending of alchemists by one court to another, and of treaties between states where monarchs traded in alchemists. Many were raised to the nobility; many worked shoulder to shoulder with their sovereigns. A number of little houses used as laboratories, situated near the beautiful castle of Emperor Rudolph II in Prague, bear testimony to that monarch's intense interest in this strange science. He neglected the affairs of state to dabble in science, and in Vienna are still displayed leaden bars which Rudolph tried to convert into gold.

Two years before Bernard Trevisan was born, England, by act of Parliament, forbade the making of gold and silver by alchemical processes. Later, however, King Henry IV granted the right to make gold to certain persons, and at the same time appointed a committee of ten learned men to investigate the possibilities of transmutation. Henry VI went further. He encouraged both the nobility and the clergy to study the science of alchemy, in the hope that they might help him pay the debts of the State. Two soldiers, Edmund de Trafford and Thomas Asheton, organized a company which was granted the privilege in 1445 to make the yellow metal and actually produced a product from which coins were minted. When the Scots heard of this English gold, their Parliament refused to allow it to enter their country. Upon analysis, they found it to be an alloy of mercury, copper and gold.

While among the alchemists there were some genuine enthusiasts like Bernard Trevisan, the annals of this queer practice are filled with accounts of charlatans and spurious adepts who, with a deluge of glib words but with only a drop of truth, turned alchemy into one of the greatest popular frauds in history. The writings of these avaricious devils and honest fools are a meaningless jargon of cryptic terms and strange symbols. Their public demonstrations of transmutation were often clever enough to fool the most cautious. Many came to witness the making of gold from lead and iron, convinced that it could be done. For had they not seen iron vessels, plunged into certain natural springs containing copper salts, emerge covered with the red metal? It was a matter of common knowledge that a dark dirty ore could be heated until all its impurities were destroyed and a bright shiny metal was obtained. Traces of silver and gold had been found in many ores. Then why could not the further heating of these ores yield larger quantities of the precious metals? In fact, with sufficient treatment, it ought to be possible to change the ore entirely into lustrous gold. Simple enough questions in the light of their ignorance of chemical facts. Besides, nature was performing marvelous transmutations every minute of the day as food was changed into blood, and sugar into alcohol. And there were mystics who saw in the change of bread and wine into the body and blood of Christ at the consecration of the elements in the Eucharist a hope that, by the help of God, a similar transmutation could be effected of the baser metals into gold.

In many of the museums of Europe we can still see shiny yellow metals reputed to be gold—products of the deceptions and trickery of the gold cooks of European courts. The Hessian thalers of 1717 were struck from alchemical gold and silver. Some of these samples came from the false bottom of a crucible whose true bottom had been publicly filled with a mysterious mixture which furnace heat was to turn into gold. Other nuggets of gold were gathered from the inside of hollow nails which had been used by impostors to produce gold by "transmutation."

The penalty for failure to produce the philosopher's stone was heavy. For Bernard Trevisan, it meant the loss of an immense fortune, the discouragement of seventy and more years of futile, tireless labor, until death finally came. For many others it was premature death. History records the exposure and punishment of more than one impostor. Marco Bragadino, the gold maker, was hanged by the Elector of Bavaria. William de Krohnemann met the same fate at the hands of the Margrave of Beyreuth. David Benther cheated the Elector Augustus of Saxony by killing himself. And in 1575 Marie Ziglerin, a female alchemist, was burnt at the stake by Duke Julius of Brunswick. Frederick of Wurtzburg maintained a special gallows, ironically painted in gold, used solely for those unfortunate alchemists who could not fulfill their promises to make real gold. On the gibbet, an inscription had been posted by the hangman for the entertainment of its victim: "I once knew how to fix mercury and now I am fixed myself."

During the summer of 1867 three clever rogues met in Paris: Romualdo Roccatani, a Roman archpriest, Don José Maroto Conde de Fresno y Landres, a Spanish grandee, and Colonel Don Antonio Jimenez de la Rosa, a Neapolitan chevalier. The possessors of these sonorous names had a secret process for turning silver into gold. They were shrewd enough to realize that Emperor Francis Joseph of Austria was, by dynastic tradition at least, keenly interested in alchemy.

Arriving in Vienna they cleverly obtained an audience with the monarch and offered him the most momentous discovery of all time. In Mariposa, California, they told His Majesty, were natural deposits of white nuggets which contained gold formed from silver by the action of mercury and the heat of the sun. They continued: "This same process of transmutation may be brought about much more quickly by artificial methods, through giving the amalgam a specific gravity of 15.47. Thereby a process of nature is imitated when the silver amalgam is exposed to a greatly increased temperature."

Francis Joseph made an initial payment of $10,000 for the secret, and assigned Professor Schrötter, discoverer of red phosphorus, to supervise a small-scale experiment in the laboratory of the Polytechnic. On October 17, 1867, two iron pots and two glass retorts were filled with silver amalgam and heated for four months. The vessels then cracked. No gold was found. Then, opportunely, the adventurers disappeared, thus cheating the gibbet of three distinguished victims.

In the New World, too, soon after the founding of the Republic, alchemy had its adepts. "Father" George Rapp, a Pietist priest, in 1804 brought over five hundred of his followers from Württemberg, Germany, in search of religious freedom. They first settled in western Pennsylvania and later in the village of Harmonie, called Economy, which they founded on the Ohio River, 18 miles from Pittsburgh, Pennsylvania. They shared their work and possessions and at first the sect prospered. When Rapp died in 1847 the Utopian Society weakened and finally dissolved in 1905.

Rapp conducted a secret school of alchemy consisting of a few disciples including three women. They worked in a well-equipped alchemical laboratory which even included a fume hood. Rapp was searching not for gold even though he used large quantities of cinnabar, the chief ore of mercury. Influenced by the writings of Paracelsus and Jacob Boehme, an alchemical mystic, Rapp invoked the aid of God in his religious quest for an elixir that would bring its adepts ultimate purity of mind. Old Economy is now a Pennsylvania State Museum.

About the time that the Harmony Society was dissolved, a more venturesome alchemical enterprise was started by a Dr. Stephen H. Emmens. This English poet, novelist, logician, chemist, and metallurgist claimed to have discovered "argentaurum," a modern philosopher's stone which could augment the amount of gold in an alloy of gold and silver. Many fanciful stories about this undertaking appeared in the press, even though Dr. Emmens tried earnestly to surround his experiments with strictest secrecy. Much of what appeared in print was deceptive, but this we know—the syndicate formed by the English adventurer sold to the United States Assay Office six ingots of an alloy weighing ten pounds which upon analysis showed the presence of gold and silver. The government paid him the sum of $954 for the metals, and Dr. Emmens straightway advanced this payment as proof of his astonishing success. For a moment the affair seemed to promise a recrudescence of alchemy. The first dividends were paid, and Emmens even promised a public demonstration at the World Fair in 1900, which, however, never materialized. The whole scheme was a fraud and before long the name of Dr. Emmens was added to that long list of men and women who have gone down in the limbo of the past among the spectacular failures of history.

Alchemy, nourished in superstition and chicanery, still has its adepts and believers. In France, there exists an Alchemical Society for the study of alchemic processes of transmutation. August Strindberg, one of Sweden's great modern literary figures, was a firm believer in transmutation. "People ask me if I can make gold," he wrote, "and I reply, 'to draw the genealogical chart of the ancestors of a cat, I do not have to know how to make a cat.' " He knew, he believed, the secrets of the great riddle, but he never professed to make gold.

In Germany, Franz Tausend, a former plumber, was arrested in 1929 on a charge of having swindled a number of prominent financiers including General von Ludendorff of more than $100,000 through asserting that he could make gold synthetically from lead.

From his cell in a jail in Munich, Tausend insisted his discovery was based on modern scientific principles and demanded a chance to show his methods. He was finally taken to the State Mint and in the presence of its Director, two detectives, the State's attorney and the examining judge he began his experiments. Every precaution was taken to prevent fraud: he was minutely searched, all apparatus and chemicals were supplied him from a carefully guarded safe.

The judges watched vigilantly while Tausend worked cleverly with his "tincture of tinctures" made up of lead chloride, and calcium hydroxide smelted with mercury, potassium and sodium. In his second method he used potassium hydroxide, rock flint and ferric oxide.

Then came the official announcement: "After experimenting for two hours Tausend produced a grain of the purest gold weighing one-tenth of a gram which was smelted from 1.67 grams of lead. Experts described the result as surprisingly favorable and contradictory to scientific knowledge." The cost of production of this alleged synthetic gold was estimated at $5 per kilogram as against $700 in the world market at that time.

The world might have accepted this man and his methods, but the following day the Director of the Mint discovered that, in spite of all the precautions that had been taken, gold which came from the ashes of a cigarette, had been smuggled in to Tausend, and had accounted for the remarkable result. Tausend was sent back to his prison cell challenging the world to try his process. No doubt the credulous did, for the spell of frenzied alchemy still persists.

What was the significance and value of this strange search for the philosopher's stone? Was it just a meaningless, childish reaching for the moon? Was alchemy really chemistry as Liebig, one of the world's greatest chemists, believed? Was this long tragedy and farce of alchemy all in vain?

Surely it was not in vain. Francis Bacon compared alchemy "to the man who told his sons he had left them gold buried somewhere in his vineyard; where they by digging found no gold, but by turning up the mould about the roots of the vines, procured a plentiful vintage." In this fanatical search a great mass of valuable discoveries was made, and many chemical facts were learned. Nitric, hydrochloric and sulfuric acids, the three most important acids employed by the modern chemist, and aqua regia, the powerful solvent for gold formed by mixing the first two of these acids, were introduced by these early gold searchers. In their quest for the seed of gold in the dirt and dross of centuries, new elements like antimony, arsenic, bismuth and phosphorus were unearthed. Many of the common chemicals of today owe their discovery to those early days—alum, borax, cream of tartar, ether, fulminating gold, plaster of Paris, red lead, iron and silver salts and heavy barium sulphide, the first substance known to glow in the dark after exposure to sunlight, stumbled upon by Cascariolo, a cobbler of Bologna.

Some of the apparatus and utensils which are the tools of the chemist of our scientific laboratories were first introduced by alchemists—cupel, distilling flask, retort, water bath and even the balance in its crude form. The extraction of gold by amalgamation with mercury, the preparation of caustic alkali from the ashes of plants, and other new processes of manipulation and methods of manufacture were developed by the gold cooks in their manifold operations.

This heritage is indeed a rich one, for in their blind groping for a new process to make gold these adepts of alchemy paved the way for the more fruitful science of chemistry. Synthetic gold, however, never came through the work of an alchemist. And as Bernard Trevisan lay dying on the Isle of Rhodes almost five hundred years ago, he uttered with his last breath his conviction: "To make gold, one must start with gold."

II

PARACELSUS

A CHEMICAL LUTHER FEEDS A BONFIRE

BERNARD TREVISAN was dead. Almost fifty years passed and in front of the University of Basel its students had lit a huge bonfire to celebrate the feast of St. John. Suddenly and unexpectedly appeared Philippus Aurelius Theophrastus Bombast von Hohenheim, lecturer in medicine and chemistry. Under his arm was a copy of Avicenna's *Canon of Medicine.* He turned to his students and ordered brought to him all the books of the old masters of alchemy and medicine which clutched the thoughts of men in a paralyzing grip. "You shall follow me," he shouted, "you Avicenna, Galen, Rhazes, you gentlemen of Paris, Cologne, Vienna, and whomsoever the Rhine and Danube nourish; you likewise Athenians, Arabs, Greeks and Jews, all shall follow me. The latchets of my shoes are better instructed than you all. All the universities, and all the old writers put together are less gifted than the hairs of my beard and the crown of my head."

Then into the flames of the roaring fire he threw the books of the masters, and as the fire consumed these evil scrolls he cried out to his students, "All misery shall be carried away in this smoke."

The world of authority stood aghast. A bonfire had just been kindled by Luther to swallow up the bull of Pope Leo X. And here was another fellow who burned the sacred works of these masters and trampled underfoot every precept they had taught! With the zeal of a religious fanatic and the courage of a crusader, he ran amuck among the treasured beliefs of his day and shattered them in bits. Had they not expected something of the sort from that crazy Paracelsus? Had he not scoffed at their veneration of the Latin tongue, and lectured his students and the crowd of barbers, bathmen and alchemists whom he invited contrary to all precedent, in a racy German? The dons had been horrified. They had warned him about this breach, but he would not be intimidated. What was to be

13

done with such a heretic? He was strongly intrenched at Basel through the influence of Johan Frobenius, the distinguished book publisher of that city to whom he had ministered in sickness, and whose right leg he had actually saved from amputation. Desiderius Erasmus, the great scholar of Rotterdam, was living with Frobenius at the time, and he, too, received medical attention from the Swiss iconoclast who cured him of the gout and kidney trouble. "I cannot offer thee a fee equal to thy art and learning," he wrote to Paracelsus. "Thou hast recalled from the shades Frobenius which is my other half; if thou restorest me also, thou restorest each through the other. May fortune favor that thou remain in Basel."

It was hard to dislodge a man with the spirit of Paracelsus. More than three hundred years later Robert Browning revealed in a poem the soul of this sane fanatic. Paracelsus speaks to his friend:

> Festus, from childhood I have been possessed
> By a fire—by a true fire, or faint or fierce,
> As from without some master, so it seemed,
> Repressed or urged its current; this but ill
> Expresses what I would convey.[3]

The authorities were afraid of this man who believed himself chosen by God. They waited for a chance to get rid of him. Besides, Paracelsus had made other enemies. The local doctors hated him. He had denounced them publicly as a "misbegotten crew of approved asses" for their practices of bleeding, bathing and torturing the sick. "The doctors who have got themselves made doctors with money go about the town as if it were a crime for the sick to contradict them," he had told the people. And he sneeringly added, "These calves think themselves great masters, for did they not go through the examination at Nuremberg?"

The apothecaries, too, were enraged against this iconoclast. For had he not, official town physician, demanded the right to inspect their stocks and rule over their prescriptions which he denounced as "foul broths"? These apothecaries had grown fat on the barbarous prescriptions of the local doctors. "The physician's duty is to heal the sick, not to enrich the apothecaries," he had warned them, and refused to send his patients to them to have prescriptions compounded. He made his own medicines instead, and gave them free to his patients.

All joined in an effort to rid themselves of this firebrand. He was peremptorily ordered to appear before the medical faculty of the University to show cause why he should be permitted to continue to practice medicine in the city of Basel. Frobenius, his patient, had died suddenly and they blamed him for his early death. But Paracelsus knew that death was caused by a stroke due to a strenuous horseback ride to Frankfort which Frobenius had undertaken against his advice. Paracelsus refused to present himself.

Then they hatched a plot and before long Basel lost Paracelsus, ostensibly because of the meanness of a wealthy citizen. Paracelsus had sued Canon Lichtenfels for failure to pay him one hundred guldens promised for a cure. The patient had offered only six guldens, and the fiery Paracelsus, when the court deliberately handed in a verdict against him, rebuked it in such terms that his life was in imminent danger. In the dead of night, he was persuaded by his friends to leave secretly the city where he had hurled defiance at the pseudo-medicos of the world.

Europe at the time was in the throes of a great intellectual upheaval. Over in Mainz, Johannes Gutenberg introduced printing from movable type. In Eisleben, in the heart of Germany, Martin Luther was born to question the orthodox religion of the day and usher in the Protestant Reformation. Columbus, seeking a westward passage to the Indies, discovered a new world. The Polish astronomer, Nicolaus Copernicus, for lack of a telescope, cut slits in the walls of his room in the University of Padua, watched through them the passage of the stars, and then revised man's idea of his place in the universe. Fifteen centuries of Ptolemaic teaching that the earth was the center of the universe were overthrown. And in Madrid, Andreas Vesalius, another graduate of the University of Padua, raised a storm as he introduced human dissection into the study of anatomy, thereby risking death at the hands of the Spanish Inquisition.

In spite of this renaissance, medicine was still a pseudoscience based on the teachings of Hippocrates of Cos, Avicenna the Persian Prince of Physicians, and Galen of Pergamos, gilder of pills and dissector of swine and apes. Superstition, mysticism and false theories were the cornerstones of its structure. Yet so intrenched were these authorities that, even when Paracelsus was still a student at Basel, a certain Dr. Geynes was refused a fellowship at the newly incorporated College of Physicians in England until he had publicly recanted of his error in having doubted the infallibility of Galen the Greek.

Most alchemists and physicians, having failed in their quest for the philosopher's stone, were now passionately hunting for the universal medicine, a panacea for all the ills of man. "There is nothing which might deliver the body from mortal death," they admitted, "but there is one thing which might postpone decay, renew youth, and prolong human life—the *elixir vitae*." What greater incentive to research could be offered than that strange fluid potent enough to ward off the dreaded encroachments of old age?

Many claimed to have discovered the Grand Catholicon. It is recorded that one man anointed his entire body with it, and lived four hundred years. But failing to anoint his soles he was forced to ride and never walk lest his feet, subject to decay, might bring him premature death. The elixir held forth an alluring promise of perpetual youth and happiness. Need we wonder that the eternal search went on in every corner of the world? Juan Ponce de Leon, landing at Porto Rico soon after the discovery of America, followed the hopeful tales of the Indians and sought the Fountain of Youth only to discover Florida.

This was the world into which Paracelsus was born three years after the death of Trevisan. His father was William Bombast von Hohenheim, a celebrated physician of the little village of Einsiedeln in the Swiss canton of Schwyz. For a while it seemed the child could not survive the weakness of its body. Small, frail and rickety, only the constant care of his mother, in charge of the village hospital, carried him through the dangerous period of infancy.

After attending school in a small lead-mining region where his father was now physician and teacher of alchemy, he was sent for higher instruction to the University of Basel where he adopted the name Paracelsus after Celsius, a famous Roman physician. Here he came across the writings of Abbott Hans Trithemius, celebrated astrologer, alchemist and inventor of a scheme of shorthand. Influenced by his writings Paracelsus went to Wurtzburg to study under him many books besides the Bible.

Then for a year he worked in the silver mines of Schwatz in the Tyrol. In 1516, at twenty-three, Paracelsus started to "transport himself into a new garden." For nearly ten years he roamed over Europe, matriculating and studying in every famous university. In Paris he met Ambroise Paré, who was learning to tie human arteries, an art which was to bring him fame as the father of modern surgery. This surgeon of King Charles IX was later spared during the slaughter of the Huguenots on the eve of St. Bartholomew's Day. He, like Paracelsus, was the first surgeon to write in his native tongue, nor did he forget to acknowledge his indebtedness to the Swiss reformer.

At Montpellier Paracelsus studied the Moorish system of medicine. He also hovered around Bologna and Padua. Spain and Portugal were included in his itinerary, before he travelled to Eng-

land. Then the restless spirit of Theophrastus Paracelsus brought
him to the Netherlands where war had broken out. He was offered
and accepted the post of barber-surgeon in the Dutch Army. Later
he served in the same capacity in the Neapolitan Wars during
which he came in possession of his famous long sword. He took
advantage of his position to attend the Diet of Worms, where he
heard his kindred spirit, Luther, make his memorable defense of his
doctrines. In Sweden he investigated the causes of mine disasters
and studied diseases of miners. The treatment of diseases of horses,
goats and cattle occupied much of his time in Russia.

Paracelsus, the crusader, did not stop with Europe. Like Bernard
Trevisan, he went to the East and visited Constantinople, the seat of
a world-famous medical practice. Trevisan had come here in search
of the secret of gold, but Paracelsus came to seek the secret of long
life. He travelled to Egypt and Tartary, and accompanied the son of
the Grand Khan in search of the tincture of life possessed by a
Greek alchemist.

These years of adventure were years of fruitful experiences. In his
passionate search for truth, Paracelsus did not hesitate to mingle
with gypsies, conjurers, charlatans, sorcerers, robbers, bandits, con-
victs, refugees from the law—all manner of rogues and honest men.
From them he gathered much curious lore about medicine and
alchemy. "My travels," he wrote, "have developed me; no man
becomes a master at home, nor finds his teacher behind the stove.
Sicknesses wander here and there the whole length of the world. If a
man wishes to understand them, he must wander too. A doctor must
be an alchemist, he must see mother earth where the minerals grow.
And as the mountains will not come to him, he must go to the
mountains. It is indeed true that those who do not roam have
greater possessions than those who do; those who sit behind the
stove eat partridge, and those who follow after knowledge eat
milkbroth. He who will serve the belly—he will not follow after
me."

During these years of travel he studied and practiced medicine
and surgery. Students flocked to him in Wurtemberg, Tübingen,
and Freiburg. The world began to hear of his wonderful cures. In
the meantime Paracelsus was filled with a realization of the wisdom
and folly of the medicine of his day. The shams of that pseudoscience kindled in him the fire of a reformer, and when at length he
came to Basel as medical lecturer, he was ready for his great battle
against the false ideas of healing—a battle which he waged until
death found him, before he was fifty, at Salzburg in Austria.

Paracelsus strove to tear away the shackles that enslaved the human mind to the ancient dogmas of the infallible Avicenna and the categorical Galen. But the authorities at Basel were too powerful. He was forced to leave that city, but not before he had struck a blow at the established leaders of alchemy and medicine from which they never recovered. When he fled from Basel he had already begun his work of destroying age-old tenets. He was still young— only thirty-five. Before leaving his students he made clear to them his plans for the future—"the restoration of true medicine."

Paracelsus, the iatrochemist, found refuge among his friends in Colmar in the province of Alsace whither he called the printer Oporinus, who brought him his chemical apparatus and notes. He set up a laboratory in a cellar and continued to work, but his enemies pursued him. They had styled him the chemical Luther, and, "Why not?" he asked. "Luther is abundantly learned, therefore you hate him and me, but we are at least a match for you."

Paracelsus was bitter. When at Basel they had pinned to the cathedral door a scurrilous attack from the shades of Galen to "Cacophrastus," he had challenged them in no uncertain terms: "Show me what kind of men you are and what strength you have. You are nothing but teachers and masters combing lice and scratching. You are not worthy that a dog shall lift his hind leg against you. Your Prince Galen is in Hell, and if you knew what he wrote me from there you would make the sign of the cross and prepare yourselves to join him. Your dissolute Avicenna, once Prime Minister, is now at the gates of Purgatory. I am preparing 'soluble gold' as a medicine for his suffering." They hated his caustic, vulgar tongue, and with reason.

For thirteen tormenting years Paracelsus led a vagabond life. Driven by poverty and reverses, he devoted his time to writing, healing and preaching with the energy and passion of one possessed. Like Bernard Trevisan, this Galahad, too, was absorbed in a great search. He was going to crush to the ground every trace of a vicious practice. He was determined to vindicate his teachings before the whole hostile world. The medical reformer met bitter opposition. Innsbruck refused him the privileges of the city. The professors were jealous and the authorities were afraid. He pleased no one but the sick whom he succeeded in healing. Tired, hungry, and in rags, he dragged his sickly body from town to town. He could gain no public hearing, no publisher would print his books; his enemies had seen to that. So for hours at a stretch he would write ceaselessly, and then, too tired to undress, throw himself upon his pauper's bed for a few hours' rest. This outcast was putting down in writing a new message which was to bring the world nearer to a clearer, saner understanding of the art of healing.

Paracelsus shouted the need for experimentation. "I admonish you not to reject the method of experiment, but according as your power permits, to follow it out without prejudice. For every experiment is like a weapon which must be used according to its peculiar power, as a spear to thrust, a club to strike, so also is it with experiments."

"Obscure diseases," he wrote, "cannot at once be recognized as colors are. What the eyes can see can be judged quickly, but what is hidden from the eyes it is vain to grasp as if it were visible. Thus it is with the obscure and tedious diseases that so hasty judgments cannot be made, though the Galenic physicians do this." And again he attacked the standpatters. "Do you think that because you have Avicenna of Bukhara, Valescus and de Vigo that you know everything? That which Pliny Secundus and the rest of them have written of herbs they have not tested, they have learned from noble persons and then with their smooth chatter have made books about it. Test it and it is true. You cannot put to proof your authors' writings. You who boast yourselves Doctors are but beginners!"

Those of his followers who believed he always carried the elixir of life with him in the pommel of his famous long sword, attributed his premature death to an overdose of his life-giving fluid or arcanum. His enemies, on the other hand, alleged that his career ended at the hands of an assassin while he was in one of his frequent drunken stupors. It is impossible to credit the first and unfair to believe the second of these stories. To-day we cannot doubt that his end came peacefully. When the time drew near for death to take him, Paracelsus cried:

> Let me weep
> My youth and its brave hopes, all dead and gone,
> In tears which burn. Would I were sure to win
> Some startling secret in their stead, a tincture
> Of force to flush old age with youth, or breed
> Gold, or imprison moonbeams till they change
> To opal shafts—only that, hurling it
> Indignant back, I might convince myself
> My aims remained supreme and pure as ever.

And through the long night, as Browning pictured it, Festus, his friend, comforted this childless man with the assurance that he had not struggled in vain.

When death retires before
Your presence—when the noblest of mankind
Broken in body or subdued in soul,
May through your skill renew their vigor, raise
The shattered frame to pristine stateliness.
When men in racking pain may purchase dreams
Of what delight them most, swooning at once
Into a sea of bliss, or rapt along
As in a flying sphere of turbulent light.
When we may look to you as one ordained
To free the flesh from fell disease, as frees
Our Luther's burning tongue the fettered soul.

Paracelsus took courage. He looked into the future.

But after, they will know me. If I stoop
Into a dark tremendous sea of cloud,
It is but for a time; I press God's lamp
Close to my breast; its splendour, soon or late,
Will pierce the gloom: I shall emerge one day.
You understand me? I have said enough.

Three days before he died, on his forty-eighth birthday, Paracelsus dictated his last will and testament. "There shall be sung in the church," he requested, "the first, seventh, and thirtieth psalms, and at all three singings a penny is to be given in hand to every poor man before the door."

Today, engraved on a broken pyramid of white marble in the cemetery of the Hospital of St. Sebastian in Salzburg, may be read: "Here is buried Philippus Theophrastus, distinguished Doctor of Medicine, who with wonderful art cured dire wounds, leprosy, gout, dropsy, and other contagious diseases of the body." No mention is made of the elixir of life.

Like Bernard Trevisan, egotistic yet earnest Paracelsus went to his grave beaten in his quest. Years before, he had realized the difficulty of his fight. The old theory of diseases accepted in those days was based upon the conception of Hippocrates of four body fluids or humors—phlegm, blood, yellow bile, and black bile, which in some mystic way were associated with the old Aristotelian elemental qualities—cold, warm, dry, and moist. Disease was caused by the improper proportions of these four fluids in the body, which also controlled the character of man. An excess of phlegm made one phlegmatic, too much blood made one sanguine, while an abundance of yellow bile produced a choleric person.

Another one of the common beliefs of his day was the doctrine of signatures which dictated the use of certain plants in medicine because their names resembled the part of the body afflicted or the disease itself. The feverwort, for instance, was used to reduce fever, and the liverwort to cure diseases of the liver.

The peculiar practice of sympathetic remedies was prevalent too. A wound was cleaned and then bandaged while the weapon which caused the wound was covered with the remedy. An axe had badly cut a butcher's hand. The bloody hand was washed and bandaged, and the axe, covered with the healing salve, was hung on a nail and carefully guarded until the hand was healed. Once, when the butcher suffered very annoying pains, it was found that the axe had fallen from the nail. It was thought that as long as the weapon was watched a magic current through the air would perform the miraculous healing.

For pains in the joints doctors prescribed an oil obtained from the bones of victims of some violent death, and for chicken pox they served their patients a soup filled with the heart and liver of vipers. Against the physicians who professed these ideas Paracelsus had fought all his life. "I shall not in my time," he had written, "be able to overthrow this structure of fables, for they are old and obstinate dogs who will learn nothing new and are ashamed to recognize their folly. That, however, does not matter much, but it does matter that, as I hope, the young men will be of a different character when the old ones have passed away, and will forsake their superstitions." That day did not come until long after Paracelsus exchanged life for death. The old order survived for decades, and the new order was ushered in only after the old dogmas were safely buried with the dead follies of the past.

The world owes bombastic Paracelsus a great debt. This revolutionist with the imagination of a poet and the fearlessness of a crusader, was much more than the bibulous braggart his enemies had called him. He was a real benefactor of mankind. His great contribution was no one epoch-making discovery, but rather the vital impetus he gave to the study of chemistry for the curing of ills of the body. He swept aside the teachings of the ancient authorities and brought alchemy to the aid of medicine.

"I praise," he told Europe, "the chemical physicians, for they do not go about gorgeous in satins, silks, and velvets, silver daggers hanging at their sides, and white gloves on their hands, but they tend their work at the fire patiently day and night. They do not go promenading, but seek their recreation in the laboratory. They thrust their fingers among the coals into dirt and rubbish and not into golden rings." Here was the true creed of the laboratory. Here alone would mankind find balm for its ills and salvation for its pains.

No longer were the rich to depend upon the playing of the flute to ward off or heal the gout as Galen had taught. Nor was man to rely any more upon the blowing of the trumpet to heal sciatica, as Æsculapius himself had prescribed. Man was no longer to remain captive to the notion that an ape's leg tied around the neck would cure the bite of this beast, nor was medical knowledge to be gained by the scanning of the heavens. And what pernicious nonsense was this singular practice among the old men of Rome of being breathed upon by young girls to prolong their lives!

Paracelsus abandoned all this witchcraft and superstition. He started the search for the potent drugs which the alchemist was to prepare or purify. Even the many herbs and extracts in common medical use were placed secondary to the value of these chemicals. There were many who gave ear to his instructions. They went back to their laboratories, threw away the crucibles filled with the strange concoctions that would not change to gold, and sought medicines to relieve human suffering. Paracelsus himself showed the way. He experimented in his laboratory, and introduced into medicine salves made from the salts of mercury. He was the first to use tincture of opium, named by him laudanum, in the treatment of disease. The present pharmacopœia includes much that Paracelsus employed— lead compounds, iron and zinc salts, arsenic preparations for skin diseases, milk of sulfur, blue vitriol, and other chemicals.

He understood the scepticism of the people about alchemy. Had they not been cheated and duped by those charlatans who claimed to possess the philosopher's stone? "Its name," he pleaded, "will no doubt prevent its being acceptable to many; but why should wise people hate without cause that which some other wantonly misuse? Why hate blue because some clumsy painter uses it? Which would Caesar order to be crucified, the thief or the thing he had stolen? No science can be deservedly held in contempt by one who knows nothing about it. Because you are ignorant of alchemy you are ignorant of the mysteries of nature." The changes which take place in the body are chemical, he said, and the ills of the body must be treated by chemicals. Life is essentially a chemical process and the body a chemical laboratory in which the principles of mercury, salt, and sulfur mingle and react to bring illness or health. Paracelsus believed that if the physician were not skilled to the highest degree in this alchemy, all his art was in vain. Here was a radical departure from the old practices. It brought to a hopeful close the age of frenzied alchemy, with its search for gold from dung.

Yet the lure of gold was still powerful. In Germany, Christian Rosenkreutz had founded the Brotherhood of the Rosy Cross, which professed to have the secret of making the yellow metal from dew. The Rosicrucians, as they were called, mixed their alchemy with a queer form of religion. Even Paracelsus, while searching for potent drugs to cure the ills of mankind, secretly sought the philosopher's stone, which might yield the cherished gold. Never did he really deny transmutation, and while he shouted from every public forum he was permitted to ascend that "the true use of chemistry is not to make gold but to prepare medicines," we find him privately attempting to prepare alchemic gold.

In his *Coleum Philosophorum* Paracelsus wrote that by the mediation of fire any metal could be generated from mercury. He considered mercury an imperfect metal; it was wanting in coagulation, which was the end of all metals. Up to the halfway point of their generation all metals were liquid mercury, he believed, and gold was simply mercury which had lost its mercurial nature by coagulation. Hence if he could but coagulate mercury sufficiently, he could make gold. And while he tried hundreds of different methods to bring this about, in the end he admitted failure. "From the seed of an onion, an onion springs up, not a rose, a nut, or a lettuce," he declared. The end of his life's journey had brought him to the same secret that Bernard Trevisan had found.

Strange that this modern Paracelsus, called by his followers Life's Dispenser, should have left among his writings such cabalistic directions as these:

> Some one may ask, what, then, is the short and easy way whereby Sol and Luna can be made? The answer is this: After you have made heaven, or the sphere of Saturn, with its life to run over the earth, place it on all the planets so that the portion of Luna may be the smallest. Let all run until heaven or Saturn has entirely disappeared. Then all those planets will remain dead with their old corruptible bodies, having meanwhile obtained another new, perfect and incorruptible body. That body is the spirit of heaven. From it these planets again receive a body and life and live as before. Take this body from the life and the earth. Keep it. It is Sol and Luna. Here you have the Art, clear and entire. If you do not understand it it is well. It is better that it should be kept concealed and not made public.

If he intended to conceal his art by clothing it in mystic words, Paracelsus did well, for even if we substitute for Sol and Luna the elements gold and silver, for Saturn lead, and for the other planets the metals mercury, tin, iron and copper, this alchemical jargon becomes hardly more intelligible.

Robert Boyle, one of the founders of modern chemistry, likened such alchemists as Paracelsus to the Argonauts of Solomon's Tarshish who brought home from their long and perilous voyage not only gold, silver, and ivory, but apes and peacocks, "for so the writings present us, together with diverse substantial and noble experiments, theories which either like peacock's feathers make a great show, but are neither solid nor useful, or else like apes, if they have some appearance of being rational, are blemished with some absurdity or other that, when they are attentively considered, makes them appear ridiculous."

And as his writings, so was Paracelsus—strange mixture of honest, fitful, fearless crusader, and mystic, cowardly seeker after gold.

III

BECHER

FIRE IS NOTHING, SOMETHING, LESS THAN NOTHING

ALMOST a century and a half had passed, witnessing the blind efforts of the followers of Paracelsus to bring order out of the chemical chaos. The quest for gold and the elixir of life still occupied the minds and hands of men.

The world wanted gold. Wars had sapped the coffers of the royal families of Europe. A wandering ambitious chemist, physician, economist, and inventor made a tempting proposal to Prince Herman of Baden, who happened to be in Holland. This practical dreamer would make gold from good Dutch sand, and sand was plentiful here. He would make gold, not like the alchemists, by cryptic formulas, but by a practical chemical method which could not fail. He needed hundreds of pounds of silver to change hundreds of tons of sand into mountains of glittering gold. Life had been hard to him and he could not finance this scheme himself. He had already experimented on a small scale in his laboratory in Vienna, whence he had fled after incurring the enmity of the prime minister of Emperor Leopold I. Now he was ready to proceed with his process on a factory scale.

It was 1673 and France was at war with Holland. Prince Herman of Baden could not help him just then, but asked him to return when hostilities had ceased. Becher could not wait, and sought aid from the Dutch Government. His plans sounded plausible enough for the States General to appoint a committee of The Hague to witness a test. He promised the state an income of millions of thalers from this project. These Dutchmen were wary of Becher. They knew the frauds of alchemy. And they did not fail to remind him of the gibbet.

One thirtieth of a million marks was Becher's first request to build a great overshot water-wheel which he had himself invented and patented. This was essential to the success of his plan, Becher

25

insisted, and the committee voted him the equivalent of twelve hundred dollars to start his experiments. The construction of this water-wheel delayed him until December, when he was at last prepared for the decisive trial.

Suddenly he was called away to Mecklenburg at the urgent request of the Grand Duke. This trip was so unexpected and so secretive that it was thought Becher had taken a discreet departure because he knew he could not fulfill his promise. So incensed were the Dutch that one newspaper suggested that he be caught and imprisoned for his frauds. Some of the members of the Committee were ready to withdraw their support.

But Becher returned to silence his enemies. A mind conscious of righteousness, he knew, laughs at the lies of rumor. He started his great experiment at once. A huge furnace was built, and on the 14th of February, 1679, he began the preliminary test in the presence of only one person, Lorenz Keerwolff, who attested to the success of the initial experiment. He had actually changed the white granular sand of Holland into the yellow metal! A month later he repeated his tests in the presence of the select commission of the government and no less a personage than the Mayor of Amsterdam. So cleverly did he work that these experiments, too, were reported to be successful, and he was authorized to continue his work on a larger scale with the assurance that all the necessary funds would be forthcoming. Becher tells us, and this is his own story, that his enemies at Vienna, where at one time he had been councillor of the Chamber of Commerce, jealous of his epoch-making successes in Holland, intrigued against him. They fomented such trouble among the Dutch Commission that he was in danger of his life, and to save himself he fled hastily, completely abandoning his great work. Kopp, one of the greatest historians of chemistry, says that Becher sought here to make gold "not from greed but as a scientific problem."

At first he found refuge in England, but his enemies discovered his whereabouts and persecuted him. He returned to the Continent, and later Prince Rupert, a nephew of Charles I of England, sent him to the Scottish mines to seek information on a new chemical project in connection with his research on the art of engraving in mezzotint. A terrific storm broke out over the sea, causing his voyage to be held up for four weeks, during which time he wrote *Foolish Wisdom and Wise Folly.* Intermingled with the story of his life, Becher described in this book some of his queer discoveries. He tells us he has seen and handled a heavy stone which rendered its possessor invisible. He relates how Count von Zinzendorf showed him his recipe for making gold. In Scotland he saw geese that grew and lived on trees, and hatched their eggs with their feet. He watched glass being rolled and flattened while cold. His description of a stentrophone conceived by a Mr. Moorland was certainly prophetic of our modern phonograph. This instrument could bottle "several words as an echo through a wire or spiral line, in a flask, so that they could carry this flask about for an hour and when they opened the flask could hear the words."

It was this strange adventurer, John Joachim Becher, born in Speyer, the son of a Lutheran minister, who took a real step forward in the development of modern chemistry by the introduction of the first important theory or generalization. Erroneous and deluding though it was, it still held the front of chemical thought for a hundred years.

The ancients considered fire the purest and most perfect of the elements of nature. The flames of fire rose ever upward reaching finally to the highest of the seven heavens—the heaven of fire and light, the empyreum, as they named it. Man's greatest and earliest achievement was the discovery of the use of this fire. The gods had not given this tool to man willingly. It had to be stolen by the demi-god Prometheus who, with the aid of Minerva, ascended to heaven and lit his torch at the chariot of the sun. Fire raised man above the brute. So highly did the ancients regard this luminous spirit that the Egyptians burned a perpetual fire in every temple, and the Greeks and Persians kindled them in every village and town. The Romans consecrated Vestal virgins to watch the sacred fire on their altars.

What is this strange phenomenon called fire? Aristotle considered it one of the great principles of all things. Heraclitus of Ephesus, a Greek philosopher, regarded it not only as the elementary principle of all things, but also as the universal force of creation. In every century men had pondered over the mystery of the flame. The fiery force that could engulf life and bring utter destruction was first regarded with fear and reverence, and later subjected to serious study. Among the ancients, Plato had assumed all burnable bodies to contain some inflammable principle, which, however, he failed to identify. The alchemists of later centuries had considered either some vague spirit of sulfur, or yellow, brittle, solid sulfur itself as the cause of fire. These vaporous concepts they used and discarded at will. Paracelsus explained the burning of wood on the basis of three elements or principles possessed by the wood. It burned because it contained sulfur, it gave off a flame because it had mercury in it, and left an ash because of the salt that was present in all wood. This explanation persisted for hundreds of years.

Then came Becher, who first enunciated a definite, though pseudo-scientific theory of the nature of fire. Crude though his idea was, it served as a bases of explanation for a great number of chemical phenomena, and gripped the scientific world for more than a century. Priestley, discoverer of oxygen; Scheele, the Swedish apothecary who discovered half a dozen other elements; Cavendish, one of the greatest chemical experimenters of England, all were ardent believers in Becher's conception, not to mention the hundreds of lesser lights in the chemical firmament of the seventeenth and eighteenth centuries. So simple was this theory and so easy to grasp and apply, that it dominated all chemical thought.

Becher's wild and imaginative nature led him into speculations which recall the palmy days of the Middle Ages. Far from being faithful to the spirit of experiment, he arrived at his conclusions outside of his laboratory. Becher threw aside the sulfur principle of his predecessors, for he had found substances which contained no sulfur and yet could burn vigorously. He postulated a "terra pinguis," that is, a fatty or inflammable earth possessed by all substances which could burn. For this fatty inflammable earth he later derived the term "phlogiston," from the Greek, "to set on fire." Phlogiston was fire itself. It was a definite chemical entity of an earthy nature, dry and adapted to solid combination. And it was not to be cast aside wherever it seemed to conflict with observed phenomena. It was to him and his followers the touchstone which explained those great chemical reactions of burning, oxidation, and calcination. The vital processes of breathing could likewise be explained by it, for did not the lungs constantly exhale phlogiston as food was consumed during digestion in human and animal bodies? When a substance burned, explained Becher, its phlogiston was given off violently in the form of the flame. Weigh the burned body, he said, and you will find that it has actually lost weight in the process, as phlogiston escaped. To him phlogiston was not merely an idea. It was a chemical substance with a definite color, odor, and weight.

But there were doubters even in those days, and some of them experimented while Becher wrote and explained. If phlogiston has weight and leaves the burning body, then the ashes of the burned body should always weigh less than the original substance. So in their crude scales they weighed the ashes of zinc and lead, and

found them to have increased in weight! Perhaps these were exceptions, they thought. They repeated their experiments with other
metals, and again found that the resulting powders weighed more
than the lustrous metals. Here were facts to stump even the wily
Becher. They came to him and faced him with the incontrovertible
proofs of their crucibles and balances.

But Becher would not acknowledge that his phlogiston had failed
as an explanation of burning. "Yes, you fools," he bellowed, "but
you do not know that my phlogiston may sometimes possess the
power of levity—it weighs less than nothing. Naturally, then, the
ash of your metals weighs more than the metals you burned.
Something, minus another thing which weighs less than nothing,
weighs more than that original something." Cardan, physician and
mathematician of Padua, a century before, had attributed the gain
in weight of lead to the loss of a similar celestial fire. We laugh at
these explanations today, but they were made before the end of the
seventeenth century. Chemical science was still struggling to see the
light, and natural philosophers were groping in abysmal darkness.

Phlogiston "explained" many facts. Phlogiston was the same in
one metal as in another, as well as in all burnable bodies. When
metals were heated or burned they changed into powders or calces,
as they were called. Why? Because they surrendered their fire, or
phlogiston, to the air. When, however, charcoal, an inflammable
substance rich in phlogiston, was added to the calx, the metallic
properties were instantly restored. In a similar way phlogiston
explained one of the most common of all chemical changes—the
rusting of iron. For what was rust, if not iron minus its phlogiston?
Add phlogiston to the rust in the form of charcoal, and lustrous
metallic iron is re-formed. A simple enough explanation of the
metallurgy of iron. Similarly could not the white ash of pure tin be
made to yield silvery tin again when phlogiston-rich coal was
heated with it? The dead calx of any metal could be instantly
restored to life by the addition of Becher's all-powerful phlogiston.
Paracelsus, himself, had written, "Dead metals may be revived or
reduced (*reduzieren*) to the state of metals by means of soot."

Yet phlogiston was not altogether so simple. There were facts that
could not be so neatly explained. For a time it seemed that
phlogiston would be discredited. But phlogiston suddenly became
chameleon-like, changing its nature to suit each perplexing problem. Lavoisier, who came a hundred years later, spoke thus of
phlogiston: "They have turned phlogiston into a vague principle
which consequently adapts itself to all the explanations for which it

may be required. Sometimes this principle has weight, and sometimes it has not; sometimes it is free and sometimes it is fire
combined with the earthly element; sometimes it passes through the
pores of vessels and sometimes they are impervious to it. It is a
veritable Proteus changing in form at each principle."

Goethe says somewhere that "a pompous word will stand you
instead for that which will not go into the head," and so this
mysterious phlogiston, child of the fanciful Becher, nearly destroyed
the progress of chemistry when the world of science accepted it as a
creed. A number of natural philosophers during the period following Becher caught a glimpse of the true meaning of fire and
burning, but Becher's creation had formed a mental smoke screen in
front of chemical progress. For almost a century the world struggled
through the chemical wilderness, like the Israelites of old before
they could see the Promised Land. Had it not been for this blight a
Joshua might have risen in the likeness of Boyle, Mayow, or Rey, all
of whom had glimpsed the truth behind the veil of fire.

Yet phlogiston did valiant service. It effectively turned the attention of chemists away from the elixir of life and the philosopher's
stone to a new field, one which yielded a rich harvest. The challenge of phlogiston led to a more serious study of simple chemical
reactions, such as combustion and oxidation, and stimulated the
development of the analysis of substances. In this way, "A vigorous
error vigorously pursued kept the embryos of truth a-breathing and
the body of chemistry was prepared for its soul."[4]

Alchemy was not altogether dead. A thin feeble blood stream still
coursed through its veins. Becher's fertile brain began to wrestle
with the mysteries of transmutation, and he proved a worthy and
enthusiastic adept. He revelled in its strange secrets.

"The chemists," he wrote, "are a strange class of mortals who
seek their pleasures among soot and flame, poisons and poverty, yet
among all these evils I seem to live so sweetly that may I die if I
would change places with the Persian King." A year after his
disastrous efforts in Holland, he published a curious account of the
philosopher's stone or "Magnalia Nature," which he claimed had
been lately exposed to public sight and sale. This was supposed to
be a true and accurate story of the manner in which Wenceslaus
Seilerus, a famous gold cook of the Emperor's court at Vienna,
obtained a very great quantity of powder of projection, an esoteric
substance akin to the philosopher's stone. The account was published by Becher "at the request and for the satisfaction of several
curious persons, especially Mr. Boyle, the eminent English scientist."

The frailties of mankind were the weaknesses of this versatile Becher. When he felt that he was no longer a young man, and fearful that his old age might be a barrier to his advancement, he cut off ten years from his life, and announced his birth date as 1635. Becher, ingenious rogue that he was, went to the extent of secretly inserting in his writings passages to falsify his age. So cleverly did he manipulate this falsehood that twenty of the foremost authorities of the eighteenth century were duped, until it was learned that Da Hayn, who attended the funeral of Becher in 1682, had frequently asked Becher why he claimed to be so young, even though everyone knew he was close to sixty.

Becher inherited from his father some of those qualities which later brought him power. The elder Becher was a teacher at Strassburg and later pastor at Speyer, the seat of the Imperial German Supreme Court and capital of the Bavarian Palatinate. He was a well-educated, forceful man, who, had he not died so early, might have properly directed young John Joachim. At the age of twenty-eight he spoke ten languages well, and wrote some works which unfortunately have not come down to us. If the cultured father wrote well, as we have reason to believe, his son John did not inherit this faculty, for his writings are barbarous both in style and content.

Young Becher's mother remarried, and his stepfather never spared him the rod. Whatever education he managed to get was obtained with great difficulty. Borrowing or stealing books for which he could not pay, he studied far into the nights, for during the day he worked to help support not only his mother, but his two brothers as well. He managed to earn a little by teaching, and with this he bought other books and acquired a smattering of theology, medicine, chemistry, politics, and law. So far as records show, he spent the equivalent of about eight dollars on elementary education from his teacher Debus—this being his only formal education.

Yet, by dint of a forceful personality, and no doubt, a little rascality, Becher soon began to attract attention. When he met the daughter of the privy-councillor, William von Hornigk, he exclaimed, "The beautiful women have bewitched Samson and Solomon. Why not me also?" After a hasty courtship he was converted to Catholicism and married her. Nothing was to be permitted to stand between him and fame. His marriage and conversion raised him in the estimation of the ruler of Mainz, who made him his personal physician and professor of medicine at the University of Mainz. Becher undoubtedly received his medical degree at the marriage altar rather than at any university—such was the spirit of the times. Drunk with the eminence to which he had been suddenly raised, he was conceited enough to declare that he could learn nothing from the university, since he had learned more by himself. Verily another Paracelsus! Becher kept pushing forward. Charmed by his delightful personality, and awed by his cleverly flaunted erudition, the Elector of Mainz, who had also befriended Leibnitz, the mathematical genius, selected him to compose a universal language for which he was promised one thousand ducats.

Feverishly he worked upon this subject, and the ingenious brain of Joachim did not rest until he had handed to the Elector a richly bound volume of his universal language, containing no less than ten thousand words. But what had come over the Elector? Someone must have been talking to him about this young upstart. For when Becher handed the manuscript to his patron, instead of the jingle of ducats and the shower of praises, all he could hear was an almost inaudible "Thank you." So embittered was Becher that he was tempted, he tells us, "to send his invention to Pekin, China." So much energy did he put into this work and so disappointed was he at its reception that he developed a high fever, and was so sick at Frankfort that he almost passed away with his universal language. But powerful Becher survived this shock; he was destined to outlive many more.

The problem of perpetual motion was again puzzling the scientists of Germany. Becher heard of it. He was going to put a stop to the perpetual talk of the impossibility of perpetual motion. He, Becher, the brainiest man in all Christendom, was going to prove to the world that a perpetual motion machine could be built. Perhaps the world was not ready for a universal language, he consoled himself. But it could not sit back and scoff at the finished product of a genuine perpetually moving clock. The world was bound to make a beaten path to his door. And while the world waited breathlessly, Becher built his clock.

This clock, which would run on forever without being rewound, needed a firm foundation upon which to rest. Becher prevailed upon the Elector to build a great tower. But a certain employee persuaded Jacob Britzly, the Swiss clockmaker, to tamper with it, so that this costly and artistic work designed for perpetual motion actually resulted in perpetual non-motion or rest. Even the prodigious brain and superhuman skill of a Becher could not induce a disemboweled clock to run forever. This failure had the desired result. His enemies at the Elector's court, and they were many, had been waiting to expose him and the time had at last arrived. The

Elector refused any longer to remain the stupid tool of a lunatic, and Becher was summarily dismissed from his court.

But was he altogether discouraged? Not Becher! A hundred and one more schemes revolved in his restless brain. He would not give up until he had beaten the sceptical world to its knees. He introduced the potato into Germany to feed its swine and cattle and to prevent famine among the poor. He patented a method of distilling coal. He started work on a lamp that would burn forever. He invented weaving and spinning machines. Various mining and metallurgical processes were devised by him. They all ended in commercial disaster. But in spite of his many bankruptcies, there was something about Becher that kept the believing world ready to listen to him once more. His fame as an economist had spread. In 1664 he was called to Mannheim to introduce new manufactures. He was to be given a free hand in the development and enrichment of the city. He planned a new era in the industrial life of that city. He outlined projects for the introduction of the manufacture of glass, paper, and even silk. Silk for the looms was to be obtained from a silk-raising industry which he was to inaugurate. It was an ambitious undertaking.

Before he set his plans in motion, however, he suddenly left for Bavaria. Contracts meant nothing to him. He visioned a greater opportunity. He was to be employed as financial and economic adviser at a very tempting salary, but misfortune still shadowed him. The Bavarian clergy was hostile. Becher had been forced some years before to leave Wurtzburg for performing an illegal operation which resulted in the woman's death; he could not hide that. The merchants hated him. They could not forget how he had planned to ruin them by building up a government monopoly of cloth manufacture. The nobility, too, was enraged. His schemes seemed to favor the peasant class. His life was secretly threatened if he failed to leave the kingdom. Becher compromised enough to quiet the most powerful of his enemies, and was finally permitted to stay. He founded a silk manufacturing company with money borrowed among the merchants. Soon, however, he left Bavaria to enter the service of Emperor Leopold I. Here at Vienna he prepared to take revenge. He would teach those Bavarians a costly lesson for the way they had plagued him. He built a large silk factory in Vienna and neglected the one in Bavaria, from which he had withdrawn his own funds. Becher chuckled when the news reached him that sericulture in Bavaria had failed dismally.

This restless, ambitious, conceited spirit who had raised himself from poverty and obscurity to the position of Medical Doctor, Privy and Commercial Advisor to His Majesty the Holy Roman Emperor, now turned his hand to another venture. He had written many economic treatises on socialistic and other schemes which gave him a wide reputation. On the strength of this he established an institution for commercial instruction at Vienna, from which he received a salary of the equivalent of ten thousand dollars a year. The four years between 1666 and 1670 were comparatively quiet ones for Becher. While over in England Sir Isaac Newton was discovering the calculus, Becher was writing his *Physica Subterranea,* by which he is best known in science. In ten days he completed another book, *Methodus Didactica.* His wonderfully equipped laboratory at Munich, the finest in Europe, helped him in his scientific writings. So successful was his *Physica Subterranea* that it was reissued by his most famous follower, George Ernest Stahl, Professor of Medicine at Halle, and private physician to the King of Prussia.

The spirit of Becher could not long remain tranquil. There were new worlds to conquer. He became involved in an ambitious scheme to unite the whole of devastated Germany commercially by breaking down the numerous tariff walls between the many independent states. Furthermore, various colonization plans were outlined. The Count of Hanau had received three thousand square miles of territory far away in South America from the Holland West India Company. Here, from a tropical jungle between the Amazon and Orinoco Rivers, Becher planned to build up a great fertile plain peopled by his countrymen. The products of this rich soil were to help feed and enrich exhausted Germany. Amid grand festivities and the firing of one hundred cannon, the Hanau West India Company was officially launched—without a single ship. Becher, exultant over his new creation, exclaimed, "We now have a New Germany alongside of New Spain, New France, and New England."

But Germany was to wait for other more successful colonization plans. Becher's brain child proved an abortion; the whole dismal failure was laid at his door. He was accused of mismanagement, theft, and even atheism! Another bubble had burst.

And now Becher was getting old. Fate had dealt harshly with him —everything he touched seemed cursed. The ten years he had tried to slice away from his life hung heavily on his weakened body, and he could no longer hide his fifty-seven years of earthly struggle. He had just returned from the rich mines of Cornwall and the Isle of Wight. While still dreaming of a new world along the Orinoco River and planning great projects which would finally free him from the evil star under which he was born, suddenly in 1682 he died in poverty in London. Reverting to Protestantism before he died, he was buried in the chancel of St. James-in-the-Fields, in London.

Becher's span of life covered a period of intense intellectual activity. Harvey, in England, was demonstrating with remarkable precision the circulation of the blood. Hooke was writing *Micrographia,* a book on microscopy which Samuel Pepys, famous diarist and President of the newly founded Royal Society of England, read with great delight. Roemer, a Dane, was measuring the speed of light. Torricelli, in Italy, was demonstrating atmospheric pressure with his newly invented barometer. Von Guericke, burgomaster of Magdeburg, was introducing the air-pump and astonishing the Emperor and his court as sixteen pairs of horses struggled in vain to pull apart two hollow iron hemispheres from which the air had been exhausted. Anthony van Leeuwenhoek, in Holland, was gazing upon red blood corpuscles and minute plants and animals under microscopes which he himself constructed, while Huygens, his compatriot, was inventing a pendulum clock, and Spinoza, the lens grinder of Amsterdam, was writing his *Ethics.* Descartes, the French philosopher, was enriching the sciences of mathematics and physics in many of their branches.

In the field of chemistry, too, there was great animation. Nicholas Lemery was publishing his *Cours de Chimie.* Johann Glauber in Bavaria was discovering important chemical salts and publishing an encyclopedia of chemical processes. Rey, a French physician, was writing a curious account of experiments on the rusts of tin and lead. John Mayow, of Cornwall, another physician, was investigating the respiration of animals in air and its relation to the burning of metals. And in the same year that Becher published his *Physica Subterranea,* Brandt, a Hamburg merchant and alchemist, accidentally discovered fiery phosphorus which shone in the dark, and awakened the curiosity of the world with his cold fire. While Krafft, who had learned the secret from Brandt, made a tour of the Continent exhibiting glowing phosphorus to the crowned heads of Europe, immortal Newton sat before a large tub of soap suds, blowing bubbles through a common clay pipe and watching their colors in the bright sun—the same Isaac Newton who was giving to the world one of the greatest scientific contributions of all times, the theory of gravitation.

Becher had travelled widely through Sweden, Italy, Holland, France, and other countries. He knew many of these illustrious scientists and philosophers. He read voluminously, was familiar with the literature of chemistry, and never failed in his writings to quote the opinions of others. His only lasting contribution to applied chemistry, however, was his experimental discovery of the gas ethylene, introduced in 1922 by Dr. Lockhardt as an anesthetic, which Becher prepared by the action of common alcohol on sulfuric acid.

Stripped of the many-sided ambitious projects which occupied the greater part of his tempestuous life, Becher stands out as the pioneer of a new path over which chemistry was thenceforth to travel. Chemistry was again at the crossroads, and it was Becher who pointed the way. Although phlogiston contained the germ of the old fire principle of Zoroaster, and, like the subtle ether of today, was based upon conjecture rather than experimentation, still it drew men to more practical endeavors. Becher gave phlogiston to the world "clothed in such a material garb that it required two centuries to unwrap the truth. Still the sparkle of the gem was there, and men followed it until it led them into a clearer day."[5]

A hundred years after Becher's death, Madame Lavoisier, robed as a priestess, and surrounded by the scientific celebrities of Paris, burned his writings with those of his illustrious follower, Stahl, upon an altar. And, as a solemn requiem was chanted, the theory of phlogiston perished in France, while out of its ashes, like the Phœnix of old, sprang up a new chemistry.

IV

PRIESTLEY

A MINISTER FINDS THE PABULUM OF LIFE

BECHER lies buried in his grave for a century. Great political upheavals have shaken the foundation of Europe and institutions have gone tumbling to the ground. The French have stormed their Bastille in Paris while eager, greedy, curious men pottered around in smoky laboratories ever seeking to unravel some of the secrets of nature.

The anniversary of the storming of the Bastille is approaching. Over in Birmingham, England, liberal men are planning to celebrate this historic day. Modestly, quietly, without drumbeat or torch-light, they gather in the meeting-house of the town. Among these lovers of human freedom is a dissenting minister, named Joseph Priestley, who, too, has joined this group to commemorate the emancipation of a neighboring nation from tyranny.

It is July 14, 1791. Outside the meeting-place two men on horseback are stationed in front of a wild mob. One of them is reading a long document, prepared by an agent of the King: "The Presbyterians intend to rise. They are planning to burn down the Church. They will blow up the Parliament. They are planning a great insurrection like that in France. The King's head will be cut off, and dangled before you. Damn it! you see they will destroy us! We must ourselves crush them before it is too late." The cry of Church and King goes up and a thousand men break loose. And as the magistrates of the city look on and applaud, Priestley's meeting-house is burned to the ground.

The clergy all over England had inflamed the people against the Dissenters. Priestley, an activist, was engaged in the Liberal party's struggle for civil and religious liberty, and was also an enemy of the government party. He had been a thorn in its side for years. Openly siding with the American colonists in their struggle for independence, he had brazenly broadcast letters like the following which

37

Benjamin Franklin had sent him. "Britain, at the expense of three millions," wrote the candle-maker's son from Philadelphia, "has killed one hundred and fifty Yankees this campaign, which is twenty thousand pounds a head; and at Bunker's Hill she gained a mile of ground, half of which she lost again by our taking post on Ploughed Hill. During the same time sixty thousand children have been born in America. From these data your mathematical head will easily calculate the time and expense necessary to kill us all, and conquer the whole of our territory."

He was fearless in espousing any cause which seemed just to him. He had just been made a citizen of the French Republic for publishing a caustic reply to Burke's attack on the French Revolution. To this dangerous agitator's home the crowd rushed, demolished his library, smashed to bits all his scientific apparatus, and burned his manuscripts. Priestley was not the only victim. The residence of Dr. Withering was also attacked and the homes of others of Priestley's friends were pillaged and burned, while some of the Dissenters, to escape the terror, scrawled "No Philosophers" on their doorsteps. Still the fury of the mob was not abated. "Let's shake some powder out of Priestley's wig," yelled one rioter, and away they went to hunt him out. But the stuttering minister had been warned, and he fled to London, while for three days the riot continued, encouraged by some members of the court of King George III, who thought to intimidate the friends of liberty by this means.

William Pitt, the Prime Minister, did not show any sign of regret after this infamous riot, and Edmund Burke "could scarcely contain his joy" when he received the news. The year before, he had declared, "We are resolved to keep an established Church, an established monarchy, an established autocracy, and an established democracy, each in the degree it exists, and in no greater."

After his flight to London, Priestley found himself much restricted with respect to philosophical acquaintances. In Birmingham since 1780 he had been the center of a stimulating intellectual circle. He had infused new vigor into this small group of men who called themselves the Lunar Society because they were accustomed to dine once a month near the time of the full moon, "so as to have the benefit of the light on returning home" as Priestley explained. Erasmus Darwin, grandfather of Charles Darwin, was its patriarch. As this portly gentleman with his scratch-wigged head buried in his massive shoulders stammered out his lively anecdotes, the room fairly rocked with laughter. On one occasion, finding himself unable to attend a meeting he wrote: "Lord, what inventions, what wit, what rhetoric, metaphysical, mechanical and pyrotechnical will be on the wing—while I, I by myself I, am joggled along the King's highway to make war upon a stomach-ache."

James Watt, celebrated Scotch engineer and perfecter of the first practical steam-engine, sat there with his business partner, Boulton, while Samuel Galton, wealthy man of letters, exchanged views on science, literature and politics with Dr. Withering, physician and chemist. Captain James Kerr, commercial chemist and author; Collins, an American rebel, and Dr. Henry Moyes, a blind lecturer in chemistry, completed this brilliant gathering among which Joseph Priestley "seemed present with God by recollection and with man by cheerfulness."

Priestley missed this social and intellectual life deeply, for most of the members of the Royal Society shunned him either for religious or for political reasons. When London's natural philosophers met once a week at Jacob's Coffee House with Sir Joseph Banks, Sir Charles Blagden, Captain Cook and Dr. George Fordyce, Priestley was not a welcome visitor. Cavendish avoided him even as "the chased deer is avoided by all the herd." Finally he resigned from that scientific body. More than a century later during the first World War, on similar grounds, the scientists of Germany struck the names of England's most eminent chemists from their list of foreign honorary members, and a generation later, during the Nazi domination of Germany similar and even worse actions were taken against scientists of Jewish extraction and others who opposed Hitler. Such is the madness of men, even among scientists, in times of stress.

And, while over in France the Department of Orne was electing this son of a poor dresser of woolen cloth a member of its National Convention, he brought action for damages to the extent of four thousand pounds against the city of Birmingham. King George wrote to Secretary Dundee: "I cannot but feel pleased that Priestley is the sufferer for the doctrines he and his party have instilled . . . yet I cannot approve of their having employed such atrocious means of showing their contempt." The case went to a jury, and after nine hours Priestley triumphed. The great wrong done was in part righted, and Priestley was enabled again to give himself to the world of science.

Born in 1733 at Fieldhead, near Leeds, of staunch Calvinists, Priestley was prepared for the ministry. At fifteen he learned Hebrew, and developed an intense distaste for dogma. At the age of twenty-two, after having been rejected because of his views on original sin and eternal damnation, he was appointed pastor of a small chapel in Suffolk, earning thirty pounds a year. Much as he was averse to teaching, he was compelled by his meager salary to do so. This master of French, Italian, German, Hebrew, Arabic, Syriac, and even Chaldee, between seven in the morning and four in the afternoon taught school; between four and seven, he gave private lessons; and then, whatever time he could snatch from his clerical duties he devoted to the writing of an English grammar. Two years later he attended some lectures given by a Mr. Haggerstone from whom he learned various branches of mathematics, and at the same time he read W.J.S. Gravesend's *Elements of Natural Philosophy*. He learned the importance of careful observation and experiments as confirmation of the principles of mechanics and astronomy as well. The book, however, contained no chemistry.

From 1752 to 1755 Priestley attended the dissenting academy at Daventry. Dissenting academies offered the most complete formal instruction in contemporary science available at that time in England. As part of his studies here he read *Observations on Man* by a Dr. Hartley from which he obtained his firm belief in mechanistic and strictly causal relationships in science.

We next find Priestley at Needham Market, Suffolk, where he ministered to a small dissenting congregation and applied himself "with great assiduity to my studies which were classical, mathematical, and theological." In 1758 he left Needham Market for Nantwich in Cheshire. Here he opened a school and taught among other things the air pump, electrical machines but no chemistry. He gave no lectures and made no original experiments.

Three years later Priestley was invited to become tutor in languages and belles lettres in another dissenting academy at Warrington. John Sedden was the chief dissenting minister in Warrington and helped to establish this academy. Together with Priestley they induced Dr. Matthew Turner, atheist, scholar, wit and surgeon to give a course of twenty lectures in chemistry here. Turner was known also as an excellent lecturer and skillful demonstrator of experiments. Priestley himself attended these lectures and assisted him in making some nitric acid.

Then at the age of thirty-four, Priestley went to take charge of the Mill Hill Chapel at Leeds about 200 miles north of London. Here he declared himself a "believer in the doctrine that Jesus was in nature solely and truly a man, however highly exalted by God." Poor, struggling to support a family on the scantiest of means, unpopular because of his religious views, and like Demosthenes, battling a serious defect in his speech, this many-sided Englishman found time between his theological duties and metaphysical dreamings for more worldly matters. During one of his occasional trips to London he met Benjamin Franklin, who stirred in him a deeper interest in electricity. Interrupting his chemical studies, Priestley planned to write a history of this subject with books and pamphlets which Franklin undertook to supply. This was the beginning of his serious career as a scientist. "I was led in the course of my writing this history," he tells us, "to endeavor to ascertain several facts which were disputed, and this led me by degrees into a larger field of original experiments in which I spared no expense that I could possibly furnish."

Some part of the fame which is Priestley's was due to the public brewery which adjoined his home here in Leeds. In this smelly factory he busied himself in his spare moments, experimenting with the gas which bubbles off in the huge vats during the process of beer-making. He lighted chips of wood and brought them near these bubbles of colorless gas as they burst over the fermenting beer. It was a queer business for a minister, and the factory hands shook their heads as they watched him bending over the bubbling cisterns that hot midsummer. Priestley was too absorbed to pay any attention to their stifled laughs. He knew little chemistry, but he was a careful observer. He noticed that this colorless gas had the power of extinguishing his burning chips of wood. He suspected that it might be the same "fixed air" which, fifteen years before, Joseph Black, son of a Scotch wine merchant, had obtained by heating limestone while on the trail of a secret remedy of calcined snails by means of which a Mrs. Joanna Stephens had cured the gout of Robert Walpole, England's Prime Minister. Could this be true? Unable to obtain sufficiently large quantities of this gas from the brewery, he learned to prepare it at home. He tried dissolving the gas in water. It was not very soluble, but some of it did mix with the water. In this manner in the space of two or three minutes he made, as he related, "a glass of exceedingly pleasant sparkling water which could hardly be distinguished from Seltzer water." Appearing before the Royal Society he told that learned body of his discovery of what we now know as soda water—a very weak acid solution of carbon dioxide gas in water. The Royal Society was intensely interested, and he was asked to repeat his experiments before the members of the College of Physicians. He jumped at the opportunity, and as he bubbled the gas through water, he asked some of those present to taste the solution. They were very much impressed, and recommended it to the Lords of the Admiralty as a possible cure for sea scurvy. Priestley received the Society's gold medal for this discovery, the first triumph of this amateur chemist in science.

Priestley, the dilettante scientist, was happy. He busied himself with other chemical experiments. What a relief to get away from his ministerial duties and lose himself in this great hobby! He tried heating common salt with vitriolic acid, and obtained a product which others had missed, for Priestley collected the resulting gas over liquid mercury rather than over water, as his predecessors had done. The colorless gas which he obtained had a pungent, irritating odor. He tried to dissolve it in water. Hundreds of volumes of this gas were easily dissolved—the water sucked it up greedily. No wonder the gas had not been collected before! It had dissolved in the water over which they had tried to imprison it. This gas dissolved in water is the hydrochloric acid used extensively today as muriatic acid (Priestley's name) for cleaning metals and in the manufacture of glue and gelatine. Here was another great contribution to chemistry by this mere amateur.

The Rev. Joseph Priestley's congregation was puzzled at his abiding interest in bottles and flasks. He seemed to be serving two altars. There was some grumbling, but the English minister was too excited to listen. He was now heating ammonia water and collecting another colorless gas over mercury. This gas, too, had a characteristic irritating odor. The fumes filled his room as he bent over the logs of the open fireplace, stirring the embers into greater activity. He was giving science its first accurate knowledge of the preparation and properties of pure ammonia—the gas which has been so successfully employed in cleaning agents, fertilizers, and refrigeration. What if the vapors did make his eyes tear until he was almost blinded, and drove the occupants of his humble house into the open to catch their breaths? This thrilled him more than any passage in the Scriptures. Then Priestley brought these two dry, colorless, disagreeable gases, hydrogen chloride and ammonia, together. He was amazed at the result. The gases suddenly disappeared and in their place was formed a beautiful white cloud which gradually settled out as a fine white powder. A great chemical change had taken place—a deep-seated change. Two pungent gases had united to form an odorless white powder—ammonium chloride, now used as an electrolyte in the dry battery, and in many other ways.

Thus in the space of a few years Priestley, eager devotee of science, made a number of significant discoveries. He began to spend more and more time in his makeshift laboratory. Chemistry had completely captivated him. And as he spread the word of God among his worshipers in Leeds, the world of science, too, began to hear of this preacher-chemist. Soon a proposal came to him to accompany Captain James Cook on his second voyage to the South Seas. He was tempted but, fortunately, another clergyman objected to him because of his religious principles. He stayed behind, and continued the great experiment that was to bring him lasting fame.

Priestley's experiments with the different kinds of gases or airs, as he called them, had made him very proficient in the preparation and collection of these elastic fluids. Until his time, the various gases had been studied by collecting them first in balloon-like bladders. This was a clumsy method, and besides, the bladders were not transparent. Priestley introduced and developed the simple modern method of collecting gases. He filled a glass bottle with liquid mercury, and inverted it over a larger vessel of mercury, so that the mouth of the bottle was below the quicksilver in the vessel. A tube was connected, by means of a cork, to the gas generator, and the end of this delivery tube was placed under the mouth of the bottle of mercury. The escaping gas displaced the mercury in the bottle, and was thus imprisoned in a strong transparent container. For those gases which are insoluble in water, Priestley used water, lighter and much less expensive than mercury, in the glass bottle and trough, even as Hales and Mayow had done before him. Here was a decided advance in the methods of studying gases.[6]

Priestley had heated a large number of solid substances in the flames of his furnace. Now he tried utilizing the heat of the sun by means of a sun-glass. By concentrating the sun's rays through a burning lens, he found that he could obtain a sufficient heat to burn wood and other solid materials. Finally he procured a very large lens, a foot in diameter, and proceeded with great alacrity to heat a great variety of substances both natural and artificial. He placed the solid substances in a bell-jar arranged so that any gas which might be formed inside it would pass out and be collected in a bottle placed over a trough of mercury. The burning lens was so placed outside the bell-jar that the heat of the sun was concentrated upon the solid to be tested. With this apparatus, he endeavored, on Sunday the first of August, 1774, "to extract air from *mercurius calcinatus per se*," a red powder known to Geber and made by heating mercury in the air. "I presently found," he reported, "that air was expelled from it readily."

But there was nothing startling about this. Others before him had obtained gases by heating solids. Scheele, the great Swedish apothecary chemist, had obtained the same result three years before by collecting "empyreal" air. Robert Boyle, a hundred years back, had heated the same red powder and obtained the same mercury.

Stephen Hales, too, had liberated a gas from saltpetre but saw no connection between it and air. Eck of Salzbach, an alchemist, had likewise performed this experiment three centuries before in Germany, and yet the world had not been aroused, for nothing further had been discovered about the gas.

A lighted candle was burning in Priestley's laboratory near him. He wondered what effect the gas might have on the flame of this candle. Merely as a chance experiment, he placed the candle in a bottle of the gas. The flame was not extinguished. On the contrary, it burned larger and with greater splendor. He was thrilled with excitement but was utterly at a loss to account for this phenomenon. He inserted a piece of glowing charcoal in another bottle of this gas, and saw it sparkle and crackle exactly like paper dipped in a solution of nitre. The burning charcoal was quickly consumed. He was astounded. He inserted a red hot iron wire. The heated metal glowed and blazed like a spirit possessed. The preacher's agitation knew no bounds.

This chance insertion of the lighted candle ushered in a revolution in chemistry. Speaking about this memorable occasion some years later, Priestley said, "I cannot at this distance of time recollect what it was that I had in view in making this experiment; but I had no expectation of the real issue of it. If I had not happened to have a lighted candle before me, I should probably never have made the trial, and the whole train of my future experiments relating to this kind of air might have been prevented . . . . More is owing to what we call chance than to any proper design or preconceived theory."

At this time Priestley had no notion of the real nature of this air. He was steeped in the fire principle of Becher and believed the gas to be, not the simple substance we know today as oxygen, but some strange compound of phlogiston, earth, and nitric acid—so completely had phlogiston befuddled him. But he kept studying this mysteriously active gas which had been driven out of his red powder. Fumbling along as best he could, hampered by meager funds, a poor foundation in chemistry, and no clear goal before him, he continued to investigate the properties of this gas. Once before, he had accidentally prepared it from saltpetre, but had neglected to carry out any further experiments with it. The perfect scientist would have probed into the character of this gas as soon as he had prepared it, but Priestley was not the perfect experimenter.

At that time, the atmosphere we breathe was thought to be a pure, simple, elementary substance like gold or mercury. Priestley himself had first conjectured that volcanoes had given birth to this atmosphere by supplying the earth with a permanent air, first inflammable, later deprived of its inflammability by agitation in water, and finally purified by the growth of vegetation. He had concluded that the vegetable world was nature's supreme restorative, for when plants were placed in sealed bottles in which animals had breathed or candles had been burned, he had noticed that the air within them was again fit for respiration. The phlogiston, he thought, which had been added to the atmosphere by burning bodies, was taken up by plants, thus helping to keep the atmosphere pure. But just about this time Daniel Rutherford, a medical man who occupied the chair of botany at the University of Edinburgh, had found two substances in the air. He had absorbed a small amount of carbon dioxide from the air, by means of lime-water, which turned milky white. Then, by allowing a small animal to breathe in a limited supply of air, he found that after the carbon dioxide had been absorbed, about four-fifths of the volume was left in the form of an inert gas. This inactive gas of the air was named by Chaptal *nitrogen*, because of its presence in nitre.

Priestley had read of these experiments. He began to suspect something. He heated some lead very strongly in the air and watched it gradually turn red. This red powder he treated in exactly the same way that he had heated the red powder of mercury. Priestley danced with glee, for he had obtained the same oxygen! He was confirmed in his suspicions that this oxygen, which he had obtained both from the red powder of mercury and the red lead, must have originally come from the atmosphere. "Perhaps it is this air which accounts for the vital powers of the atmosphere," thought Priestley. "I shall find out how wholesome is this dephlogisticated air."

On the eighth of March, 1775, we find this honest, religious heretic working on a queer experiment in the large castle of Lord Shelburne in Bowood near Calne. The night before he had set traps for mice in small wire cages from which the animals could be easily removed alive. But what is this moulder of the souls of men to do with mice? They are going to unravel a mystery for him. He takes two identical glass vessels, fills one with oxygen and the other with ordinary common air, and sets them aside over water.

The next morning he removes one of his captive mice from the trap, takes it by the back of the neck and quickly passes it up into the vessel of common air inverted over water. He sets the mouse on a raised platform within the vessel, out of reach of the water. The little beast must not drown. Then under the second vessel, filled with oxygen, he places an equally vivacious mouse with the same care.

Seated on a chair, Priestley amuses himself by playing the flute as he watches his curious experiments. He has no idea how long he will have to wait. Suddenly he stops playing. The mouse entrapped in the glass vessel containing common air begins to show signs of uneasiness and fatigue. Priestley puts away his flute and looks at his clock. Within fifteen minutes the mouse is unconscious. Priestley seizes its tail, and quickly yet carefully pulls it out of its prison. Too late—the mouse is dead. He peers into the second vessel containing his oxygen. What is happening to its tiny inmate? Nothing alarming. It keeps moving about quite actively. Ten minutes more pass. Priestley is still watching the animal. It begins to show unmistakable signs of fatigue. Its movements become sluggish—a stupor comes over it. The minister rushes to set it free, and takes it out of its tomb apparently dead. It is exceedingly chilled, but its heart is still beating. Priestley is happy. He rushes to the fire, holds the little mouse to the heat, and watches it slowly revive. In a few minutes it is as active as ever. He is unable to believe his senses. For thirty minutes this animal has remained in his oxygen and survived, while the first mouse, confined in common air, had died in half that time!

What can account for this? It is possible that his oxygen is purer than common air, or does common air contain some constituent which is deadly to life? Perhaps it is all an accident. That night Priestley keeps pondering over the mice and his oxygen. He begins to suspect that his oxygen is at least as good as common air, but he does "not certainly conclude that it was any better, because though one mouse might live only a quarter of an hour in a given quantity of air, I knew," he told himself, "it was not impossible but that another mouse might have lived in it half an hour." And the next morning finds Priestley experimenting with more mice to probe this mystery of the air.

He looks for the glass vessel in which a mouse had survived fully thirty minutes the day before. He is in luck. The vessel still contains oxygen. He is going to use this air over again, even though it has been rendered impure by the breathing of the mouse. He thinks of putting two or three mice in this vessel but abandons the idea. He has read of an instance of a mouse tearing another almost to pieces, in spite of the presence of plenty of provisions for both. So he takes a single mouse and passes it up on to its floating platform. He watches it intently for thirty minutes while it remains perfectly at ease. But slowly it passes into a slumber, and, "not having taken care to set the vessel in a warm place, the mouse died of cold. However, as it had lived three times as long as it could probably have lived in the same quantity of common air, I did not think it necessary," wrote Priestley, "to make any more experiments with mice."

Priestley was now convinced of the wholesomeness of his oxygen. The mice had proved this to him beyond doubt. He might have ended his experiments at this point, but he had the curiosity of the true natural philosopher. He decided to substitute himself for his humble mice, and partake of this gaseous pabulum of life. Breathing strange gases was a dangerous business but Dr. Mayow, a hundred years before him, had found that a certain gas (nitro-aerial spirit obtained by him from nitre) when breathed in the lungs gave the red color to arterial blood. Priestley wondered if his oxygen would be just as effective. He inhaled some freshly prepared oxygen through a glass tube, and found to his astonishment that the feeling in his lungs was not sensibly different from that of common air, "but I fancied," he noted, "that my breath felt peculiarly light and easy for some time afterward. Who can tell but that in time this pure air may become a fashionable article in luxury. Hitherto only my mice and myself have had the privilege of breathing it." Priestley foresaw many practical applications of this very active gas—"it may be peculiarly salutary to the lungs in certain morbid cases when" (as he explained it in his terms of phlogiston) "the common air would not be sufficient to carry off the phlogistic putrid effluvium fast enough." Today oxygen is, in fact, administered in cases of pneumonia where the lungs have been reduced in size and the patient cannot breathe sufficient oxygen from the air. Firemen fighting suffocating fumes, rescue parties entering mines, aviators and mountain climbers, who reach altitudes where the air is very rare, carry tanks of oxygen.

Priestley, the tyro, more than a century and a half ago had dreamed of these modern practical uses of oxygen. Priestley, the minister, also saw a possible danger of using this gas constantly instead of common air, "For as a candle burns out much faster in this air than in common air, so we might *live out too fast*. A moralist at least may say that the air which nature has provided for us is as good as we deserve."

Priestley kept testing the purity of his newly discovered gas. He found it to be "even between five and six times as good as the best common air" that he had ever handled. His imaginative mind was often very practical, and again he thought of a possible application of this oxygen. He saw in it a means of augmenting the force of fire to a prodigious degree by blowing it with his pure oxygen instead of common air. He tried this in the presence of his friend Jean H. de Magellan by filling a bladder with oxygen and puffing it through a small glass tube upon a piece of lighted wood. The feeble flame burst at once into a vigorous fire. Here was the germ of the modern blow-pipe which uses yearly billions of cubic feet of oxygen for cutting and welding. He even suggested that it would be easy to supply a pair of bellows with it from a large reservoir, but left to Robert Hare, of Philadelphia, the actual invention of the oxy-hydrogen torch.

The results of his experiments set Priestley all a-quiver. A few weeks later Lord Shelburne, who had shared his views regarding the American colonists, took a trip to the Continent. This scholarly statesman had offered Priestley an annuity of two hundred and fifty pounds, a summer residence at Calne, and a winter home in London, to live with him as his librarian and literary companion. For eight years this beautiful relationship lasted, and it was during these years that Priestley performed his most productive experiments. On this trip to the Continent, Priestley accompanied his patron. In Paris, Priestley was introduced by Magellan, a Portuguese descendant of the circumnavigator of the globe, to the famous chemists of France. In Lavoisier's laboratory, in the presence of a number of natural philosophers, he mentioned some of the startling results of his experiments. Lavoisier himself honored him with his notice, and while dining with him Priestley made no secret of anything he had observed during his years of experimentation, "having no idea at that time to what these remarkable facts would lead." Lavoisier listened to every word of this Englishman, and when Priestley left to visit Mr. Cadet, from whom he was to secure a very pure sample of the red mercury powder, Lavoisier went back to his laboratory, lit the fire of his furnace, and repeated the experiments of the minister.

Now Priestley was back in England, little dreaming to what his meeting with Lavoisier was to lead. To Priestley the atmosphere was no longer a simple elementary substance. The riddle of the air was already on the threshold of solution when Priestley was born. The Chinese, many centuries before, had written of "yin," the active component of the air which combined with sulfur and some metals. Leonardo da Vinci, that versatile genius of Italy, had been convinced back in the fifteenth century of two substances in the air. Others, too, had caught faint glimpses of the true nature of the atmosphere. Yet it was Priestley who, by the magic of chemistry, called up invisible oxygen from the air and first solved, by his discovery of this most abundant element of the earth, the profound enigma of the atmosphere. This puzzle, so simple today that few cannot answer it, so important that its mystery impeded the progress of chemistry for centuries, was finally solved by this man who typifies the intellectual energy of his century. To this heretic of the church, chemistry was but a hobby, a plaything that filled the spare moments of his varied life. Out of this almost juvenile pursuit came the unravelling of one of the world's great mysteries. Priestley's discovery of oxygen marked a turning point in the history of chemistry.

On August 1, 1874, there was celebrated in Birmingham, England, the centennial of this great discovery. A statue of Priestley was to be unveiled. Three thousand miles away, in America, a cablegram was dispatched by a group of American chemists gathered in a little graveyard in Northumberland, Pennsylvania, overlooking the north branch of the Susquehanna River. Dr. Joseph Priestley, a great-grandson of the English scientist, was present to witness the ceremonies in honor of his illustrious ancestor. For Priestley had been buried in America.

He had come to the New World when conditions in England became unbearable for him. The press had attacked him, and Edmund Burke had assailed him on the floor of the House of Commons for championing the cause of the French revolutionists. Finally, when his scientific friends began to snub him, Priestley though past sixty, decided to come to America.

Priestley and his wife left England in April and spent most of their two months at sea suffering from seasickness. In June they landed in New York. While they were at sea the great Lavoisier was guillotined. Priestley's landing in New York was like the arrival of a conquering hero. His fame as theologian, scientist, and liberal had spread to the Colonies. Governor George Clinton and Dr. Samuel L. Mitchill, professor of chemistry at Columbia University and a former pupil of the celebrated Dr. Black, of Edinburgh, were among the distinguished citizens who met him at the pier. The Tammany Society of New York, "a numerous body of freemen who associate to cultivate among them the love of liberty," sent a committee to express their pleasure and congratulations on his safe arrival in this country. "Our venerable ancestors," they told him, "escaped as you have done from the persecution of intolerance, bigotry, and despotism. You have fled from the rude arm of violence, from the flames of bigotry, from the rod of lawless power, and you shall find refuge in the bosom of freedom, of peace, and of Americans."

When Priestley left for America on the *Sansom* on the 7th of April, 1794, with one hundred others there were many Englishmen who realized their country's loss. The Rev. Robert Garnham expressed this misfortune in verse:

> The savage, slavish Britain now no more
> Deserves this patriot's steps to print her shore.

Despots, and leagues, and armies overthrown,
France would exult to claim her for her own.
Yet no! America, whose soul aspires
To warm her sons with Europe's brightest fires,
Whose virtue, science, scorns a second prize,
Asks and obtains our Priestley from the skies.

Three young men of the University of Cambridge presented him with a silver inkstand with the inscription, "To Joseph Priestley LLD on his departure into Exile, from a few members of the University of Cambridge, who regret that expression of their Esteem should be occasioned by the ingratitude of their Country." The Society of the United Irishmen of Dublin sent him a valedictory message: "Be cheerful, dear Sir, you are going to a happier world—the world of Washington and Franklin."

Here the *American Daily Advertiser* printed an editorial: "It must afford the most sincere gratification to every well wisher to the rights of man that the U.S.A., the land of freedom and independence, has become the asylum of the greatest characters of the present age, who have been persecuted in Europe merely because they have defended the rights of the enslaved nations."

America did more than greet this slender, active man with flattering phrases. The Unitarian Church offered him its ministry. The University of Pennsylvania was ready to make him professor of chemistry. Other offers of speaking tours and the like came to him. He accepted none. Benjamin Franklin had made great efforts to have him settle in Philadelphia, but Priestley preferred the serenity and wild seclusion of Northumberland, where his three sons and other English emigrants had attempted to found a settlement for the friends of liberty. The scheme had been abandoned, but Priestley's children stayed on. Here he hoped to find support for the Unitarian parish he hoped to establish. Here the amateur chemist built himself a home and a laboratory, and settled down to writing and experimenting. Thomas Jefferson came to consult him in regard to scientific matters, and to education and the founding of the University of Virginia. Occasionally he left Northumberland to attend the meetings of the American Philosophical Society at Philadelphia before which he read several scientific papers, or to take tea with George Washington, who had invited him to come at any time without ceremony. He also met the distinguished American astronomer, David Rittenhouse, and became a close friend of the celebrated Dr. Benjamin Rush, who attended him in 1796 when he was stricken with pleurisy.

Toward the end of 1797 Priestley's house and laboratory were completed, and before the close of the century he performed his greatest chemical experiment in America. Still working with gases, he passed steam over glowing charcoal and collected a small volume of a new gas, "burning with a lambent flame," now known as carbon monoxide. The discovery of this colorless gas explained for the first time the light blue flickering flame seen over a furnace fire. Today some of the gas used in our homes for cooking and heating is manufactured in essentially the same way originated by Priestley in 1799.

He continued to communicate with his friends of the Lunar Society to whom he sent accounts of his scientific discoveries. They in turn did not forget him, and, as late as 1801 Watt and Boulton presented him with "furnace and other apparatus for making large quantities of air."

Priestley was interested in a great deal more than science. Religious philosophy was his first interest and his major preoccupation. While in the United States he spent much time on his *Church History*. He completed the first three volumes and was working on the fourth. He lectured regularly in his home on theology and philosophy to a class of fourteen young men who adopted his Unitarian ideas. In July, 1796, he wrote to his friend Lindsey in England, "I do not know that I have more satisfaction from anything I ever did than from the lay Unitarian congregation I have been the means of establishing in Philadelphia."

He attempted to build a bridge between various sects of Christianity as well as one including Judaism. In 1787, while still in England, he had invited Jews to an "Amiable Discussion of the Evidences of Christianity." His purpose was not conversion but rather his concern with the restoration of the Jews to their ancestral homeland in Palestine. This invitation was taken up by David Levi, an Anglo-Jewish scholar who lived in London. Several letters were exchanged. Jefferson became interested in them but nothing came of Priestley's efforts.

When he went to Philadelphia for his fourth and last time he had his little volume *Socrates and Jesus Compared* printed there. He sent a copy to Jefferson who was delighted with it. Priestley wrote to some friends in England, "Jefferson is generally regarded as an unbeliever. If so, however, he cannot be far from us, and I hope in the way to be not only *almost* but *altogether* what we are."

Priestley's interest in politics never waned. In July, 1798, while John Adams was President, the Alien and Sedition Acts were passed by the Federalists. The Alien Act was introduced to get rid of foreign "agitators," and the Sedition Act was aimed at the Republicans who were the radicals of that day. Priestley was threatened as was his very close friend Thomas Cooper, chemist and political refugee from England who was agitating for social changes, insisting that "the right of exercising political power is derived solely from the people."

The following year Priestley published his *Letters to the Inhabitants of Northumberland and Vicinity* in which he again clarified his political views. President Adams was ready to prosecute Cooper for libel but did not want to expel Priestley from the country because he considered him harmless. Cooper was sent to jail for six months. Jefferson defended Priestley and when he took office as the third President of the United States on March 4, 1801, Priestley rejoiced, adding, "I trust that *Politics* will not make you forget what is due to *Science.*"

Priestley's long years of preaching and experimenting were now drawing to a close. Had he not been hampered by his deep-rooted belief in the phlogiston of Becher, his contributions in the field of chemistry would undoubtedly have been greater. Much that he discovered was not very clear to him, for he saw those things in the false light of the phlogiston theory. He had called an hypothesis a cheap commodity, yet Becher's hypothesis held him in its power, and clouded almost every great conclusion he had drawn. Across the sea a chemical revolution was taking place. Phlogiston as a working foundation was being annihilated. One by one its believers were forsaking it for a newer explanation born in the chemical balance. The great protagonists of science were gradually being won over to the new chemistry. Priestley alone, of the eminent chemists of the time, clung tenaciously to Becher. So thick-ribbed a believer was he in this theory that, when his health began to fail him, and he was no longer strong enough to light the fire in his laboratory, Priestley sat down in the quiet and tranquillity of his study to throw the last spear in defense of phlogiston. "As a friend of the weak," he wrote to Berthollet in France, "I have endeavored to give the doctrine of phlogiston a little assistance."

In this document, the last defense of phlogiston, Priestley honestly and courageously stated his beliefs. He was not altogether blind to the apparent weaknesses of the theory which he still championed. "The phlogistic theory," he wrote, "is not without its difficulties. The chief of them is that we are not able to ascertain the weight of phlogiston. But neither do any of us pretend to have weighed light or the element of heat." He had followed the fight very closely. Here in America his friends were helping in the destruction of the phlogiston hypothesis. Within the pages of Dr. Mitchill's *Medical Repository* many had discussed the fire principle. James Woodhouse, professor of chemistry at the University of Pennsylvania, Pierre Adet, French Minister to the United States and devotee of chemistry, and John MacClean of Princeton University, besides Dr. Mitchill and Priestley, had threshed out the matter in a friendly spirit.

Priestley felt keenly the overthrow of this doctrine. It had served men of science for a century and had pointed out a way. "The refutation of a fallacious hypothesis," he declared, "especially one that is so fundamental as this, cannot but be of great importance to the future progress of science. It is like taking down a false light which misleads the mariner, and removing a great obstacle in the path of knowledge. And there is not perhaps any example of a philosophical hypothesis more generally received or maintained by persons of greater eminence than this of the rejection of phlogiston. In this country I have not heard of a single advocate of phlogiston." And yet, in spite of this, he was not a mental hermit. He honestly believed in phlogiston—he had been brought up in it; yet he was open-minded. "Though I have endeavored to keep my eyes open, I may have overlooked some circumstances which have impressed the minds of others, and their sagacity," he added, "is at least equal to mine." His was not the stupid, obstinate clinging to an old hypothesis simply because it had been handed down. He sincerely believed in its truth. "Yet." he wrote, "I shall still be ready publicly to adopt those views of my opponents, if it appears to me they are able to support them."

Priestley was now past seventy. Mentally he was still very alert; physically his tired body was beginning to show signs of weakness. "I have lived a little beyond the usual term of human life," he told his friends. "Few persons, I believe, have enjoyed life more than I have. Tell Mr. Jefferson that I think myself happy to have lived so long under his excellent administration, and that I have a prospect of dying in it. It is, I am confident, the best on the face of the earth, and yet, I hope to rise to something more excellent still." Death did not crush him. A year after his arrival in America he had lost his son Henry, after only a few days' illness, and within a few months his wife, too, was taken from him. But he hoped soon to meet them again, for he awaited a real material return of Christ upon earth.

At eight o'clock, Monday morning, February 6, 1804, the old minister lay in bed knowing the end was very near. He called for three pamphlets on which he had lately been at work. Always a careful writer, clearly and distinctly he dictated several changes to be made before they were sent to the printer. He asked his secretary to repeat the instructions he had given him. The dying man was dissatisfied: "Sir, you have put it in your own language; I wish it to be in mine." He then repeated his instructions almost word for word, and when it was read to him again, he was contented. "That is right," he said, "I have done now." Half an hour later, at 11 A.M., he was dead.

He was buried in Northumberland. On a simple upright flat stone one can still read his epitaph:

> Return unto thy rest, O my soul, for the
> Lord hath dealt bountifully with thee.
> I will lay me down in peace and sleep till I
> wake in the morning of the resurrection.

Priestley's house in Northumberland still stands at Priestley Avenue and Hanover Street. His laboratory is at the north end and the kitchen at the southern end. Each is 22 feet by 22 feet. A small brick building on the grounds serves as a museum for Priestley's apparatus—flasks, gun barrels, glass tubes, vials, corks, bottles, balance, crucibles, pneumatic trough—chiefly the work of his own hands.

In 1956 title of the Priestley home passed from Pennsylvania State University to the Borough of Northumberland. Four years later the Pennsylvania Historical and Museum Commission assumed administration of the house. On August 1, 1974, chemists met on this site to celebrate Priestley's discovery of oxygen and the "Second Centennial of Chemistry." It was marked among other events by the unveiling of a plaque by the President of the American Chemical Society commemorating the 100th Anniversary of the Society's origin at the same site.

Among another collection at Dickinson College in Carlisle, Pennsylvania, presented by his lifelong friend Thomas Cooper, is a large compound burning-glass similar to the one with which he prepared the gas that has placed the name of Joseph Priestley among the immortals of chemistry.

V

CAVENDISH

A MILLIONAIRE MISANTHROPE TURNS
TO THE ELEMENTS

In 1366 King Edward III of England raised John de Cavendish to the exalted office of Lord Chief Justice of the King's Bench. Sir John could trace his ancestry back to Robert de Gernon, a famous Norman who aided William the Conqueror. This same Cavendish was later murdered for revenge, because his son was accused of slaying Wat Tyler, leader of an insurrection. Two centuries later the name of Cavendish was again glorified by the noted freebooter Thomas Cavendish, the second Englishman to circumnavigate the globe.

On October 10, 1731, at Nice, a son was born to Lady Anne Cavendish, who had gone to France in search of health. This Cavendish was not destined to wield power in public life, as his parents had hoped. Rather did he devote his long life to the cultivation of science purely for its own sake. In him the pioneer spirit was to push back the frontiers of chemical knowledge.

Here was a singular character who played with chemical apparatus and weighed the earth, while more than a million pounds deposited in his name in the Bank of England remained untouched. His bankers had been warned by this eccentric man not to come and plague him about his wealth, or he would immediately take it out of their hands.

Gripped by an almost insane interest in the secrets of nature, this man worked alone, giving not a moment's thought to his health or appearance. Those who could not understand the curiosity of this intellectual giant laughed at the richest man in England, who never owned but one suit of clothes at a time and continued to dress in the habiliments of a previous century, and shabby ones, to boot. This man could have led the normal life of an active nobleman. His family wanted him to enter politics, but instead he lived as a recluse, and devoted his life to scientific research. While other

55

natural philosophers wasted time and energy squabbling over the priority of this or that discovery, or arguing one theory or another, Cavendish could be found among his flasks and tubes, probing, experimenting, discovering—altogether unconcerned about the plaudits and honors of his contemporaries.

An immense fortune, inherited after he was forty, gave him that material independence so helpful to the research worker. A temperament that knew neither jealousy nor ambition gave him the freedom of mind so vital to the clear and unemotional consideration of theoretical problems. It is no wonder that he was able to accomplish so much in his long life.

A mind so free of dogma could not stand the strict religious tests applied to candidates for degrees at the universities. After spending four years at Cambridge, where he knew the poet Gray as a classmate, Cavendish left without taking a degree, and went to London.

Unlike Priestley, when the phlogiston theory began to crumble, he did not cling to it to the last, even though he did not openly accept the newer chemistry of Lavoisier, believing it at best "nearly as good" as phlogistonism. Elusive phlogiston still remained only a word, while all the natural philosophers of Europe and America went hunting for it in every school and private laboratory. When, in 1772, Priestley was being honored with a medal for his discovery of soda water, the President of the Royal Society, Sir John Pringle, remarked: "I must earnestly request you to continue those liberal and valuable inquiries. You will remember that fire, the great instrument of the chemist, is but little known even to themselves, and that it remains a query whether there be not a certain fluid which is the cause of this phenomenon." Here was the biggest single problem in chemistry. If this principle of fire could only be trapped —if it could be captured between the sealed walls of a bottle to be shown to every sceptical chemist, then Becher and his followers would be vindicated. To identify it with heat or light as Scheele and Macquer had done was not sufficient. It must be ponderable and possess all the other properties of real matter.

In the sixteenth century the Swiss medicine man, Theophrastus Paracelsus, had noticed bubbles of air rising from sulfuric acid when pieces of iron were thrown into it. He had also discovered that this gas could burn, but that was the limit of his investigation. Later Jan Van Helmont, a Flemish physician, made a similar observation, but he, too, neglected to continue the study of this gas.

Then came Cavendish, to whom the pursuit of truth in nature was a thing almost ordained. He, likewise, had noticed the evolution of a gas when zinc or iron was dropped into an acid. He went cautiously to work to investigate this phenomenon. He hated errors and half truths, and while the instruments which he constructed for his experiments were crudely fashioned, they were made accurately and painstakingly. This eccentric mortal, who could make the half mythical calendar of the Hindoos yield consistently numerical results, proposed to investigate this mysterious gas which burned with a light blue flame. Perhaps here he would find the key to phlogiston. Perhaps this gas was phlogiston itself!

He took a flask and poured sulfuric acid into it. Then into the acid he threw some bits of zinc. Through a cork which sealed the mouth of the flask, he attached a glass tube to the end of which a bladder was tied. Slowly at first, and then more rapidly, bubbles of a colorless gas began to rise from the surface of the metal to find their way into the bladder. Then, when the bladder was full, Cavendish sealed it and set it aside. He repeated this experiment, using iron instead of zinc, and again collected a bladderful of gas. Still another metal he tried—this time tin, and now a third bladder of gas was collected. Cavendish must make sure of his conclusions. He repeated these three experiments using hydrochloric acid instead of sulfuric, and three more sacs of gases were prepared.

The experimenter now brought a lighted taper to his six samples of gas. He watched each specimen of gas burn with the same pale blue flame. Strange that the same gas should be evolved in each case! What else could this inflammable air be, but that elusive phlogiston? For had not Becher taught that metals were compounds of phlogiston and some peculiar earths? Surely Cavendish had proved that the gas came, not from the acids or water in the bottles, but from the metals themselves! But he must not announce this until he had investigated further—it would not do to startle the world before he had made certain he was right.

With the crude instruments at his disposal, he passed the gases through drying tubes to free them of all moisture, and then he weighed the pure imprisoned "phlogiston." Though extremely light, he found it actually had weight. It was ponderable. He had nailed phlogiston itself! Now, at the age of thirty-five, he published an account of this work on *Factitious Airs* in the *Transactions of the Royal Society*.

Priestley, accepting these results, discussed them with the members of the Lunar Society and the "Lunatics," as they were called, agreed with him. Boulton especially was enthusiastic. "We have long talked of phlogiston," he declared, "without knowing what we talked about, but now that Dr. Priestley brought the matter to light we can pour that element out of one vessel into another. This Goddess of levity can be measured and weighed like other matter."

So immersed was Cavendish in the phlogiston of Becher that he did not know he had isolated, not the principle of fire, but pure, colorless, hydrogen gas.

When the daring Frenchman, Pilatre de Rozier, heard of this invisible combustible gas, he tried some queer experiments to startle the Parisians. He inhaled the gas until he filled his lungs, and then, as the gas issued from his mouth, set fire to it. Paris held its sides as it watched this Luciferous devil spitting fire. When, however, he endeavored to set fire in the same way to a mixture of this gas and common air, "the consequence was an explosion so dreadful that I imagined my teeth were all blown out," and he turned to other applications of the gas. Dr. Jacques Charles of Paris constructed the first large hydrogen-filled balloon, and in the presence of three hundred thousand spectators de Rozier bravely climbed inside the bag and started on the first aerial voyage in history.

There were many who would not accept this inflammable hydrogen as the real phlogiston. Even England's literary genius, Samuel Johnson, busied himself with chemical experiments—Boswell tells us: "a life-long interest." Now past sixty-three, he found running around London increasingly arduous. Boswell relates that he sent Mr. Peyton to Temple Bar with definite instructions: "You will there see a chemist's shop at which you will be pleased to buy for me an ounce of oil of vitriol, not spirits of vitriol. It will cost three halfpence." He, too, was going to investigate.

Cavendish now continued to pry into the problem of what really happens when a substance burns in the air. He was true scientist enough to consider what others had already done about this problem. He set feverishly to work to read some pamphlets.

In Dean Street, Soho Square, the quietness of which Dickens so well described in his *Tale of Two Cities,* Cavendish had filled a London mansion with his library, and during his long continued researches in the field of science he had occasion to refer to many of its volumes. Dressed as a gentleman of the previous half century, this shabby, awkward, nervous philosopher would come here to draw his books. His soiled, yet frilled shirt, his cocked hat, buckled shoes, and high coat collar pulled up over his neck, made this pernickety eccentric a ludicrous figure. Advancing towards the librarian, the fair-complexioned man would talk into space while asking for his books. He would sign a formal receipt for the volumes he was borrowing—this he insisted upon—and then walk slowly home, always taking the same path. He would thrust his walking stick in the same boot and always hang his hat on the same peg. He was a creature of habit, rigidly self-imposed, and seldom did he vary his daily routine.

Here was a lively account of an electrical machine which Pieter van Musschenbroek, a Dutch physicist, had accidentally discovered in 1746 while attempting to electrify water in a bottle. This Leyden jar, as it was called, produced sparks of electricity at the operator's will. It was a curious instrument and a powerful one whose shocks were claimed to work miraculous cures. It was shown to gaping crowds throughout rural England and on the Continent. Nine hundred monks at a monastery in Paris, formed in a single line linked to one another by iron wire, gave a sudden and tremendous jump as the discharge of this mighty device was sent through them. They would not take another shock for the Kingdom of France!

Cavendish was fascinated by such stories. He read also about Franklin's experiments with atmospheric electricity—how he had flown a kite in the summer of 1752 and felt the electric shock of the thunderstorm. This force must be a powerful weapon, thought Cavendish, for a year later Dr. Georg Richmann who tried the same experiment had been killed. Here was a potent instrument which the chemist might use to solve great mysteries.

He read in another pamphlet of an experiment performed about ten years after Franklin's. Giovanni Beccaria, an Italian, had passed some electric sparks through water, and had noticed a gas issuing from the water. But he missed discovering a great truth. Cavendish, the acute, saw something significant behind this ingenious experiment. He read on. The year which marked the beginning of the American Revolution witnessed an experiment by an Englishman, John Warltire. This natural philosopher who helped Priestley in the discovery of oxygen, was trying to determine whether heat had weight or not. In a closed three-pint copper flask, weighing about a pound, he mixed some common air and hydrogen, and set fire to the mixture by means of an electric spark. An explosion took place inside the flask, and, upon examination, Warltire detected a loss in weight of the gases, and incidentally the formation of some dew. Cavendish saw in this another clue to a great discovery which had just been missed by inches.

Now he came across another natural philosopher, Pierre Joseph Macquer, a scientist of the Jardin des Plantes, who described an experiment he had performed that same year. He, too, set fire to hydrogen in common air, and as the gas burned he placed a white porcelain saucer in the flame of the inflammable gas. The flame was accompanied by no smoke—the part of the saucer touched by the flame remained particularly white, "only it was *wetted by drops of a liquid like water,* which indeed appeared to be nothing else but pure water."

Cavendish heard from his friend Priestley, working away in his laboratory in Birmingham. On April 18, 1781, this preacher-scientist, using the spark of an electric machine, fired a mixture of common air and hydrogen in a closed thick glass vessel. He was working on a different problem at this time so that his observations were not very pertinent when he wrote, "Little is to be expected from the firing of inflammable air in comparison with the effects of gunpowder." Cavendish's suspicions became more and more confirmed.

The facts seemed to be as clear as daylight. He went to his bottles and his bladders, his gases and his electrical machine to probe a great secret. The way had been shown him—this fact Cavendish, like Priestley, never denied. He sought no fame in the pursuit of truth. Not that anything mattered to this misanthrope, yet he could not help peeping into nature's secrets. He was a machine, working to unfold hidden truths—not because they were useful to mankind, but because he delighted in the hunt.

Suddenly the voice of his housekeeper was heard through the door which separated his laboratory from the rest of the house. "I found your note on the hall table this morning, Sir. You have ordered one leg of mutton for dinner." "So I have," cried Cavendish gruffly. He was not to be disturbed. He had more important things to think about than his stomach. "But, Sir," ventured the maid, "some of your friends from the Royal Society are expected here for dinner." "Well, what of it?" stammered Cavendish. "But," she pleaded, "one leg of mutton will not be enough for five." "Well, then, get two legs," came the final reply. She dared not risk another question. She knew how strange and frugal was her master.

Cavendish was busy repeating the experiments of Warltire, Macquer, and Priestley. He performed them with greater skill and care, and with a clearer understanding of what was before him. He had cut down the underbrush and headed straight for his goal. Day after day, week after week, this "wisest of all rich men and richest of all wise men," hit nearer and nearer to his target. And as he worked, the solution of his problem grew clearer. He did not jump to hasty conclusions. Instead of common air, which his predecessors had used, Cavendish employed the newly discovered oxygen. He broke many a flask as he sparked this explosive mixture of oxygen and hydrogen. A great number of measurements and weighings had to be repeated. He had the patience of an unconquerable spirit. Had he not read of Boerhaave, the Dutchman whose fame as physician had spread so far that a Chinese mandarin seeking medical aid had sent a letter addressed: "Boerhaave, celebrated physician, Europe"? Boerhaave, in an endeavor to discover a chemical fact, had heated mercury in open vessels day and night for fifteen successive years. Cavendish could be just as persevering.

Here was an error in his figures which he had not noticed before. He must dry his gases to remove every trace of water. And there was another matter he had failed to take into account in measuring the volumes of his gases. He proceeded to change the volumes of his gases to conform to standard conditions. Where the ordinary experimenter detected one flaw, this recluse saw two and sometimes many more. As his calculations filled page after page, his results began to verify one another. Now, after more than ten years of labor, Cavendish was almost ready to make public his proofs. Had he not, like his contemporaries, delayed the publication of these results, he would not have started a controversy which lasted half a century.

Before March, 1783, he made known his experiments to Priestley. Then his friend Blagden was informed of his work, and the following June, Blagden notified Lavoisier. The year 1783 passed and Cavendish had not yet published the result of his work. He never displayed that keen desire to rush into print which so generally ensues an important discovery.[7] He was interested in experimentation—not publicity through publication. Not until the following January did he read his memoir on *Experiments on Air* before the Royal Society of England.

And this is what he told them: "Water consists of dephlogisticated air united with phlogiston." Translated into the language of modern chemistry, Cavendish informed his hearers that water was really a compound of two gases, hydrogen and oxygen, in the proportion of two volumes of hydrogen to one volume of oxygen. That clear, life-sustaining, limpid liquid was not the simple elementary substance all the savants of the world thought it to be. Not at all. The crowning wonder of chemistry had formed it out of two invisible gases.

What a startling announcement! Water a compound of two tasteless vapors! Where were his proofs? Cavendish told them quietly and without emotion. He had introduced into a glass cylinder, arranged so that its contents could be sparked without unsealing the vessel, four hundred and twenty-three measures of hydrogen gas and one thousand parts of common air. When they were sparked "all the hydrogen and about one-fifth of the common air lost their elasticity and condensed into a dew which lined the glass." Hydrogen and oxygen had combined to form pure potable water.

But how could he be sure that this dew was really water? They were certain to ask this question. He had to prove it for them. He collected very large volumes of the gases—500,000 grain measures of hydrogen and 1,250,000 grain measures of common air, and burned the mixture slowly. "The burnt air was made to pass through a glass cylinder, eight feet long and three-quarters of an inch in diameter. The two airs were conveyed slowly into this cylinder by separate copper pipes, passing through a brass plate which stopped up the end of the cylinder." He thus condensed "upwards of one hundred and thirty-five grains of water which had no taste or smell and left no sensible sediment when evaporated to dryness, neither did it yield any pungent smell during the evaporation. In short, it seemed pure water." Positive enough experiments—tests that were infallible, and yet Cavendish said "it seemed." He suspected his listeners would not be convinced. Water a compound of two gases—incredible!

Cavendish went further. "If it is only the oxygen of the common air which combines with the hydrogen," he argued, "there should be left behind in the cylinder four-fifths of the atmosphere, as a colorless gas in which mice die and wood will not burn." He tested the remnant of the air left in the cylinders and found that to be the case. The nitrogen gas was colorless and mephitic. He weighed all the gases and all the apparatus before and after sparking, and found that nothing had been added or lost. Only oxygen and twice its volume of hydrogen had disappeared, and in their place he always found water of the same weight.

To convince the sceptics, Cavendish varied his experiments once more. Now he used only pure gases, not common air but pure oxygen obtained, as Priestley had shown him, by heating the red powder of mercury. He took a glass globe (still preserved in the University of Manchester), holding 8800 grain measures, furnished with a brass stop-cock, and an apparatus for firing air by electricity. The globe was well exhausted by an air-pump, and then filled with a mixture of pure hydrogen and oxygen. Then the gases were fired by electricity as before. The same liquid water resulted and the same gases disappeared. Again he weighed the gases and their product as well as the glass globe, before and after combining them. Again the same remarkable result—two volumes of hydrogen always united with one volume of oxygen to form a weight of water equal to the weights of the gases. He had proved it conclusively.

A few years later Deiman and Paets van Troostwijk passed electric sparks from a frictional machine through water and decomposed it into hydrogen and oxygen. Fourcroy, in France, left burning 37,500 cubic inches of hydrogen and oxygen continuously for a week, and got nothing else but water. There could no longer be any question about the nature of water.

In the history of science is now recorded the story of a great controversy. It stands beside the great discussion of 1845 between the friends of John Adams, an Englishman, and Urbain Le Verrier, a young Frenchman, as to the real discoverer of Neptune, the ninth planet of our solar system, and the equally vehement controversy between the friends of Crawford Long and William Morton as to the true pioneer in the use of ether as an anesthetic.

Three men were claimants to the discovery of the composition of water. Two of them claimed the discovery for themselves, the third for Cavendish. Priestley, who too might have sought credit for this discovery, or who might at least have helped settle the discussion, remained for a time on the sidelines, watching the great verbal battle.

Two months after Cavendish read his paper to the Royal Society, Le Duc communicated the contents of this same discovery to James Watt, the inventor, who had likewise been interested in experiments on the nature of water. In consequence of this communication, Watt transmitted a report to the same Society, claiming its discovery as early as April of the preceding year. Lavoisier laid claim to its discovery on the basis of an oral report submitted in conjunction with Laplace to the French Academy in June, 1783. In this report he announced the composition of water without acknowledging any indebtedness to other scientists, even though he had by that time been informed by Blagden of the work of Cavendish.

Cavendish was not interested in such squabbles. When, in August, 1785, the shy, unsocial chemist visited Birmingham, where Watt was living, he met the Scotch engineer and spent some time with him discussing their researches. Watt, too, was not looking for notoriety, and while they said not a word about the priority of the discovery of water, both felt that Lavoisier might have been gracious enough to have acknowledged that his work on water was based on their previous work. Ten years later came Lavoisier's tragic end, and by 1819 the last of the figures directly concerned in the water controversy had died.

Another twenty years passed, and little was mentioned of this matter. Then Dominique Arago, celebrated astronomer and Secretary of the French Academy, came to England to gather material for a eulogy on James Watt. He made what seemed to him a thorough examination of the water controversy, and came to the conclusion that James Watt was the first to discover the composition of water, and that Cavendish had later learned of it from a letter written by Watt to Priestley. And while the principals of these wranglings lay in their graves, their friends started a turmoil which did not subside for ten years. The friends of Watt accused Cavendish of deliberate plagiarism. To vindicate Cavendish, the President of the British Association for the Advancement of Science published a lithographed facsimile of Cavendish's original notebook, and today the world gives credit for the discovery of the nature of water to him who sought this honor least.

The more Cavendish frowned upon fame the more fame wooed him. At twenty-nine he had been elected a Fellow of the Royal Society (F.R.S.) following in the footsteps of his father who had been honored with that society's Copley Medal for inventing the maximum and minimum thermometers. Every Thursday this awkward, gruff-speaking philosopher attended its meetings to keep in close touch with the progress of science. He seldom missed a meeting, and while he kept a good deal to himself, his ear was always cocked for new developments in science. He was appointed member of a committee to consider the best means of protecting a powder magazine against lightning, and the following year was placed in charge of a meteorological bureau which was to make and record daily observations of temperature, pressure, moisture and wind velocity around the building of the Royal Society.

Cavendish was even persuaded now and then to attend a soirée of the Society held at the home of its president, Sir Joseph Banks. He would be seen standing on the landing outside, wanting courage to open the door and face the people assembled, until the sound of stair-mounting footsteps forced him to go in. On one such occasion this tall, thin, timid man was seen in the center of a group of distinguished people. His eyes downcast, he was visibly nervous and uncomfortable. Suddenly he flew panic-stricken from the group and rushed out of the building. He had been talking with an acquaintance when John Ingenhousz, Dutch physician to Maria Theresa, appeared. Cavendish recognized this scientist by his queer habit of wearing a coat boasting buttons made of the recently discovered metal platinum. With Ingenhousz was a gentleman who had heard of Cavendish and wanted to be introduced to the illustrious philosopher. Cavendish was annoyed almost to frenzy, but managed to control his temper. But when the dignified Austrian visitor began to laud him as a famous and most distinguished man of science, then Cavendish, with a queer cry like that of a frightened animal, bolted from the room.

Cavendish had turned the family residence, a beautiful villa at Clapham, into a workshop and laboratory. The upper rooms became his astronomical laboratory, for he was interested in every phase of natural phenomena. On the spacious lawn he had built a large wooden stage which led to a very high tree. When he was sure not to be seen, he would climb this tree to make observations of the atmosphere. Often, in the dusk of the evening, Cavendish would walk down Nightingale Lane from Clapham Common to Wandsworth Common. He took this walk alone, rambling along in the middle of the road, performing queer antics with his walking stick, and uttering strange, subdued noises. Once when, to his utter horror, he was observed climbing over a stile by two ladies, he forsook that road forever, and thenceforth took his solitary walks long after sundown.

There is only one likeness of Cavendish in existence—a watercolor sketch which hangs in the British Museum. It was impossible to make him sit for his portrait. The painter Alexander had to sketch this one piece-meal while Cavendish was completely unaware that he was being memorialized.

Cavendish was a confirmed woman-hater. He never married—he could not even look at a woman. Returning home one day, he happened to meet a female servant with broom and pail on the stairway. So annoyed was he at seeing her that he immediately ordered a back staircase to be built. He had already dismissed a number of maids who had crossed his path in the house. On another occasion, he was sitting one evening with a group of natural philosophers at dinner, when there was a sudden rush to the windows overlooking the street. Cavendish, the scientist, was curious. He, too, walked over to gaze, as he expected, at some spectacular heavenly phenomenon. Pshaw! he grunted in disgust. It was only a pretty girl flirting from across the street!

Although a misanthrope, Cavendish was, strangely enough, charitable. His unworldliness made him an easy mark for unscrupulous beggars and borrowers, and he was even addicted to handing out blank checks. He naïvely believed every charity monger who accosted him. One of his librarians became ill, and Cavendish was approached for help—a hundred pounds would have more than sufficed. But Cavendish, too impatient to listen to the verbose details of the plea, asked if ten thousand pounds would do. It did!

As an experimenter Cavendish was superb—to him science was measurement. In 1781 he had collected, on sixty successive days, hundreds of samples of air, gathering them in all sorts of ingenious ways, and from as many different places as he could possibly reach. He subjected these samples to innumerable experiments, weighings, and calculations. He was repeating the work of Priestley and others, which was to lead him to the conclusion that the atmosphere had an almost uniform composition in spite of its complex nature. He was the first accurate analyst of the air. He had found air to contain twenty per cent of oxygen by firing it with pure hydrogen gas in a glass tube. During these experiments, a small quantity of an acid had found its way into the water in the eudiometer. He was not the first to detect this impurity; Priestley, Watt, and Lavoisier had all observed it, but they were at a loss to explain its formation. Cavendish, however, was not satisfied to leave this observation without a reasonable explanation. Again he showed his powers as an original researcher. By a series of carefully planned and skillfully executed experiments he tracked this minute quantity of acid to its source. He found it to be the result of a chemical reaction between the nitrogen and oxygen of the air, during the passage of the electric spark through the eudiometer. This he demonstrated privately to some friends. Nitrogen and oxygen had united to form oxides of nitrogen which Priestley had already prepared. This discovery was the basis of the first process used in the commercial fixation of nitrogen utilized in the manufacture of fertilizers and high explosives.

Cavendish determined to change all the nitrogen of the air into nitrous acid by repeated sparking of the air in an enclosed vessel. During these experiments he left records in his notebooks of the crowning achievement which stamped him as one of the outstanding scientific experimenters among the early chemists. It had taken a hundred years to discover a gas which Cavendish during these experiments had isolated from the air. What every investigator before him, and for a century after him, had either missed entirely or ignored, Cavendish noticed and recorded.

A hundred years of chemical progress passed. Lord Rayleigh and Sir William Ramsay, two of his compatriots, while searching for a suspected element in the air, turned over the pages of Cavendish's memoirs, at Dewar's suggestion, and read this statement: "I made an experiment to determine whether the whole or a given portion of the nitrogen of the atmosphere could be reduced to nitrous acid . . . Having condensed as much as I could of the nitrogen I absorbed the oxygen, after which *only a small bubble of air remained unabsorbed,* which certainly was not more than 1/120 of the bulk of nitrogen, so that if there is any part of the nitrogen of our atmosphere which differs from the rest, and cannot be reduced to nitrous acid, we may safely conclude that it is not more than 1/120 part of the whole."

Here was a clue to their search. They repeated the experiments of Cavendish and isolated a small volume of gas from the nitrogen of the air. They subjected it to every test for an unknown, and identified a new element. Small wonder that this colorless, odorless, insoluble gas would not form nitrous acid, as Cavendish had remarked. This idle gas, *argon*, was found to be incapable of combining with even the most active element. It was present in the atmosphere to the extent of one part in 107 by volume. Henry Cavendish had recorded one part in 120—remarkable accuracy in the light of a century of experimental advance.

From this clue came also the later discovery of three other inert elements of the air. From liquid argon, the same scientists separated new "neon," hidden "krypton," and "xenon" (the stranger) present to the extent of one part in eighty thousand, twenty million, and one hundred and seventy million parts of air respectively. With modern apparatus at his disposal it is not difficult to believe that Cavendish might have been the discoverer of these *noble gases* one hundred years before they were given to the world.

Cavendish's writings were rendered somewhat obscure by the verbiage of phlogiston. He knew no other chemical language. When the flood of the new chemistry began to rise in France, when the chemical revolution which followed the French Revolution began to question and destroy the beliefs in which he had been reared, Cavendish changed to a new field of scientific research. And while the world of science was set agog by the new developments in chemistry, Cavendish was busy measuring the force with which two large leaden balls attracted two small leaden balls. He was finding the weight of the earth. He would rather do this than be embroiled in the heat and fury of foolish discussions over new theories.

Cavendish left London on very rare occasions. He visited Sir Humphry Davy a number of times to watch him experiment on the alkalis in which he used some pieces of platinum which Cavendish had given him. During these meetings his conversation could not have proved very agreeable. The utterance of unnecessary words he regarded as criminal. Once, while staying in a hotel at Calais with his younger brother Frederick, whom he saw seldom, they happened to pass a room through the open door of which they could see a body laid out for burial. Henry was much attached to his brother, yet not a single word passed between them until the following

morning, when, on the road to Paris, the following lengthy conversation broke their silence:

Frederick to Henry: "Did you see the corpse?"

Henry to his brother: "I did."

This man never wasted a single word, spoken or written, on the
beauties of natural scenery, even through he had spent his whole life
engrossed in the study of nature. In the diary of his travels we may
come, with surprise, upon the following: "At——I observed——".
What?—a piece of sculpture or a beautiful sunset? No! only the
readings of a barometer or thermometer. He inherited from his
father an intense interest in mathematical measurements. On those
rare occasions when he travelled in his carriage, he attached to the
wheels an antique wooden instrument, called a "way-wiser," to
show him how far he was travelling. His biographer has summed up
his life thus: "Such was he in life, a wonderful piece of intellectual
clockwork, and as he lived by rule he died by it, predicting his
death as if it had been the eclipse of a great luminary."

One evening Cavendish returned as usual from the Royal Society
and went quietly to his study. He was ill, but this non-religious man
told no one. Soon growing worse, he rang the bell and summoned
his servant. "Mind what I say," he told him, "I am going to die.
When I am dead, *but not till then,* go to my brother, Frederick, and
tell him of the event. Go." An hour passed. Cavendish was growing
weaker. Again he rang for his valet. "Repeat to me what I have
ordered you to do," he demanded. This was done. "Give me the
lavender water. Go."

Another half-hour passed, and the servant, returning, found his
master a corpse. Thus passed England's great chemical luminary,
leaving part of his fortune to science, and his fame to be commem-
orated in the Cavendish Laboratory for Experimental Research at
Cambridge, where today other oracles are travelling the path he
helped illuminate.

VI

LAVOISIER

THE GUILLOTINE ROBS THE CHEMICAL BALANCE

DURING the frenzy of the French Revolution, when the King and Queen were guillotined for conspiracy against the liberty of the nation, and a dozen men sitting in the Palace of the Tuilleries were sending thousands to their death, a scientist was quietly working in a chemical laboratory in Paris.

This scientist was a marked man. He had given much of his energy and wealth to the service of France, but hatreds were bitter in those days and he had many enemies. Yet, while the streets of the city were seething with excitement, and his foes were planning to destroy him, he stood over his associate, Seguin, and slowly dictated notes to his young wife beside him.

Seguin was seated in a chair in the laboratory. He was hermetically enclosed in a varnished silk bag, rendered perfectly airtight except for a slit over his mouth left open for breathing. The edges of this hole were carefully cemented around his mouth with a mixture of pitch and turpentine. Everything emitted by the body of Seguin was to be retained in the silken bag except what escaped from his lungs during respiration. This respired air was passed into various flasks and bottles, finally to be subjected to an accurate and complete analysis. Whatever escaped from Seguin's body in the form of perspiration or other waste material was to remain sealed in the silken covering.

Lavoisier was investigating the processes of respiration and perspiration of the human bady. Weighings of Seguin, the silk bag, the inhaled air, and the respired air, and determinations of the gain in weight of the bag and loss in weight of his associate, were made on the most accurate balances in all France. Lavoisier trusted his scales implicitly. But these experiments were never to be completed by him. The door of his laboratory was pushed open with sudden

69

violence. A pompous leader, wearing the liberty cap of the revolutionists, entered the room, followed by the soldiers of the Revolutionary Tribunal and an uncontrollable mob.

Marat, member of the National Assembly and self-styled Friend
of the People, had attacked the scientist in bitter, dangerous terms:

> I denounce to you this master of charlatans, Monsieur Lavoisier,
> son of a rent collector, apprentice chemist, tax collector, steward of
> ammunition and saltpetre, administrator of discount funds, secretary
> to the King, member of the Academy of Sciences. Just think of it, this
> little gentleman enjoyed an income of forty thousand livres and has
> no other claim to public gratitude than to have put Paris in prison by
> intercepting the circulation of air through it by means of a wall
> which cost us poor people thirty-three million francs, and to have
> transferred the gunpowder from the Arsenal to the Bastille the night
> of the 12th or 13th of July, a devil's intrigue to get himself elected
> administrator of the Department of Paris. Would to heaven he had
> been hanged from the lamp post!

Lavoisier had offended this man years before. He had exposed
Marat as a very poor chemist when the latter had tried to gain
election to the Academy of Sciences. The future revolutionary had
struck back and denounced Lavoisier as "the putative father of all
the discoveries that are noised about, who having no ideas of his
own snatches at those of others, but having no ability to appreciate
them, rapidly abandons them and changes his theories as he does
his shoes." The learned societies of France had been suppressed for
harboring disloyal citizens. Even among his scientific collaborators
Lavoisier had enemies. Fourcroy and De Morveau, scientists and
members of the Assembly and Convention, loathed the old government, and Lavoisier, aristocrat and appointee of the King, became
an object of their hate.

Paris was ready to listen to such inflammatory words. The
conflict of the privileged classes and the third estate had culminated
in the Reign of Terror, during which a Committee of Public Safety
sent traitors, conspirators, and suspects to a quick doom. The deluge
had come. Lavoisier had been, until very recently, a member of the
Fermes Générales, a sort of Department of Internal Revenue made
up of aristocrats. It was essentially a financial company whose
members paid the government a nominal sum for the privilege of
collecting taxes which they themselves kept. They had been guilty
of outrageous abuses and were finally ordered disbanded.

As the document for his arrest was read, Lavoisier serenely and bravely made ready to obey the order. Saying goodbye to his wife, he entrusted his unfinished manuscript to Seguin and left his laboratory for the last time. In May, 1794, he was called by the Committee of Finance before the Revolutionary Tribunal. He was tried and falsely convicted on the grounds that he had plotted against the government by watering the soldiers' tobacco, and had appropriated revenue that belonged to the State. Others before him had been condemned for less. In spite of the petitions of his friends in the Bureau of Consultation, who reminded the judge of the greatness of this man of science, in spite of Lavoisier's years of unselfish devotion to his country, Coffinhal, president of the Tribunal, would not relent. "The Republic has no use for savants." The sentence was death, and no appeal could be taken. Carried in a cart to the Place de la Révolution, he and twenty-seven others were to be decapitated. The third to be executed was his father-in-law, and then the head of Lavoisier fell into the insatiable basket of the guillotine. "It took but a moment to cut off that head, though a hundred years perhaps will be required to produce another like it." This was the verdict of the great mathematician Lagrange, then living in Paris. Truer words were seldom uttered. Thus died France's great chemical revolutionist. His burial place has never been found for the body was lost in that mad upheaval.

Just a month before, Priestley had fled from the religious bigotry of England. His great work had already been done. But Lavoisier was cut off in the midst of productive investigations, and who can say what might have come from this genius? "Until it is realized that the gravest crime of the French Revolution was not the execution of the King, but of Lavoisier, there is no right measure of values; for Lavoisier was one of the three or four greatest men France has produced." This is the judgment of posterity.

The eighteenth century witnessed the efforts of other chemists besides Priestley and Cavendish. Hundreds were working with the flask, the crucible, and the balance. And while the great oracles of chemistry were discovering new truths or unmasking old errors, these lesser lights kept plodding away, building up a storehouse of chemical facts which soon cried out for order. Every bit of chemical information dug out of the fruitful mines of Europe's laboratories was put to the test of phlogiston. Phlogiston was the all-explaining touchstone. If this universal principle seemed unable to fit a new discovery into the structure of chemistry, then those ingenious creatures of the crucible could twist it into a form which would fit.

Scheele's chlorine, that yellowish greenish gas which both kills and purifies, and which the Swedish apothecary had torn out of muriatic acid, was explained by the phlogistonists as being oxymuriatic acid. Water was a compound of the phlogisticated air of Cavendish and the dephlogisticated air of Priestley. Rutherford's nitrogen was mephitic air devoid of phlogiston. The language of chemistry, too, was stagnant; it had not been revised or rejuvenated since the ancient days of alchemy, and its literature was filled with such barbarous expressions as phagadenic water, pomphlix, oil of tartar per deliquim, butter of antimony, calcothar and materia perlata of Kerkringius. Yet in spite of this confusion of terms and explanation, the facts kept piling up, waiting only for someone to dispel the mist that enshrouded and enveloped chemistry. It is truly remarkable that, working in such a wilderness, those early researchers were able to extricate so much of permanent value.

Lavoisier's appearance at this juncture was timely. Chemistry was in dire need of such a figure. Here was a man of influence whose voice was not lost. His were the words of power and position, not only in the councils of natural philosophers, where he had no peer, but also in the assemblies of politics, where he played a leading part. Lavoisier was heard, and science profited by the tactics of the publicity agent. Liebig said of him, "He discovered no new body, no new property, no natural phenomenon previously unknown. His immortal glory consists in this—he infused into the body of science a new spirit."

Lavoisier's mind was clear. He had been trained in mathematics and physics. Few possessed better foundations for the pursuit of the science of chemistry. His well-to-do parents had sent this imaginative boy to the Collège Mazarin, where at first he intended to study law. But he soon turned to science. He was greatly influenced by Guillaume Rouelle who held the position of "Demonstrator" at the Jardin des Plantes. For more than a century and a half it was the custom here for the Professor of Chemistry to lecture on the theories and principles of science. He performed no experiments and never soiled his fingers with chemicals. His realm was theory.

Bourdelain was Professor at the time. Concluding his discourse he would wind up with "Such, gentlemen, are the principles and the theory of this operation. The Demonstrator will now prove them to you by his experiments." And as Bourdelain stepped out of the room, Rouelle appeared, greeted with loud applause. Fashionable audiences came to listen to him. Lavoisier sat spellbound as Rouelle, instead of proving all the theory of the Professor, would, with his skillful experiments, destroy it. The young student never forgot how Rouelle one day became excited and waxed eloquent.

Removing his wig which he hung on a retort, and throwing off his waistcoat, he suddenly rushed out of the lecture hall, in search of some chemical apparatus, still absentmindedly continuing to lecture while out of sight and hearing of his audience.

On one of his scientific excursions Lavoisier met Linnæus, the great Swedish naturalist and botanist, who, too, captivated his interest. He definitely decided to devote his life to science.

Young Lavoisier's activities soon became so varied that he had scarcely time to eat. He started to write a drama, *La Nouvelle Héloise,* which was never completed. One full day each week he lived in his laboratory—never leaving it for a moment. Besides this he worked at his furnace every day from six to nine in the morning and from seven to ten at night. He would not allow himself the luxury of leisurely eating. To save time, he put himself on a bread and milk diet. One of his friends felt the need of warning Antoine. "I beseech you," he wrote, "to arrange your studies on the basis that one additional year on earth is of more value to you than a hundred years in the memory of man." Accompanying this letter was a package containing a bowl of thin, milky porridge. Lavoisier, however, did not adopt this suggestion. Before he was twenty-five, the French Academy of Sciences had already heard from him on such diverse subjects as the divining rod, hypnotism, and the construction of chairs for invalids. He soon gained recognition, and was elected a member of this body. Young as he was, he directed an active discussion about a wholesome drinking water supply for the city of Paris, and his practical mind led him to advocate fire hydrants as a protection against great conflagrations in crowded communities.

In the year following his admission to the Academy, Lavoisier became associated with the Fermes Générales, and made the acquaintance of Jacques Paulze de Chastenolles. Monsieur Paulze, member of the Fermes Générales, was an aristocrat at whose home gathered many men prominent in the social and political life of France—Turgot, Comptroller General of France; Laplace, greatest of French astronomers; Franklin, the American; Condorcet, mathematician and humanitarian; and Pierre Du Pont de Nemours, who later, marked for destruction, emigrated to America with his sons, to found the great industrial institution that still bears his name. To Paulze's home came also Antoine Laurent Lavoisier, young, good-looking, keen-minded, a good conversationalist, and eager to mix with the intellectual élite of France. Lavoisier soon became interested, not so much in the distinguished guests, but in a petite, blue-eyed brunette, the daughter of Paulze. Lovable little Marie Anne Pierretti became very fond of the tall, gray-eyed,

simple-mannered scientist. Her father noticed this and encouraged
the lovers. Antoine was eligible! Soon the busy man found time to
walk with Marie Anne, and he would talk to this fourteen-year-old
girl about love, and his career in the field of science. She understood. She was going to study English, Latin and even science so
that she could help him in his work. Besides, she had a talent for
drawing and they planned to have Marie do the drawings and
plates for his scientific memoirs. The courtship was a short one, and
when they were married that year they were given a beautiful home
at 17 Boulevard de la Madelaine with a salon over which Mme.
Lavoisier was to preside. It was a happy marriage, and Marie never
showed that violence of temper which she displayed years later
when she remarried. During a stormy domestic quarrel she is said to
have ordered her second husband, Count Rumford, out of the house
with the warning never to return.

Lavoisier's first research in chemistry was a simple analysis of
gypsum. Then this son of a wealthy Parisian merchant directed all
his skill toward an attack upon the old notion that water could be
converted into earth and rocks. Ever since Thales of Miletus,
worshiping the Nile, had attributed the origin of all things to water,
science had believed that water became stone and earth by evaporation. For twenty centuries this had been taught. Men had taken
flasks of water and heated them over fires until all the water had
boiled out. Inside the flasks they had found dull, earthy substances
which must have come from the water. Van Helmont had planted a
small willow tree weighing five pounds in a pot of two hundred
pounds of earth that had been thoroughly dried and weighed. He
had nourished the plant for fifteen years with nothing but water,
and the tree had increased in weight to one hundred and sixty-nine
pounds. The soil having in the meantime lost but two ounces, he
had "proved" that water had been converted into one hundred and
sixty-four pounds of solid material in the tree! Lavosier saw the
obvious fallacy of this demonstration.

"As the usefulness and accuracy of chemistry," he held, "depend
entirely upon the determination of the weights of the ingredients
and products, too much precision cannot be employed in this part of
the subject, and for this purpose we must be provided with good
instruments." Borrowing the most sensitive balance of the French
Mint, he weighed a round-bottomed flask which he had cleaned
until it glistened in the sunlight. Into this flask he poured a
measured volume of drinking water, which he distilled into another
carefully weighed flask. Just as he expected, a gray, earthy material
clung to the bottom of the empty flask. He weighed the flask and its

earthy impurity and subtracted from this the weight of the flask. He thus obtained the weight of the earth. He compared this weight of earth with the loss in weight sustained by the drinking water during distillation. The weights were identical! This earth must have come from the drinking water! But he had still to answer this question: Was this solid impurity which clung to the glass *dissolved* in the drinking water, or had the water *changed* into an earthy material?

He took a pelican, an alchemical flask shaped so that a boiling liquid would drop back again into the same flask. Into this pelican he poured a definite weight of pure sparkling rain-water and boiled the liquid over a low even fire. For one hundred consecutive days he distilled this rain-water, never allowing the fire beneath the flask to go out. When he finally stopped the distillation, he noticed a few specks of solid material floating in the water. They had not been there before. He weighed the pelican and its contents. There was no loss in weight. The distilled water, too, had remained constant in weight during the long boiling. Then he placed the pelican on a balance and found it had lost weight equal to that of the solid material in the flask. These seventeen grains of mud, he concluded, must have come from the glass of the pelican. There was no other explanation. The water itself had remained unchanged. Water could never be transmuted into earth. With the aid of his balance Lavoisier had destroyed another false heritage of antiquity.

Lavoisier was a careful worker with an idea at the back of his head which grew clearer as he read or repeated the experiments of his predecessors and contemporaries. Slowly he began to weed out the faulty explanations and weak theories that had crept into chemistry. Phlogiston did not fit into his scheme of chemistry. While the rest of Europe clung to it tenaciously he could see through it. To him it was a myth, an idle mischievous theory with neither foundation nor substance. There must be a simpler and more logical explanation of burning than Becher's phlogiston. With the coolness and dexterity of a skilled surgeon, he began to dissect the old idea. The creature was rotten to the core.

With scientific intuition he rejected this theory before he had thought of a substitute, but he was going to find an alternative. This practical Lavoisier who, at twenty-two, received a gold medal from the Academy of Sciences for working out the best method of lighting the streets of Paris; this same Lavoisier who, before submitting his essay, had worked for months on this problem, shutting himself up in a dark room for six weeks to render his eyes more sensitive to different lights; he was going to find the true explanation of burning! Phlogiston would not do.

He quickly dropped phlogiston and jumped to "caloric," or heat. Half a century before, the French Academy had offered a prize for an essay on the nature of heat. All the three winners favored a materialistic theory. It was not strange, therefore, that Lavoisier accepted the explanation that heat was a subtle fluid which penetrated the pores of all known substances. He frankly admitted, however, that he had no very clear conception of the real nature of this caloric. "Since there are no vessels which are capable of retaining it," he wrote, "we can only come at the knowledge of its properties by effects which are fleeting and difficultly ascertainable."

In avoiding the pitfall of one monstrosity, Lavoisier fell into the snare of caloric, the imbecile heir of phlogiston. It is difficult to explain this widespread acceptance of caloric. There were some, however, who recognized the evil kinship of phlogiston and caloric, among them Benjamin Thompson, the first great chemist of American birth. This adventurer had left Massachusetts to fight on the side of the English during our War for Independence. In Bavaria, as Count Rumford, he had a model law passed to put a stop to mendicancy. Problems of science also interested him. He made a study of foods and promptly tested his pet theories while feeding the troops of the Elector of Bavaria. While in charge of the military foundry in Munich, he bored through a cannon surrounded by a wooden box containing two gallons of water, which in two hours began to boil. The astonishment of the bystanders was indescribable. Water boiling without fire! He had transformed the mechanical force of a horse-driven boring machine into the energy of heat. To Count Rumford heat was a form of energy, the energy of particles of matter in motion as Newton and Lomonossov, a Russian, had held—not a ponderable fluid. He knew that caloric would soon perish. To a friend he wrote, "I am persuaded that I shall live a sufficiently long time to have the satisfaction of seeing caloric interred with phlogiston in the same tomb."

But caloric was not quite so vicious a theory. Here was the great difference between the myth of phlogiston and the fiction of caloric. Lavoisier did not depend upon caloric to explain the facts of chemical changes. His chemistry was not *based* upon vaporous caloric, while Becher's phlogiston was the actual foundation of the structure of chemistry. Lavoisier wanted to crush phlogiston. To appease those chemists who demanded a substitute, he gave them the comparatively harmless prescription of caloric. Believe in it or not, caloric would do no harm either way. It served as a vicarious palliative to save chemistry from the lethal dose of phlogiston.

But even Lavoisier was not satisfied with caloric as an explanation of burning. The phenomenon of burning still puzzled him. He was determined to solve it scientifically. Neither the fetish of phlogiston nor the belief in caloric was going to decide it. "We must trust in nothing but facts. These are presented to us by nature and cannot deceive. We ought in every instance to submit our reasoning to the test of experiment. It is in those things which we neither see or feel that it is especially necessary to guard against the extravagances of imagination which forever incline to step beyond the bounds of truth." Rich enough to secure the best in apparatus and chemicals, he spared neither wealth nor effort. As he worked, he kept building chemical structures in his mind, rejecting one after another as his furnace brought cogent objections.

Lavoisier worked tirelessly. He was bound to conquer the mystery of burning. After years of experimentation he reached a conclusion. He went to his desk and penned to the French Academy a memoir to be kept hidden and unread until he had completed further experiments. In this sealed note he wrote: "A week ago I discovered that sulfur on being heated gained weight. It is the same with phosphorus. This increase in weight comes from an immense quantity of air. I am persuaded that the increase in weight of metal calces is due to the same cause. Since this discovery seemed to be one of the most interesting which had been made since the time of Becher, I have felt it my duty to place this communication in the hands of the secretary of the Academy, to remain a secret until I can publish my experiments." Always shrewd, Lavoisier made sure that no one would snatch away from him the credit for the discovery of a great truth. By entrusting his secret memoir to the Academy he established his priority to the discovery of the nature of burning.

This was November 1, 1772. Priestley had not yet concentrated the heat of the sun's rays upon his red mercury; oxygen was still undiscovered. For three years more Lavoisier labored to unravel further the meaning of fire.

In October 1774, Priestley visited his fellow scientist in his laboratory in Paris, and gave him an account of his experiments on the preparation of oxygen. Macquer was present and helped to correct Priestley's imperfect French. Lavoisier, armed with this information, immediately performed his classic Twelve Day Experiment.

> I took a matrass (a glass retort), [he wrote] of about thirty-six cubic inches capacity, and having bent the neck so as to allow its being placed in the furnace in such a manner that the extremity of its neck

might be inserted under a bell glass placed in a trough of quicksilver, I introduced four ounces of pure mercury into the matrass. I lighted a fire in the furnace which I kept up almost continually during twelve days, so as to keep the quicksilver always almost at its boiling point. Nothing remarkable took place during the first day. On the second day, small red particles began to appear on the surface of the mercury: these during the four or five following days gradually increased in size and number, after which they ceased to increase in either respect. At the end of twelve days, I extinguished the fire.

He examined the air which was left in the matrass. It amounted to about five-sixths of its former bulk, and was no longer fit for respiration or combustion. Animals were suffocated in it in a few seconds, and it immediately extinguished a lighted taper. This remaining gas was, of course, nitrogen. He then took the forty-five grains of red powder which were formed, and heated them over a furnace. From these he collected about forty-one and a half grains of pure mercury and about eight cubic inches of a gas "greatly more capable of supporting both respiration and combustion than atmospherical air." He had prepared a pure gas which he later named oxygen or "acid former" thinking it to be a constituent of *all* acids.

Lavoisier came forward with an explanation of burning which completely rejected the old notion of phlogiston. That air was necessary for combustion and breathing was known. Leonardo da Vinci during the fifteenth century believed "fire destroyed without intermission the air which supports it and would produce a vacuum if other air did not come to supply it." Paracelsus back in 1535 wrote that "man dies like a fire when deprived of air." Robert Boyle, too, was "prone to suspect that there may be dispersed through the rest of the atmosphere some odd substance on whose account the air is so necessary to the subsistence of flame." But what function did this air play? Jean Rey had, years before, curiously explained that the increase in weight of a burning object came from the air "which has been condensed and rendered adhesive by the heat, which air mixes with the calces not otherwise than water makes sand heavy by moistening and adhering to the smallest of its grains." But no sensible scientist could accept such an explanation.

Lavoisier described this experiment to the French Academy a few months later, mentioning not a word of the work of Priestley. In a letter to his friend, Dr. Henry, written on the last day of that memorable year, the English minister felt that Lavoisier "ought to have acknowledged that my giving him an account of the air I had got from mercurius calcinatus led him to try what air it yielded, which he did presently after I left." It is difficult to explain this omission, for Lavoisier later acknowledged his indebtedness to Priestley for his work on the composition of nitric acid.

Lavoisier was the first to interpret the facts clearly. Burning, he said, was the union of the burning substance with oxygen, the name he gave to the dephlogisticated air discovered by Priestley. The product formed during burning weighed more than the original substance, by a weight equal to the weight of the air which combined with the burning body. Simple enough. No mysterious phlogiston, not even caloric—and the testimony of the most sensitive balances in Europe to support his reasoning.

Everything was accounted for by his three delicate balances. His most sensitive one, for weighing about a fifth of an ounce, was affected by the five-hundredth part of a grain. To Lavoisier, the balance was indispensable. It allowed nothing to escape his attention. "One may take it for granted that in every reaction there is an equal quantity of matter before and after the operation. Thus, since wort of grapes gives carbonic gas and alcohol, I can say wort of grapes equals carbonic acid and alcohol." All chemical changes obeyed the law of the indestructibility of matter. Likewise, in this chemical change of burning, nothing was gained or lost. Even the vaporous air was weighed and made to give consistent results. There was no intangible ghost mixed up in his explanation. Here was a new, unorthodox idea—an exposition that ushered in a revolution in chemical thought.

The world did not accept Lavoisier's explanation at once. But he kept on working. Emperor Francis I had heated three thousand dollars' worth of pure diamonds for twenty-four hours. The diamonds disappeared. They had volatilized, or changed to vapor, he thought. Lavoisier saw the error. He heated a diamond away from air and it lost no weight. But when he subjected it, inside a jar of oxygen, to the heat of the sun's rays, it disappeared and changed into carbon dioxide. Carbon had burned or oxidized into carbon dioxide gas. In the meantime, Cavendish had proved the composition of water. Lavoisier brilliantly repeated the work of this Englishman and introduced an ingenious experiment to verify the composition of water from the standpoint of his new theory of combustion. These experiments were conclusive. French scientists began to rally around him—Fourcroy, De Morveau, Berthollet, and others.

Outside France, opposition was still strong, especially in England where William Ford Stevenson F.R.S., in an exposé of the "deception" of Lavoisier, declared: "This arch-magician so far imposed upon our credulity as to persuade us that water, the most powerful natural antiphlogistic we possess is a compound of two gases, one of

which surpasses all other substances in its inflammability." Cavendish, discoverer of the composition of water, never accepted the new
explanation. As late as 1803 Priestley wrote from Pennsylvania, "I
should have greater pride in acknowledging myself convinced if I
saw reasons to be, than in victory, and shall surrender my arms with
pleasure. I trust that your political revolution will be more stable
than this chemical one."

Yet Lavoisier's contribution triumphed. In Edinburgh, Dr. Black
accepted his explanation and passed it on to his students. Italy and
Holland fell into line at about the same time. From Sweden,
Bergman wrote to Lavoisier offering him his support. The Berlin
Academy of Sciences, urged by Martin Klaproth, ratified Lavoisier's views in 1792. American scientists rallied to him almost to
a man. Even Russia endorsed the new system, for it boasted of
a forerunner of Lavoisier in the person of Michael Vasilievic
Lomonossov, vodka-loving poet and scientist who a generation back
had "conducted experiments in air-tight vessels to ascertain whether
the weight of a metal increased on account of the heat," and
"showed that without the admission of external air the weight of the
metal remained the same."

Then Lavoisier delivered a master stroke. He realized the importance of language to a science. In 1789, while the Bastille was
being stormed, he published his *Traité Elémentaire de Chimie,* which
helped destroy another citadel of error. This book was written in the
new language of chemistry. For the first time a textbook spoke the
language of the people. Lavoisier took chemistry away from the
mystics and the obscurantists, and gave its knowledge to every man
who would learn. Too long had this science been burdened and
obscured by cryptic words and pompous phrases. Uncouth and
barbarous terms were to be banished forever. Secret "terra foliata
tartari of Muller" became potash. The new nomenclature coupled
with a scientific explanation of the process of combustion gave
chemistry a new birth.

The new terminology had not sprung up overnight. As early as
1782, four men began to meet regularly in "the little Arsenal," the
chemical laboratory of Lavoisier on Rue Neuve-des-Bons Enfants,
in Paris. There were Guyton de Morveau, a lawyer who had come
to Paris to suggest the simplified nomenclature to Lavoisier;
Berthollet, personal instructor of chemistry to Napoleon; and
Antoine François Fourcroy, dramatist and relentless orator of the
Reign of Terror, all seated around Lavoisier. A herculean task was
before them. What a jumble of names, what a mess of alchemical
débris had to be sorted out and organized! Lavoisier spoke calmly to

his collaborators: "We must clean house thoroughly, for they have made use of an enigmatical language peculiar to themselves, which in general presents one meaning for the adepts and another meaning for the vulgar, and at the same time contains nothing that is rationally intelligible either for the one or for the other." "But," ventured the mild Berthollet, "there might be objections to a radical change." Some had raised the cry of ancestor worship. "The establishment of a new nomenclature in any science ought to be considered as high treason against our ancestors, as it is nothing else than an attempt to render their writings unintelligible, to annihilate their discoveries and to claim the whole as their own property." This accusation had come later from Dr. Thomas Thomson, who reproached the French scientists for their presumption in daring to change the language spoken and written by their masters. Others resented the effort to interfere with the "genius of the language." But Lavoisier answered, "Those who reproach us on this ground have forgotten that Bergman and Macquer urged us to make the reformation." De Morveau upheld his leader: "In a letter which the learned Professor of Upsala, M. Bergman, wrote a short time before he died he bids us spare no improper names; those who are learned will always be learned, and those who are ignorant will learn sooner."

The four kept working, and in May, 1787, a treatise on the new nomenclature of chemistry was proposed before the French Academy. In Ireland that odd chemist, Kirwan, lying on his belly on a hot summer's day before a blazing fire, and eating ham and milk, received the new language of chemistry with disdain. "So Lavoisier has substituted the word 'oxide' for the calx of a metal," he sneered. "I tell you it is preposterous. In pronouncing this word it cannot be distinguished from the 'hide of an ox.' How impossible! Why not use Oxat?" He refused to agree to the new changes "merely to gratify the indolence of beginners." But Lavoisier's views prevailed. Professor Thomas Hope, at the University of Edinburgh, soon after his arrival from Paris, was the first teacher to adopt the new nomenclature in his public lectures. Dr. Lyman Spalding, at Hanover, New Hampshire, published some chemical tracts in the new system, using the name "septon" for nitrogen and "septic acid" for nitric acid, on the principle that nitrogen was the basis of putrefaction.

During his lifetime, Lavoisier's name was known throughout France for his varied activities. He rivaled Franklin in his versatility. In 1778 he was named by King Louis XVI as a member of a committee to investigate the strange claims of a physician who had come to Paris from Vienna. This Dr. Friedrich Mesmer created a great deal of excitement by practicing what he called "animal magnetism." The King and Queen suspected a plot. Lavoisier and Benjamin Franklin, who was also on the committee, watched the new miracle man, at a séance, making passes over a patient and finally putting him into a trance. Mesmer then suggested to the sleeper a cure for his ailment, and, by some occult magnetic influence which passed from the doctor to the sick one, the patient was cured. Both Lavoisier and the American Ambassador vigorously denied that animal magnetism had anything to do with the trance whose reality, nevertheless, they admitted. And as the actuality of his cures remained unsettled Mesmer continued to attract disciples, among the most ardent of whom was young Lafayette.

At thirty-two, as comptroller of munitions, Lavoisier abolished the right of the State to search for saltpetre in the cellars of private houses, and by improving methods of manufacture, increased France's supply of this chemical. Later he was appointed to investigate new developments in the manufacture of ammunition. On October 27, 1788, accompanied by his wife, he went to the town of Essonnes to report on some experiments. When within a few hundred feet of the factory, they heard a terrific explosion. Rushing to the ruins Lavoisier found several mutilated bodies. He had missed death by moments. The experiments, nevertheless, were continued.

Although condemned as a "damned aristocrat," Lavoisier was by no means blind to the poverty and suffering of the lower classes. In spite of his being a staunch royalist, he urged reforms simply on humanitarian principles. He believed that in these reforms lay France's political salvation. Investigating conditions among the French farmers, he reported to the comptroller-general that "the unfortunate farmer groaned in his thatched cottage for lack of both representation and defenders." He realized they were being neglected, and tried to improve their economic status. At Fréchine, Lavoisier established a model farm, and taught improved methods of soil cultivation and other aspects of scientific farming. During a famine in 1788, he advanced his own money to buy barley for the towns of Blois and Romorantin. To avoid a recurrence of such suffering, he proposed a system of government life insurance for the poor. Blois remembered this act of kindness, and in December of that year sent him as its representative to the States General. Lavoisier, the humanitarian, also made a tour of inspection of the various prisons in Paris, and expressed his utter disgust at France's method of treating her criminals. The dungeons were foul, filthy, and damp—he recommended an immediate fumigation of all these pest-holes with hydrogen chloride gas, and the introduction of sanitation.

Today the undying fame of Lavoisier rests not upon these fleeting social palliatives, but upon the secure foundation of his explanation of burning, and the simplified chemical nomenclature we have inherited from him. Armed with these new weapons, men were equipped to storm other bulwarks of chemical obstruction.

VII

DALTON

A QUAKER BUILDS THE SMALLEST OF WORLDS

In May, 1834, there came to London from the city of Manchester, a tall, gaunt, awkward man of sixty-six years. He was dressed in Quaker costume: knee breeches, gray stockings, buckled shoes, white neckcloth, gold-topped walking stick. His friends had raised a subscription of two thousand pounds for a portrait statue of this world-famous natural philosopher. He had come to sit for Sir Francis Chantrey, the court sculptor, who was to mold his head in clay, and then model a life-sized statue to be placed in the hall of the Manchester Royal Institution. The clay model of the head of the venerable seer was soon completed. As Chantrey sat chatting with him, he carefully scrutinized his head, which looked so much like the head of Newton. He noticed that the ears of the philosopher were not both alike, while the model showed the two ears to be the same. In a moment the sculptor leaped to his feet, cut off the left ear of the bust, and proceeded to fashion another one. The old school-master-scientist was amused. How absurdly careful was this Chantrey!

Honors came pouring in on this scientist. The French Academy of Sciences elected him a corresponding member. He was made a Fellow of the Royal Society of England, and President of the Literary and Philosophical Society of Manchester. And now his friends wished to present him to the King, who, years before, had given a gold medal to be awarded to him for his great scientific contributions. Henry Brougham, the Lord Chancellor, offered to present him to His Majesty. But this could not be arranged without breaking the rules of the Court. John Dalton was a Quaker who still respected the tenets of his religion, even though forty years before, loving certain favorite airs, he had dared ask permission of the Society of Friends to use music under certain limitations. A Quaker could not wear court dress because this included the carrying of a

84

sword. A way was soon found out of the difficulty. The University of Oxford had recently conferred upon him an honorary degree. He could be properly introduced to the King in the scarlet robes of a Doctor of Laws. The old philosopher agreed. The part was carefully rehearsed. "But what of these robes?" someone pointed out. "They are scarlet, and no Quaker would wear such a colored garment." "You call it scarlet,"replied Dalton who was color-blind. "To me its color is that of nature—the color of green leaves."

The stage was all set for the momentous event. Dalton approached King William IV and kissed his hands. They stood chatting for a while. "Who the devil is that fellow whom the King keeps talking to so long?" someone asked. He had never seen John Dalton and had probably never heard of him, for Dalton led a very uneventful, contemplative life. From this studious existence, however, came one of the greatest contributions to chemistry—a contribution upon which much of the later chemistry rested.

Dalton, like Priestley, was the son of a poor English weaver. When only twelve years old he had already requested permission from the authorities of his native village of Eaglesfield to open a school. He had by this time studied mensuration, surveying, and navigation, and his scientific knowledge convinced the authorities of his competence. They remembered how, at the age of ten, he had astonished the farmers of his village by solving a problem they had discussed for hours in a hay field. He had proved to them that sixty square yards and sixty yards square were not the same. He was always solving mathematical problems for which he won many prizes. Like most boys, he would have preferred to do other things than teach, but his poor Quaker parents had five other children, and John had to help.

At first he opened his school in an old barn, and later held his classes in the meeting house of the Friends. Some of his pupils were boys and girls much older than he. He did not mind teaching them so long as he could find time after school to make weather observations. He had become deeply engrossed in the study of the atmosphere. What a hobby that was! He would rather mark down all sorts of weather observations in one of his innumerable notebooks, than hunt or fish or go swimming. He worked for hours at a time constructing crude thermometers, barometers, and even hygrometers. Between his duties as a schoolmaster and his work as farmer on his father's small patch of land, the boy found time to play with the atmosphere and dream of it.

As this lad grew older, he studied Latin, Greek, mathematics, and more natural philosophy. But his hobby of meteorology fascinated him most of all. When he was fifteen he left Eaglesfield for the village of Kendal to teach in the school which his brother Jonathan conducted. As he passed through Cockermouth, he saw an umbrella. He had never seen one before, except in prints of fine ladies and gentlemen. He bought the umbrella, feeling, as he said years later, that he was now to become a gentleman. At the Kendal school his authority was soon questioned. One of the older boys challenged the young schoolmaster to a fight in the graveyard. Dalton knew he was no match for this bully. He locked the ruffian in his room, and his classmates outside broke his windows in revenge.

In 1793, at the recommendation of his friend John Gough, a distinguished, blind, natural philosopher of Kendal, Dalton was invited to become tutor in mathematics and natural philosophy at Manchester College at an annual salary of eighty pounds. But he needed more time and freedom for his all-absorbing pursuit of aerology. At the close of the century, he resigned from the college to become a private tutor, earning his livelihood at two shillings a lesson. He might have gone on a lecture tour, but he knew he was a failure as a public lecturer. He had been convinced of this when at Kendal he had given twelve lectures on natural philosophy to the general public, charging one guinea for the entire course. These discourses included such fascinating subjects as astronomy and optics. But his deep, gruff, indistinct voice, his slow association of thoughts, his dry humor and unattractive appearance, could not draw a large audience, although he had announced that "subscribers to the whole course would have the liberty of requesting further information, also of proposing doubts or objections." Even the lure of a public forum in science had not made his lectures popular.

Dalton could devote more time now to his study of the atmosphere. He made scores of weather observations every day. Occasionally he was called away to other cities to tutor. His life became filled with such a passion for collecting data on the air that when he went to Edinburgh, London, Glasgow, or Birmingham, he never failed to spend most of his time making observations and recording results. When conditions permitted him to take a brief vacation, he travelled to the Lake District, where he added to his almost numberless records. He tramped through northern England, explored valleys, forded streams, climbed mountains, went sailing over the lakes, not for health or pleasure, but with a greater incentive— he was studying the atmosphere. He never forgot to carry his scientific apparatus. For forty-six consecutive years he kept records of the daily weather and atmospheric conditions, and there were few entries missing in this colossal record of more than two hundred thousand observations. Goethe, at sixty-eight, hearing of Dalton's passion for weather observations, took to this new science of meteorology, and made numerous cloud calendars.

Dalton never married. He said he had no time for such a luxury. Yet he enjoyed the society of beautiful and talented women. In Lancaster there was a family of Friends he never failed to visit when in the neighborhood. In writing to his brother Jonathan, who likewise remained a bachelor, John was not ashamed to admit his infatuation. "Next to Hannah," he declared, "her sister Ann takes it in my eyes before all others. She is a perfect model of personal beauty." He is even said to have composed verse to this lady. When he visited London in 1809 to attend a meeting of the Royal Society, he reported to his brother, "I see the belles of New Bond Street every day. I am more taken up with their faces than their dress. Some of the ladies seem to have their dresses so tight around them as a drum, others throw them round like a blanket. I do not know how it happens, but I fancy pretty women look well anyhow." A generous observation from a pious Quaker.

His only relaxation, besides his scientific excursions, was bowling. Every Thursday afternoon he went outside the town to the "Dog and Partridge" to indulge in a merry game of bowls. A few pence for each game were paid for the use of the green, and Dalton meticulously noted his gains and losses in his book. He could really never stop entering figures in notebooks.

As Dalton's observations on the atmosphere filled notebook after notebook, he began to wonder about a problem which no one had as yet made clear. He knew that the atmosphere was composed of four gases—oxygen, nitrogen, carbon dioxide, and water vapor. Priestley, Rutherford, Cavendish, and Lavoisier had proved that point. But how were these gases held together? Were they chemically united or were they merely mixed together, just as one mixes sand and clay? There were two theories. Berthollet believed the air to be an unstable chemical compound, while others considered it a physical mixture of gases.

Dalton's own observations led him to accept the idea that air was a mechanical mixture of gases. Yet the composition of the atmosphere was constant. His records proved that without question. He had analyzed the atmosphere taken from hundreds of different places in England—from the tops of mountains, over lakes, in valleys, in sparsely settled regions and in crowded cities. Yet the composition was the same. Gay-Lussac in France had ascended in a balloon filled with hydrogen to an altitude of 21,375 feet over Paris, and had collected samples of air at this height. And this air differed only very slightly from the air taken in the streets of the city. Why did not the heavier carbon dioxide gas settle to the bottom of the sea of air, covered in turn by the lighter oxygen, nitrogen, and water vapor? Had he not tried to mix oil and water and had not the lighter oil collected at the surface of the heavier water? Perhaps the currents of air and the constantly moving winds mixed the gases of the atmosphere and kept their composition uniform.

Dalton could not understand it. Had he gone to the laboratory, where the masters of chemistry had sought out the answers to other baffling question? He had tried, but his flasks had not helped. Dalton knew himself—he was not a careful experimenter. This problem had to be solved in the workshop of his brain.

Dalton had read Lavoisier's *Traité Élémentaire de Chimie.* The French chemist had suggested that the particles of a gas were separated from each other by an atmosphere of heat or caloric. "We may form an idea of this," he had written, "by supposing a vessel filled with small spherical leaden bullets among which a quantity of fine sand is poured. The balls are to the sand as the particles of bodies are with respect to the caloric; with this difference only, that the balls are supposed to touch each other, whereas the particles of bodies are not in contact, being retained at a small distance from each other by the caloric."

Perhaps diagrams would help. Dalton drew pictures—he was enough of the pedagogue to know how much a simple sketch had helped his students understand a hazy point.

This drawing represented the water vapor in the air.

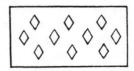

This would stand for the oxygen of the atmosphere.

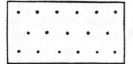 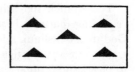

These dots were the nitrogen, and small black triangles designated the carbon dioxide of the atmosphere. Now he mixed these signs together, and drew a picture to represent how these gases were found present in the atmosphere, thus:

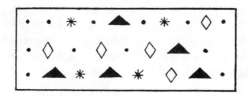

His pictorial mind began to see the particles of the different gases diffusing through each other, and thoroughly mixing, thus keeping the composition of the atmosphere uniform.

While he was interpreting this physical phenomenon of the diffusion of gases, a little word began to loom larger and clearer in his mind. He had come across that word in his readings. Kanada, the Hindu "atom eater," had centuries ago conceived matter to be discontinuous and made up of small eternal particles in perpetual motion. Leucippus, a famous scholar and teacher of Greece, had also speculated on the nature of matter twenty-four centuries before Dalton, and had concluded that everything consisted of tiny particles of various kinds, separated by space through which they travelled. Then Democritus, the "laughing philosopher" who in 500 B.C. declared, "We know nothing, not even if there is anything to know," developed his teacher's idea, and taught that matter was composed of empty space and an infinite number of invisible atoms, *i.e.*, small particles. "Why is water a liquid?" asked Democritus. "Because its atoms are smooth and round and can glide over each other." Not so with iron, however, whose atoms are very rough and hard. He constructed an entire system of atomism. Color was due to the figures of the atoms, sourness was produced by angular atoms, the body of man was composed of large sluggish atoms, the mind of small mobile atoms while the soul consisted of fine, smooth, round particles like those of fire. Even sight and hearing were explained in terms of atoms. Lucretius in his poem *De rerum natura* taught the same idea to the Romans.

The Manchester schoolmaster had also read of Newton's ideas regarding matter. "It seems probable to me," wrote Newton, "that God in the beginning formed matter in solid, massy, hard, impenetrable, movable particles . . . so very hard as never to wear or break to pieces; no ordinary power being able to divide what God himself made One, in the first creation." A beautiful idea, thought Dalton, but was this really true? He pondered over it constantly. Suddenly, after deep thought, the whole atomic theory was revealed to him. He did not wait for experimental verification. Like Galileo, he did not feel that experimental proof was always absolutely essential. Like Faraday, he possessed, to an extreme degree, a sense of physical reality. Dalton, cat on his knee, began to draw pictures of his atoms. Each atom was represented by a sphere, and since the atoms of the different elements were unlike, he varied the appearances of these tiny globes, as follows:

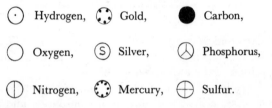

Dalton, like the ancient philosophers, could not actually see the particles which he pictured. Yet his atoms were only remotely akin to the atoms of antiquity. To Dalton, atoms were definite, concrete particles of matter, even though the most delicate instrument could not render them visible to the human eye. A hundred and forty years after Dalton formulated his Atomic Theory, the electron microscope was developed. This instrument revealed particles as small as one-four-hundredth of a millionth of an inch. Yet the atom was still hidden, even to this most sensitive eye. For even the largest atom is still smaller than the tiniest particle that even the electron microscope can reveal. The tiniest corpuscle that the Dutch lens grinder Leeuwenhoek beheld in a drop of saliva under his crude microscope was thousands of times larger than the biggest of Dalton's atoms. In every single drop of sea water there are fifty billion atoms of gold. One would have to distill two thousand tons of such water to get one single gram of gold.

And yet Dalton spoke and worked with atoms as if they were tangible. These atoms, he claimed, were indivisible—in the most violent chemical change, the atoms remained intact. Chemical change he pictured as a union of one or more atoms of one element with atoms of other elements. Thus, when mercury was heated in the air, one atom of mercury united with one atom of oxygen to form a compound particle of oxide of mercury. Billions of these particles finally appeared to the eye as a heap of red powder of mercury. He had some wooden spheres constructed by a Mr. Ewart, most of which were unfortunately lost to posterity. Some can still be seen in the Science Museum, London. These little spheres, one inch in diameter, he used for thirty years in teaching the Atomic Theory. He brought one ball representing an atom of mercury, in contact with another ball representing an atom of oxygen, and showed the formation of a particle of mercury oxide, thus:

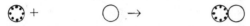

| 1 atom of silvery, liquid *mercury* | 1 atom of colorless, gaseous *oxygen* | 1 compound particle of red powder of *mercury oxide* |

Dalton asked himself another question. "Are all atoms alike in size and weight?" Here he made a distinct contribution which again stamped his theory as different from those of the ancients. Democritus had declared the atoms to be infinite in number and infinitely various in form. Dalton postulated that the atoms of the *same* element were all alike, but the atoms of *different* elements differed in both shape and weight. The weights of the atoms of each element, however, were always fixed, and never varied. Here was a bold statement. He had neither seen nor weighed an atom. Yet Dalton's Theory stood the test of almost two centuries of investigation, and today scientific evidence bears testimony to the truth of his conceptions.

In the meantime, a controversy was raging between two distinguished French chemists. Berthollet, the mild-mannered, believed that while chemical compounds showed *almost* constant composition, yet the proportions in which the elements had chemically combined were not *absolutely* rigid. Water, for example, had been proved to be a compound of oxygen and hydrogen. Berthollet insisted that, within moderate limits, the composition of this compound might vary. Usually 11.1% of hydrogen unite with 88.9% of oxygen to form water, but 11% of hydrogen might unite with 89% of oxygen, at another time, to form the same water. Strange, we might say, that so eminent a chemist as Berthollet could have championed such an absurdity. But we must remember it is still the eighteenth century, and chemistry is still in its swaddling clothes. Even at the close of the nineteenth century Ostwald, a Nobel Prize winner in chemistry, held similar views and supported Franz Wald, a Bohemian chemist and natural philosopher, who maintained that the composition of chemical compounds varied, depending upon their manner of production. Berthollet was scientist enough to experiment before making positive assertions. He made hundreds of analyses. His conclusions, as well as those of other experimenters, seemed to add force to his claims.

The contending scientist, Joseph Louis Proust, was at that time teaching chemistry in Spain. He had made numerous experiments to determine the proportions in which various compounds were formed, and had arrived at the conclusion that Berthollet was entirely mistaken. Proust repeated the experiments of his countryman. He used the purest of chemicals and the most accurate apparatus. He took every precaution to avoid error, and found mistakes in Berthollet's determinations. Besides, Berthollet had used substances like glass, alloys, and mixtures of various liquids, all of which were not true compounds. For eight years Proust tried to persuade the scientific world, and especially the followers of Berthollet, that when elements combined to form chemical compounds, the elements united in definite proportions by weight—a theory advanced as early as the fourteenth century by Jildaki, an alchemist of Cairo.

Never did this controversy become anything more than a courteous and brilliant, truth-seeking discussion. When Berthollet, discoverer of the use of chlorine as a bleaching agent, a discovery which he would not patent but gave to the world free, when this man Berthollet saw the error of his conclusions, he graciously withdrew his arguments and accepted the conclusions of Proust. And what a marvelous order had Proust found in nature! "The stones and soil beneath our feet, and the ponderable mountains, are not mere confused masses of matter; they are pervaded through their innermost constitution by the harmony of numbers."[8] Kepler, Galileo and Newton regarded nature as mathematical. Here was added testimony. The composition of every true compound never varies. This Law of Definite Composition remains a fundamental principle of the science of chemistry.

This law which, for the first time, made chemistry a mathematical science, was discovered while Dalton sat sketching figures of the atoms. Dalton's little spherical atoms could very neatly confirm this law. For, if the weight of the atom of every single element is constant, and this he had postulated in his theory, then the composition of all compounds must be definite, since all chemical union meant the combination of these minute unchangeable atoms. Here is carbon monoxide, composed of one atom of carbon and one atom

of oxygen: ●○ . And here is nitrous oxide made up of one

atom of nitrogen and one atom of oxygen: ①○ . And ⊙○

represented water, believed to contain one atom of hydrogen and one atom of oxygen. The composition of every one of these compounds must be constant, for if the eye could see beyond its limited range, it would witness single elementary atoms join hands, atom for atom, in definite combinations. How perfect, as Dalton showed, was the phenomenon of chemical union!

The ancients had speculated about the nature of matter and had written of atoms. But their atoms were not the building stones of the Manchester schoolmaster. Dalton's little figures were realities too small to be seen. He has left in his writings and charts evidences that they were, to him, concrete particles. He was not trammelled with mathematical acumen, experimental dexterity, or the wisdom of scholarly institutions. He discarded the accepted notions of scientists and contemplated nature as unbiased as a child. He would never have arrived at his immortal conception had he depended upon the results of his laboratory experiments—they were far too inaccurate. The discovery of his great generalization was based upon the imaginative boldness of a mature thinker, and the simplicity of a boy playing with a hobby.

Dalton had theorized that the atoms of the different elements had different weights. If he could only find out what these weights really were! He could not think of determining the weights of the individual atoms. They were so small and light that science was to wait a century and more before those actual weights could be determined. Decades had to pass before sufficient facts could be collected and more delicate instruments perfected, to solve this problem. But Dalton realized that chemists had to know at least the *relative* weights of the atoms, lest the progress of the science be impeded. Relative atomic weights—these he could determine.

In addition to Cavendish and Lavoisier a host of other workers had accumulated a mass of mathematical results. Wenzel had studied the effects of an acid like vinegar on a base like ammonia water. Later, Jeremias Richter found that they, like other acids and bases, combined in constant proportions and in 1794 he published his *Foundations of Stoichiometry or Art of Measuring the Chemical Elements.* It was from this hazy book that G. E. Fischer collected the data which enabled him to arrange a clear, simple table. This table was the key to Dalton's problem.

He must start with the lightest substance known—hydrogen gas. Its atomic weight he took as standard and called it one. Hence, the relative atomic weights of all the other elements must be greater than one. He knew that hydrogen and oxygen united in the ratio of one to seven by weight. Dalton believed that one atom of hydrogen united with one atom of oxygen to form water. Therefore, the relative weight of the atom of oxygen was seven. In this way he prepared the first table of relative atomic weights—a table of fourteen elements, which, though inaccurate, remains as a monument to this schoolmaster's foresight. The *Table of the Atomic Weights of the Elements—relative* atomic weights, to be sure—is today the cornerstone of chemical calculations.

While working on the relative weights of the atoms, Dalton noticed a curious mathematical simplicity. Carbon united with oxygen in the ratio of 3 to 4 to form carbon monoxide, that poisonous gas which is used as a fuel in the gas-range. Carbon also united with oxygen to form gaseous carbon dioxide in the ratio of 3 to 8. Why not 3 to 6, or 3 to 7? Why that number 8 which was a perfect multiple of 4? If that were the only example, Dalton would not have bothered his head. But he found a more striking instance among the oxides of nitrogen, which Cavendish and Davy had investigated. Here the same amount of nitrogen united with one, two and four parts of oxygen to form three distinct compounds. Why these numbers which again were multiples of each other? He had studied two other gases, ethylene and methane, and found that methane contained exactly twice as much hydrogen as ethylene. Why this mathematical simplicity?

Again Dalton made models with his atoms, and found the answer.

Carbon monoxide (CO) was ⬤◯ , while carbon dioxide (CO$_2$) was ⬤◯◯ . Nitrous oxide (N$_2$O) was ◐◐◯ , and ◐◯ was nitric oxide (NO), while nitrogen peroxide (NO$_2$) could be represented as ◐◯◯ . He had discovered

another fundamental law in chemistry! Berzelius later stated this law as follows: *In a series of compounds made up of the same elements, a simple ratio exists between the weights of one and the fixed weight of the other element.*

He wrote to Dalton to tell him that "this Law of Multiple Proportions was a mystery without the atomic hypothesis." Again Dalton's little spheres had clarified a basic truth.

On October 21, 1803, Dalton made before the Manchester Literary and Philosophical Society, of which he was secretary, his first public announcement of the relative weights of atoms. It excited the attention of natural philosophers at once. He was invited by the Royal Institution of London to lecture to a large and distinguished audience.

Dalton's atoms had started a heated discussion. His papers were translated into German. This infused a new spirit in him, and he continued to expand and clarify his theory. Then, in the spring of 1807, he made a lecture tour to Scotland to expound his theory of the atoms. Among his audience at Glasgow was Thomas Thomson, who was at work on a new textbook of chemistry. This Scotch chemist was impressed with Dalton's new conception of chemical union, and followed him to Manchester. As a result of a brief meeting, a few minutes' conversation and a short written memorandum, Thomson incorporated the Atomic Theory of Dalton in his textbook. The following year Dalton himself expounded his hypothesis in his own *New System of Chemical Philosophy*.

Before long the Atomic Theory, shuttlecock of metaphysicians for two thousand years, was finally brought to rest as an accepted and working hypothesis, eventually to be completely and experimentally proven. But not without a struggle. Those little circles of the Quaker schoolmaster were an abomination to many who would accept only what they could actually see and touch in the laboratory. They would have none of this fantastic dream.

Dr. William J. Mayo, the celebrated American surgeon, recalled that "my father was a student of Dalton and when my brother and I were small boys he told us much about this tall, gaunt, awkward scholar, and how little it was realized in his day that the atomic theory was more than the vagary of a scientist." It might be good enough for schoolboys who had to be amused in studying chemistry, or for natural philosophers who were never inside the chemist's sanctuary where the delicate balance and the glowing crucible told the whole truth. It was true that famous philosophers like Spinoza, Leibnitz, and Descartes had propounded similar ideas. But who listened to speculative philosophy? Lactantius, fifteen hundred years ago, had laughed at the idea of atoms. "Who has seen, felt, or even heard of these atoms?" he jeered. And now once more they sneered at the idea of atoms. Dalton was suffering from hallucinations, declared some. Instead of snakes he had visioned little spherical balls of atoms. How absurd! Was science again to be fettered by such scholasticism? What was this but confounded pictorial jugglery? Could any serious-minded chemist accept a theory just as baseless as the four elements of Aristotle that had chained men's minds for twenty centuries?

But here was a chemist, a natural scientist working in a chemical laboratory, who was bold enough to apply the idea of real atoms to chemical reactions. True, Thomson and William Hyde Wollaston, who confirmed the Law of Multiple Proportions experimentally, were ready to accept it. But Davy, England's most celebrated chemist, was bitterly hostile. He had been present at that meeting of the Royal Society when Dalton had first lectured about his atoms and had left the hall sceptical. But Dalton, an inveterate smoker, consoled himself. He could not see in young Davy any signs of a great natural philosopher, for, as he expressed himself, "Davy does not smoke."

Thomson had tried to convince Davy of the value of the theory. But Davy was adamant in his opposition, and caricatured Dalton's theory so skilfully that many were astonished "how any man of sense or science would be taken up with such a tissue of absurdities." Charles William Eliot, President of Harvard University, who began his career in the field of education as a teacher of chemistry, cautioned his students as late as 1868 that "the existence of atoms is itself an hypothesis and not a probable one. All dogmatic assertion upon it is to be regarded with distrust." Berthollet, too, was so sceptical of the atomic theory that as late as 1790 he still wrote the formula for water as if it were hydrogen peroxide—to him atoms were but fabrications of the mind. Wilhelm Ostwald, who did not hesitate to champion the unorthodox theories of many young chemical dreamers, wanted as recently as 1910 to do away completely with the atomic theory.

Some accepted the atomic theory with reservations. Fifty years after it was formulated, one eminent English scientist declared it "at best but a graceful, ingenious, and *in its place,* useful hypothesis." But the barriers against its acceptance were finally broken down. Even Davy was eventually converted to the abominable little atoms, and in 1818, when the Government was making ready to send Sir John Ross on a scientific exploration to the Polar regions, Davy wrote to Dalton, "It has occurred to me that if you find your engagements and your health such as to enable you to undertake the enterprise, no one will be so well qualified as yourself." Dalton appreciated this compliment, but had to refuse.

At about this time William Higgins, member of the Royal Dublin Society and F.R.S., wrote a pamphlet *Observations and Experiments on the Atomic Theory and Electrical Phenomena* in which he claimed that the atomic theory had been applied by him long before Dalton, in various abstruse researches, and that "its application by Mr. Dalton in a general and popular way gained it the name of Dalton's theory." Besides, he declared himself to be "the first who attempted to ascertain the relative weights of the ultimate particles of matter."

Here was another epic challenge in the chronicle of chemistry. Yet there was no element of attack in these statements—Higgins made no charge of plagiarism. He never even hinted at any evidence of piracy. For more than a decade he had modestly watched Dalton struggling to make the world accept his atoms, refusing to inject himself into the controversy until success had been assured for the Englishman.

Higgins was no trouble-maker. He was an eccentric Irishman of keen intellect—in fact, the first in Great Britain to see the fallacy of phlogiston. As early as 1789 when both he and Dalton were only twenty-three he had published *A Comparative View of the Phlogistic and Antiphlogistic Doctrine.* Herein he came to the defence of Lavoisier's new system of chemistry thereby daring estrangement from his irascible uncle Bryan Higgins who conducted the Greek Street School in Soho where Priestley often came for chemicals. The germ of the modern atomic theory appeared in this pamphlet. While he did not actually use the terminology of the atomic theory its method of reasoning was undoubtedly there. "Water," he wrote, "is composed of molecules formed by the union of a single particle of oxygen to a single ultimate particle of hydrogen."

Davy and Wollaston appreciated his pioneer work especially in his glimpsing of the Law of Multiple Proportions which he saw exemplified in the oxides of sulfur and of nitrogen even as Dalton years later discovered it in the oxides of carbon and nitrogen. Others, too, realized his greatness. Even Sir John Herschel in his *Familiar Lectures on Scientific Subjects* had Hermione ask: "Do tell me something about these atoms. It seems to have something to do with the atomic theory of Dalton." "Higgins, if you please," came Herschel's answer from the lips of Hermogenes.

Yet such is the not too uncommon fate of history that Higgins died in obscurity while Dalton rose to the heights of fame. In 1822 he visited Paris and received a great ovation. The most illustrious scientists of France paid homage to him. He met Laplace, seventy-three years old, who discussed with him his development of the nebular hypotheses of cosmogony. Venerable Berthollet walked arm in arm with him for the last time, for before many more weeks had passed, France's grand old man of science was dead. Cuvier, founder of the science of comparative anatomy, delighted him with his sparkling conversation. At the Arsenal, made famous by the work of Lavoisier, Dalton met Gay-Lussac, who told him of his balloon ascents. Thénard, who four years earlier had startled the scientific world with his discovery of hydrogen peroxide, amused the English schoolmaster with experiments on this strange liquid, the second compound of hydrogen and oxygen. Dalton never forgot this cordial reception by France's scientists.

Dalton's own country did not show the same reverence to this seer of Manchester. Though past sixty, he was still compelled to teach arithmetic to private students [9] in a small room of the Manchester Literary and Philosophical Society at 36 George Street. When, in 1833, his friends tried to get a pension for him from the Government, they were told by the Lord Chancellor that while he was "anxious to obtain some provision for him, it would be attended with great difficulty." Dalton's intimate friend, Dr. William Henry, made a last plea.

> It would surely be unworthy of a great nation, [he wrote] to be governed in awarding and encouraging genius by the narrow principle of a strict barter of advantages. With respect to great poets and great historians, no such parsimony has ever been exercised. They have been rewarded, and justly for the contributions they have cast into the treasure of our purely intellectual wealth. The most rigid advocate of retrenchment cannot object to the moderate provisions which shall exempt such a man in his old age from the irksome drudgery of elementary teaching. It is very desirable that the British government shall be spared the deep reproach which otherwise assuredly awaits it, of having treated with coolness and neglect one who has contributed so much to raise his country high among intellectual nations.

Lord Grey's government granted Dalton a yearly pension of a hundred and fifty pounds, later increased to three hundred. Yet he continued to teach and work in the field of science. In 1837, suffering from an attack of paralysis and unable to go to Liverpool where the British Association was meeting, he communicated a paper on the atmosphere—his first love. He had calculated that the assistance of plants in purifying the atmosphere, by absorbing carbon dioxide in starch making, was not necessary. He had figured that during the last five thousand years, animals had added only one-thousandth of one per cent of carbon dioxide to the air. When he was seventy-six, the Association met in Manchester, his home city, and Dalton was able to attend its meetings. He was still working in his laboratory. "I succeed in doing chemical experiments," he told them, "taking three or four times the usual time, and I am no longer quick in calculating."

Two years later he was still making weather observations. He made entries in his notebook of the readings of his barometer and thermometer for the morning of Friday, July 26, 1844. The figures were written in a weak, trembling hand. Over the entry "little rain" was a huge blot—he could not hold his pen firmly. This was his last entry. The next morning Dalton was dead, having passed away "without a struggle or a groan, and imperceptibly, as an infant sinks into sleep." Forty thousand people came to witness his funeral procession.

Dumas, the French savant, called theories "the crutches of science, to be thrown away at the proper time." Dalton lived to see his theories still held tenaciously by the natural philosophers of the world. For, "without it, chemistry would have continued to consist of a mass of heterogeneous observations and recipes for performing experiments, or for manufacturing metals." Dalton's Atomic Theory remains today one of the pillars of the edifice of chemistry—a monument to the genius of the modest Quaker of Manchester.

VIII

BERZELIUS

A SWEDE TEARS UP A PICTURE BOOK

ONLY the skilled adept could make sense out of the maze of strange pictures and symbols which filled the writings of the early chemistry. The alchemists had couched their ideas in an obscure sign language. Perhaps it did not require omniscience to understand that a group of dots arranged in a heap ⸫ represented sand. Maybe the connoisseur of wine knew that this symbol ⧺ meant alcohol. But who could guess that ◿ meant borax, and ◇ stood for soap, while glass was designated by two spheres joined by a bar ◯—◯ ? Clay, to be sure, must be ⎔ , and this strange sign ⊙ meant sea salt. Could ○ mean anything but a day, and its inverted image ○ a night? And what of these other strange markings which filled many a manuscript of ancient alchemy, and even found their way into current literature?

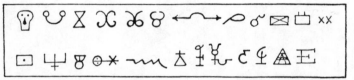

What a jumble of meaningless pictures!

The foundations of chemistry were now more or less completed. Phlogiston had been slain, and Lavoisier's theory of burning was safely established. De Morveau's new chemical nomenclature had been accepted, and Dalton had promulgated his atomic theory, which clearly explained two cornerstones of the structure of chemistry—the Laws of Constant Composition and Multiple Proportions.

But the bog of astrological and occult signs had to be cleared before an enduring edifice of chemistry could safely be raised. The muddle of arbitrary signs had to be destroyed and a more reasonable system substituted for it. The wild belief in alchemy had been scotched, but the serpent still lived, for its symbols still wriggled and twisted over the pages of chemical writings. No amateur could venture alone through its labyrinthine jungles. In one Italian manuscript of the early seventeenth century by Antonio Neri, the metal mercury was represented by no less than twenty symbols and thirty-five different names! In another book, lead was designated by fourteen symbols and sixteen names. Kunkel had rightly complained about this confusion. The old alchemists had tried to hide their pretence of knowledge in the secrets of confused hieroglyphics.

Something had to be done if chemistry was to become intelligible to everyone who wished to study it with reasonable diligence. At about the time that Priestley was discovering oxygen, Olaf Bergman of Upsala had attempted to solve the difficulty. But his figures were almost as barbarous. Still in awe of the ancient masters, he dared not forget them altogether. He continued to use for the metals the ancient symbols that had been handed down from Persia, India, and Egypt, through Greece and Rome to Europe. The number of common metals known to the ancients was seven. This was also the number of planets they had recognized and deified. The Chaldeans, believing that the metals grew by the influence of the planets, had assigned to each god and planet a metal. The Persians represented the revolution of the heavenly bodies by seven stairs leading up to seven gates—the first of lead, the second of tin, the third copper, the fourth iron, the fifth of a mixed metal, the sixth silver, and the last of gold.

To the Egyptians the circle was the symbol of divinity or perfection, hence it logically represented the sun. The circle was taken also as the symbol of gold, the perfect metal. The moon, seen as a crescent suspended in the sky, gave this planet and its metal silver

the symbol of the crescent $)\!)$. The scythe of Saturn, $\hbar$, dullest of

the gods, symbolized the character of this heavenly body, as well as

lead, dullest of the metals. $2\!\!\!\downarrow$, the thunderbolt of Jupiter, was the symbol of lustrous tin. The lance and shield of Mars, god of war, was represented by ☍ which stood appropriately for iron. The looking glass of Venus, pictured thus, ♀ , was also the symbol of copper, for Venus had risen full formed from the ocean foam on the shores of Cyprus, famous for its copper mines. Mercury, the speedy messenger of the gods, was pictured with the caduceus or wand ☿ .

Bergman clung to these old symbols and introduced a few others like:

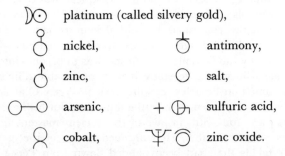

An attempt was made to change the ancient sign language. At the time when Lavoisier and his associates were reforming the nomenclature of chemistry, the Academy of Sciences at Paris selected Hassenfratz and Adet to improve the chemical ciphers. Their system, likewise, was too complex. They represented the metals as circles enclosing the Latin or Greek initials of the elements, thus: Ⓐ , Ⓟ , Ⓗ , Ⓢⓣ . Burnable bodies they represented by semicircles in four different positions, as follows:

⊃ hydrogen, ∪ sulfur, ⊂ carbon, and ⋂ phosphorus. Three short straight lines in different positions represented caloric, oxygen, and nitrogen. Compound substances whose compositions were still unknown were designated by squares standing on one point. By placing their symbols in different positions, pictures could be made for more than three hundred thousand different compounds, each consisting of three simple substances.

The result was again confusion. For not every student of chemistry was a draftsman. A simpler system was soon devised by a man whose work in the field of chemistry was so eminently successful, that for years he was respected as a lawgiver, a veritable autocrat of the chemical laboratory. In 1796 we find Berzelius at the University of Upsala in Sweden preparing for his medical degree. He was accustomed to hardship. For years this orphan boy had worked on his stepfather's farm, living in a room which, fortunately for him, was also the storehouse of a crop of potatoes. His mean, thrifty stepfather made sure that these potatoes would not freeze during the cold winter. So the warmth that protected them kept the boy, John, alive.

Four dollars and a pair of woolen stockings were his meager pay for his years of service. He had set out for the high school at Linköping near where he was born, dreaming of what he might become in ten years or so. Perhaps a clergyman. Had not his father, his grandfather, yes, and even his great-grandfather been clergymen, and why not he, John? But he was not to enter the ministry. At school he became interested in nature—more especially in the collection of flowers, insects and birds. He bought a gun and whenever the chance presented itself, he would forget the rigid rules of the school and steal off to hunt specimens of birds. His teacher had encouraged him in his love of natural history, which almost ended in disaster. To keep himself at school he managed, like other students at the gymnasium, to do some private teaching. He was tutoring the two sons of a widow, and in his zeal almost killed one of them with his gun while they were out hunting birds. Widow Elgerus complained to the rector, who instantly forbade him the use of the weapon. But John hated authority as much as he loved shooting. For using the gun on further occasions, he was almost expelled. Besides, he had cut some of his classes. During his last term he had been absent from his Hebrew classes a total of sixty-three hours! The rector did not forget, for when young Berzelius came up for his certificate of graduation he was warned that he was a young man of good abilities and doubtful ambition and had cut sixty-three hours. He would have to mend his ways if anything creditable was to become of him, for "he justified only doubtful hopes."

At the University he became interested in experimental chemistry. He had picked up the cheapest textbook he could buy— Girtanner's *Anfangs Grunde der Antiphlogistischer Chemie*, the first German book based on the antiphlogistic chemistry of Lavoisier, and had asked his teacher, Professor John Afzelius, for permission to work in the small laboratory in his spare time. Students were at liberty to work there only once a week, but that was not enough for Berzelius. He pleaded with Afzelius, who, as a test, tried to discourage him by ordering him to read several voluminous works on pharmacy. This would have checked the most ambitious college student. But Berzelius waded through the mass of involved preparations and hieroglyphics, and once more appealed to Afzelius.

"Do you know a laboratory from a kitchen?" laughed his teacher. Strange that he should have asked such a question, when his own laboratory was but a converted kitchen. Little did he realize that years later, when Berzelius was to do his classic work, his laboratory would also be his kitchen. Afzelius was adamant. "You may come only when the others work," he told him. But even school authorities were not going to stand in his way. He pleaded with the caretaker, even bribed him and soon found access to the laboratory through the back door when Afzelius was away. For some days John worked in secret excitement performing the textbook experiments and trying some of his own invention. Then one day he was caught. For a while, Afzelius stood in the darkness watching this boy carefully handling all kinds of chemical apparatus. Then he confronted the culprit. He rebuked Berzelius for daring to break the rules of the school. John made no answer. He was picturing expulsion. But Afzelius was only jesting. "Hereafter you must use the front entrance of the laboratory. And you may steal in even when I am looking."

But still Berzelius did not have enough freedom for his own work. He rented a student's room which boasted an adjoining windowless den with a fireplace. Here he spent some of the most exciting hours of his life. "One day," he wrote, "I was making fuming nitric acid and noticed some gas escaping. I collected it over water in bottles to find out what the gas was. I suspected oxygen, and seldom have I had a moment of such pure and heartfelt joy as when the glowing splint placed in the gas burst into flame and lighted up my dark laboratory."

Then, after a series of painstaking experiments, he prepared a paper on a peculiar gas called nitrous oxide. He presented it to his teacher, who shook his head and sent it first to the College of Medicine and then to the Academy of Science. To the disgust of Berzelius, the paper was refused, not because it was unworthy of an expert experimenter, but "because they did not approve the new chemical nomenclature" of Lavoisier which he had dared to use. Against such scientific inertia did Berzelius have to contend.

In the meantime, while Berzelius was completing his work at the University, Alessandro Volta, professor of physics at Pavia, invented a new machine for producing electricity. This invention like many others was the result of an accident. A few years before, his countryman, Aloisio Galvani, had left some dissected frogs hanging by a copper hook from an iron balcony. As the wind blew the bodies of the frogs against the iron the legs of the dead frogs contracted and wriggled. An almost indescribable phenomenon. Active muscular contractions from the limbs of dead frogs! Galvani was amazed. In his place, another man might have astonished the world with some mysterious explanation of life after death. But he did not explain the phenomenon on the basis of a resurrection. The days of the old alchemy were past.

He made an ingenious but erroneous explanation. Volta set to work to find the true cause of this "animal electricity." Slowly he came to believe that the electricity produced belonged to the metals and not to the frogs' legs. He proved it to the astonishment of the scientific world. Furthermore, he made good use of his discovery. By connecting a series of two dissimilar metals, zinc and silver, separated by a piece of cloth moistened in a solution of salt, he obtained a weak electric current. He joined a larger series of these metals and obtained a stronger flow of electricity. Volta had invented the "voltaic pile," forerunner of the modern storage battery.

Sir Joseph Banks, President of the Royal Society of England, received the first announcement of this discovery in a private letter (March 20, 1800) from Volta who was a Fellow of the Society. Before reading it to the Royal Society in June, Banks showed it to Sir Anthony Carlisle and William Nicholson. These men were not slow to grasp the immense possibilities of this new force. It could, perhaps, be used to disrupt hitherto unbreakable substances. They immediately sent the energy of a voltaic pile through water decomposing it into hydrogen and oxygen which formed at the two platinum poles of their electric machine. Fourcroy, friend and enemy of Lavoisier, built a large voltaic pile and ignited with it hitherto incombustible metals.

The imagination of Berzelius was at once kindled. Here was a mighty weapon for the chemist. He began to work with his oldest half-brother, Lars Ekmarck, on voltaic electricity. His thesis for his medical degree was on the action of electricity on organic bodies. The following year, with his friend, von Hisinger, he published a paper on the division of compounds by means of the voltaic pile, in which he propounded the theory that metals always went to the negative pole and non-metals to the positive pole of the electrical machine. Benjamin Franklin had introduced this idea of positive and negative electricity. He had called a body positively electrified when it could be repelled by a glass rod rubbed with silk.

The work of Berzelius, however, hardly caused a ripple in the chemical stream of progress. But four years later a young chemist in England, reading an account of his works and following them up, fired the imagination of the world. Benjamin Franklin had "disarmed the thunder of terrors and taught the fire of heaven to obey his voice," but now Humphry Davy, using Volta's electric pile and the research of Berzelius, isolated such new and strange elements as staggered men even more than the discovery of phosphorus a century before.

Potash and soda had been known to be compound in nature, but no method had been found to break them up into their component elements. Davy, in whose laboratory immortal Faraday washed bottles, built a powerful voltaic battery of copper, and in October, 1806, sent the energy of one hundred and fifty cells through some molten potash. He watched for a deep-seated decomposition. At the negative wire of platinum he soon saw globules of a silvery substance spontaneously take fire. "His joy knew no bounds, he began to dance, and it was some time before he could control himself to continue his experiments." He worked so hard that he soon became ill and all London prayed for his recovery.

Fashionable London received Davy's isolation of the metal potassium as another wonder of the world, and he was lionized. People paid twenty pounds to gain admittance to his lectures. The French Academy of Sciences awarded him a medal. Berzelius would have shared this prize had it been known that Davy's discoveries resulted from the previous work of the Swede. This was the statement of Vauquelin, discoverer of chromium.

Once before, Davy, son of a poor woodcarver of Penzance, had achieved overnight fame by his discovery of the physiological effects of laughing gas—that colorless nitrous oxide first obtained by Priestley in 1776 and later described by Berzelius to his teacher Afzelius. Distinguished people in all walks of life had come to London to inhale the gas which had raised Davy's pulse "upwards of twenty strokes and made him dance about the laboratory as a madman." Even the poet Coleridge was among those who came, but admitted that Davy's epic poem on the deliverance of the Israelites from Egypt had interested him more.

For a long time chlorine was considered to be compound in nature. Berzelius, too, believed this and disagreed with Davy, who considered it an element. Davy's illuminating experiments later convinced the Swede that chlorine was not "oxymuriatic acid," an oxygen compound of hydrochloric acid, but a simple elementary gas. When Anna, his housekeeper, complained that a dish she was cleaning "smelled of oxymuriatic acid," Berzelius now corrected her: "Listen, Anna, you must not say oxidized muriatic acid any more. Say chlorine, it is better."

After this controversy over chlorine Berzelius, now professor of chemistry, biology, and medicine at the University of Stockholm, was eager to meet Davy. However, he had previously received an invitation from Berthollet to visit Paris. While he was wavering between Paris and London war broke out between Sweden and Napoleon, and Berzelius travelled to England. He met Davy in his laboratory at the newly founded Royal Institution. They spoke about chlorine and the visitor complimented Davy on his important contributions.

He invited his visitor to his house the next morning. Berzelius was ushered into the dining room by the French butler. Davy made him wait there long enough to become fascinated by all its splendor and wealth—he was the husband of a wealthy widow. Then while they breakfasted, the Englishman and the Swedish scientist talked again about chemistry. Davy tried to impress upon his visitor his own eminence. At twenty-two he had been selected by Count Rumford as professor of chemistry at the Royal Institution. At thirty-three he had been knighted by the King. Fashionable London was at his feet. Berzelius, who was to be the teacher of kings and princes and the recipient of every honor that the chemical world had to offer, found such putting on of airs distasteful. Many years later, while travelling through Denmark and Sweden, Davy visited Berzelius, whom he considered "one of the great ornaments of the age." But a breach between the two chemists occurred soon after, due to the mischief of the secretary of the Royal Society, and they never saw each other again.

Before leaving for home, Berzelius bought much chemical apparatus, and made a trip to visit Sir William Herschel at Slough, where the erstwhile oboist and now celebrated astronomer showed him his great telescopes, whose mirrors he had stood grinding for hours with his own hands while his sister fed him. Then Berzelius visited Cambridge where he wrote, "It was with a feeling of reverence I visited the room where Newton made the greater part of his splendid discoveries." Later, at a luncheon, he spent "one of the most memorable days of my life" when he met, among other distinguished scientists of England, Thomas Young, the versatile genius who established the wave theory of light by his discovery of the interference of light. Upon his return, the King of Sweden appointed him Director of the newly established Academy of Agriculture.

Shortly afterwards, Berzelius accomplished one task which did much to make the road of chemical learning easier to travel. Quickly and decisively he abandoned the old sign language of chemistry and introduced in its place a rational system of chemical shorthand. "It is easier to write an abbreviated word than to draw a figure which has little analogy with words and which, to be legible, must be made of a larger size than our ordinary writing." This was the basis of the great change he had planned when the Swedish Government put him in charge of compiling the new Swedish Pharmacopoeia. "The chemical signs ought to be letters for the greater facility of writing, and not to disfigure a printed book. I shall therefore take for the chemical sign," he said, "the initial letter of the Latin name of each chemical element," thus:

Carbon	C	Oxygen	O
Hydrogen	H	Phosphorus	P
Nitrogen	N	Sulfur	S

"If the first two letters be common to two metals I shall use both the initial letter and the first letter they have not in common," as:

Gold (aurum)	Au	Silicon (silicum)	Si
Silver (argentum)	Ag	Antimony (stibium)	Sb
Copper (cuprum)	Cu	Tin (stannum)	Sn
Cobalt (cobaltum)	Co	Platinum	Pt
Potassium (kalium)	K (written *Po* for a while)		

A firm believer in the atomic theory of Dalton, Berzelius made his new symbols stand for the relative atomic weights of the atoms. The initial letter capitalized represented one atom of the element. These symbols stood for definite quantitative measurements and "enabled us to indicate without long periphrases the relative number of atoms of the different constituents present in each compound body." Thus they gave a clue to the chemical composition of substances. This was a tremendous step toward making chemistry a mathematical science.

True, William Higgins a generation before had introduced symbols, writing "I" for inflammable air of hydrogen, "D" for dephlogisticated air or oxygen, and "S" for sulfur. He had even suggested the use of equivalent weights of the elements (attractive forces, he called them) expressing the formula of water as "I $\dfrac{6\frac{5}{8}}{}$ D" where $6\dfrac{5}{8}$ represented the equivalent weight of oxygen to hydrogen. But his writings were unclear, his explanations hazy, and he never undertook to generalize his innovations.

Berzelius went further in his attempt to simplify the science. He joined the symbols of the elements to represent the simplest parts of compounds. Thus copper oxide was written CuO, and zinc sulfide ZnS. He had, at first, denoted the number of oxygen atoms by dots and the number of sulfur atoms by commas; thus carbon dioxide was $\overset{..}{C}$ and carbon disulfide was $\overset{..}{C}$. But he soon discarded these dots and commas, although for decades after, mineralogists utilized this method of writing the formulas of minerals.

Berzelius introduced the writing of algebraic exponents to designate more than one atom of an element present in a compound. These exponents were later changed by two German chemists, Liebig and Poggendorff, to *subscripts*. Subscripts are small numbers placed at the lower right corner of the symbols of substances where the atoms occur in the compound in numbers greater than one. Thus carbon dioxide, which contains one atom of carbon and two atoms of oxygen, is written CO_2.

These symbols and formulas were first introduced in 1814 in a table of atomic weights published in the *Annals of Philosophy*. Within a few years the literature of chemistry began to show a radical change. Dr. Edward Turner of Union College, London, in the fourth edition of his *Elements of Chemistry*, published in 1832, used these symbols with the apology that he "ventured to introduce chemical symbols as an organ of instruction." Instead of the hieroglyphics of the gold seekers, chemists used the simple system of Berzelius. And what a world of difference there was between the following symbolic language of Lavoisier:

and this translation of the above in the Berzelian system:

$$Fe + 2H_2O + 3O_2 + 4N_2O.$$

As with every great advance in science, there were objections. Dalton, himself, strangely enough thought his own picture-language superior. "Berzelius' symbols are horrifying," he wrote. "A young student might as soon learn Hebrew as make himself acquainted with them." He must have forgotten his own picture of alum which he represented thus:

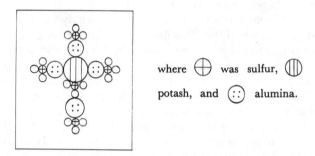

where ⊕ was sulfur, ⬭ potash, and ⊙ alumina.

The new system, however, stood the test of time. Not only were the original symbols of Berzelius accepted but they also formed the basis for the naming and writing of newly discovered elements and compounds.

And now Berzelius set himself a still greater task. While working on a new textbook, he came across the work of Richter on the proportions in which substances combine. This started him on an investigation of atomic weights. Dalton's relative weights of the atoms were both inaccurate and incomplete. The Swedish scientist realized that the chemist must have accurate relative weights if chemical manipulations were to become more than the guesswork of the old alchemists. He was going to find out the relative weights of all the different elements then known. "Without work of this kind," he declared, "no day could follow the morning dawn." At the same time he was ready to put all his indefatigable energy into the work of establishing Dalton's atomic theory by analyzing every chemical compound he could obtain.

Few would have attempted such a colossal task. Think of the time and the conditions under which he worked. In the reminiscences of his famous pupil, Woehler, is a description of the room in which Berzelius labored. "The laboratory consisted of two ordinary rooms with the very simplest arrangements; there were neither furnaces nor hoods, neither water system nor gas. Against the walls stood some closets with the chemicals, in the middle the mercury trough and the blast lamp table. Beside this was the sink consisting of a stone water holder with a stopcock and a pot standing under it. In the kitchen close by, in which Anna prepared the food, stood a small heating furnace." The chemicals he had to use in his analyses were often either unpurchasable or too impure to be used for accurate results.

Today, with every modern appliance and technique of science at his disposal, many an analyst would shrink from such a stupendous undertaking. But Berzelius did not even waver. He would weigh and measure until he had established the true relative weights of the atoms. To insure accuracy he purified every chemical reagent he used, not once but dozens of times. Even the best apparatus which the chemical world could offer him was still very crude when compared with that of today. As he had to construct his own apparatus in many cases, he took lessons from an itinerant Italian glassblower, Joshua Vacanno. He devised numerous novel instruments of precision. He invented new processes of purifying chemicals, and changed many practices of analysis then current.

Holidays, distractions, hobbies, even food meant very little to Berzelius during these months of toil. His was an indomitable spirit. Once, while he was attempting to recover gold from some fulminates, a violent explosion almost killed him. For a month he was forced to remain in a dark room to save his eyesight. When he finally emerged he went back to his laboratory.

A new observation always gave him great pleasure, and with beaming eyes he would call to his students, "Well, boys, I have found something interesting." When he had to use a platinum crucible he found there was only one in all Sweden. Fortunately von Hisinger was ready to lend it to him. Less lucky with other necessary pieces of apparatus, he had to do without them or invent some other method of analysis. For ten long years, in the midst of teaching and editorial work, he kept analyzing compound after compound, until he had studied the compositions of more than two thousand chemical substances. His blowpipe and balance, his eudiometer and crucible, finally gave him a set of atomic weights of the fifty different elements known to the scientific world of his time.

Compare his list of atomic weights with those of Dalton two decades before him and with the International Table of Atomic Weights of today, and marvel at the skill and accuracy of this giant among experimenters.

His numbers were not entirely accepted at first. The British Association for the Advancement of Science was sceptical. Turner was asked to verify the figures and found them to be correct. Later Jean Stas, a Belgian chemist, found an error in the atomic weight of carbon. Berzelius' whole list began to be questioned. Experiments were started in many of the chemical laboratories of Europe to find other errors, but the results only vindicated the experimental exactitude of Berzelius. Some have attempted to rob him of this glory by crediting the remarkable agreement of his figures with those of the present day to a fortunate balancing of experimental errors. The fact remains, however, that his Table of the Atomic Weights of the Elements still stands as a record of skilful manipulation and extraordinary perseverance.

	Dalton's Atomic Weights—1808	Berzelius' List 1826	International Table—1974
Chlorine	unknown	35.41	35.453
Copper	56	63.00	63.54
Hydrogen	1	1.00	1.008
Lead	95	207.12	207.19
Nitrogen	5	14.05	14.007
Oxygen	7	16.00	16.000
Potassium	unknown	39.19	39.102
Silver	100	108.12	107.870
Sulfur	13	32.18	32.064

In 1816 Gahn, an industrial chemist, though past seventy, persuaded Berzelius to join him in the purchase of a chemical factory at Gripsholm. In the course of this undertaking Berzelius discovered the element selenium while examining sulfuric acid. He did not remain here very long, for fire soon destroyed the factory. This was not his only industrial venture. He joined A. G. Werner, Professor of Mineralogy at Freyberg, in a mineral water business and later attempted, with borrowed capital, the commercial manufacture of vinegar. But he was no business man. His ventures all ended disastrously, and it took Berzelius ten years of tireless work to repay his enormous debts.

He undertook extensive editorial work[10] and worked long in his laboratory. Berzelius was as a rule cheerful, Woehler reports, and during his work "he used to relate all sorts of fun, and could laugh right heartily over a good story. If he was in bad humor and had red eyes, one knew that he had an attack of his periodic nervous headaches. He would shut himself up for days together, ate nothing and saw no one." Never-ending work, no relaxation, a life of solitude, had sapped his health. His pains in the head Berzelius curiously associated with the phases of the moon. He seemed to suffer most between eight in the morning and eight in the evening on the days of full and new moon. On one occasion, in Paris, he was invited to attend a dinner. Laplace, the astronomer, sceptical of the

Swedish scientist's association of headache with the moon's position in the heavens, had sent him the invitation as a test, believing that the Swede could not possibly know the day of the new moon in a foreign country. Yet Berzelius had to refuse the invitation because of a severe headache.

Before very long he was back in Stockholm, working again in his laboratory, and taking occasional excursions, until his health became so poor that he could not handle his apparatus. Berzelius was now the most eminent chemist in all the world. He was called upon to fill all kinds of honorary positions. An outbreak of cholera in Stockholm in 1834 found Berzelius chairman of a committee superintending the burial of the victims of the deadly epidemic. At five every morning, he was at the graveyard, until one morning a severe cold weakened him so that he lost all desire to live. He was aging, sick and terribly lonely.

Berzelius comes to his friend Count Trolle-Wachtmeister to talk about a subject which, until then, has seldom worried him. The Count listens tenderly to the old man and then advises him: "I suppose one can be quite happy without entering into the married state, but he who has never experienced the happiness of having a beloved wife by his side knows nothing of the finest side of life." Berzelius' eyes brighten and he asks a very personal question. "By a judicious choice it is not too late to enjoy this experience," is the answer. Berzelius is reassured but he must ask another question. "To be perfectly happy," replies his friend, "a man should have a *chez soi* and he ought not to look for it outside his own dwelling." Plain words and to the point.

Many years back, while he was still young and had just thrown himself into the fascinating work of a chemist, Berzelius had thought about marriage. One of his foreign friends, a scientist, was happily married. To him he had gone for advice. "Could a man divide his time between the strenuous work of the laboratory and the responsibilities of domestic life?" The answer he received had helped him to decide. "Although I am as happy as only the father of a family can be," his friend had told him, "I believe that if I were now unmarried, I should certainly not marry except under the influence of an unconquerable passion." Berzelius chose the path of whole-hearted devotion to science. Willingly he followed the lure of the laboratory.

But things were different now. His health was very poor and he was lonely. Count Trolle-Wachtmeister's words were balm to his aching soul. The aged chemist lost no time. He visited the town councillor, his old friend Poppius, who, he knew, had a daughter just twenty-four years old. Hesitating, and fearful of the answer, Berzelius asked him for the hand of his beautiful daughter. Great was his surprise when not only the parents showed delight, but the girl herself exhibited no displeasure at the thought. For was he not the most distinguished chemist of Europe? Students flocked to him from all over the world. Even kings came to him to learn his science. Sweden's King and Crown Prince were among his pupils, and the Czar and Prince of Russia came to visit him in his laboratory. A signal honor to be the wife of such a celebrity!

Encouraged in his suit, Berzelius must first regain his lost health. He visited Paris and was introduced to King Louis Philippe who talked to him for fully an hour, while the heir apparent, Ferdinand, Duke of Orleans, flattered him by saying that he had received his first lessons in chemistry from the pages of a French edition of Berzelius' *Lehrbuch der Chemie*. In Austria he met Metternich, and at Eger he received a luncheon invitation from the poet Goethe. More than fifty years back Goethe, as a lad of nineteen, had become interested in chemistry while at Leipzig, and had built himself a little blast furnace in which he labored to make alchemic gold and medicinal salts. He soon abandoned this futile pursuit to engage in more practical science. He bought a laboratory and undertook the analysis of well water. At thirty-five this poetic dreamer developed an interest in osteology. While at Jena in 1787, he was comparing human and animal skulls with his friend Lodi, when he hit upon the right track. "Eureka!" he cried, "I have found neither gold nor silver but the human intermaxillary jaw bone." This discovery stunned the world. This bone had heretofore been known to exist only in animals. It had distinguished man from the ape. What a flood of discussion was started by this dilettante in science! Once again science heard of him when, seven years later, he stood brazenly alone to attack the color theory of Newton. Single-handed he fought every hoary-bearded physicist of Europe with his own conception of color as a combination of light and shadow. He, too, had refused to accept the atoms of Dalton.

Goethe's interest in science was lifelong. The great poet and the great chemist Berzelius walked arm in arm to an extinct volcano whose origin and nature they discussed. Goethe had been interested in the phenomenon of volcanic eruptions and many years before had written a pamphlet on this subject, claiming that no lava would be found in the crater of the volcano. Berzelius believed that if they dug they would find lava. This turned out to be the case and the poet was pleased. Goethe asked him to stay another day, for he wanted to watch Berzelius work with his blowpipe. He marvelled at the skill of his guest, and expressed the regret that, at his age (he was now over seventy), it would be impossible to become an expert even with the help of Berzelius.

At fifty-six Berzelius, stout, middle-sized, of a pleasing personality and polished manners, married the eldest daughter of Poppius. Napoleon's former marshal, Bernadotte, then Charles XIV, King of Norway and Sweden, sent him a personal letter extolling his greatness. He was created a baron and elected member of the Upper Chamber of the Swedish Diet. In 1840 when he was sixty-one, the Diet voted him two thousand dollars as an annual pension. His marriage proved a happy one, and instead of forsaking the strenuous work of the chemist, as Davy had done, he continued to make important contributions from his laboratory. Although, toward the close of his life, some of his generalizations were discarded, yet in his two hundred and fifty papers covering every phase of chemical work he gave to posterity an abundance of facts from which the world long continued to gather rich harvests.

IX

AVOGADRO

THE SPIRIT OF A DEAD MAN LEADS A BATTLE

"The devil may write chemical textbooks," Berzelius had remarked, "because every few years the whole thing changes." Chemical books once more had to undergo a radical revision, for down in Turin, Italy, an unknown professor, thinking about the atom of John Dalton and certain behaviors of gases, had seen a great truth behind them all. His brilliant conception became the starting point of a new development in chemistry and, in a way, bears the same relation to chemistry that Newton's Law of Gravitation does to astronomy.

This hypothesis, fully explained, he had published in 1811 in a scientific magazine. Within the leaves of this *Journal de Physique* it remained for half a century while verbal controversies raged over dubious points, and the dust of the disturbances settled thickly upon the volume that held a hidden gem.

Had the scientific world seized upon this conception and understood its full meaning, chemistry might have been advanced by decades. But its formulator was an unknown, and he had introduced a new word which chemists were afraid to accept. Had not Dalton stated that the atom was the simplest particle of matter? Then why listen to the dreams of this Professor Lorenzo Romano Amedeo Carlo Avogadro di Queregna e di Cerreto who spoke of little masses of matter which he called *Molecules!*

With the passing of the years, atomic weights and formulas became more confusing than ever. Now and then the results of quantitative experiments seemed to belie the accepted atomic weights of the elements. To this teacher in Turin, however, everything was clear. His theory seemed to explain all the apparent inconsistencies. And it was so simple! No wild speculations. No involved formulas. No incredible dogmas. Amedeo, the beloved,

116

kept teaching his classes about his molecules. He was not a Berzelius, czar of all the chemical world, nor was he a crusader. He would not storm the scientific world to accept his theory. His students understood him clearly. That satisfied this unaspiring teacher. Others would find their way to his conception soon enough. So why not wait? And the chemical world waited.

In 1860 chemical science was in a turmoil. Berzelius, the venerable lawgiver of chemistry, was dead. Chaos prevailed. Karl Gerhardt, breaking with his father who wanted him to work in Strassburg at the manufacture of white lead, came to Paris and wrote a book explaining apparent inconsistencies in the established atomic weights. A radical young Frenchman, Auguste Laurent, assayer of the mint of Paris, attempted a new classification. Professor Petit, another Frenchman, dying at the age of twenty-nine, left a research on the specific heats and atomic weights of the elements which startled the scientific world. Atomic weights, molecular weights and equivalent weights of the elements were all in a muddle. Dalton had said one thing and Berzelius another. Whom was one to believe?

The cry for order was finally lifted above the clash of contending theories and explanations. In September, 1860, there gathered at Karlsruhe the lights of the chemical firmament. From all corners of the world they came. The crisis had to be met, and the clearest minds must attempt to settle it. From France came Béchamp and Wurtz, then professor of chemistry at the Sorbonne. Anderson, Frankland and Roscoe represented England. Germany, too, sent her chemical mentors—Liebig, Woehler, Mitscherlich, Erdmann, Erlenmeyer and Bunsen. Russia delegated her Mendeléeff. And from Italy journeyed one who had sat in the classroom of Karl Gerhardt listening to his new system of chemistry. He had come not to speak for Gerhardt, but to bring the message of his countryman, the great, modest teacher, Amedeo Avogadro, who four years before had passed away in Turin.

The hundred and forty chemists who had gathered fraternized first at a banquet laid in the large hall of the museum. Chemistry was forgotten for the moment, as they sat eating and drinking, and listening to the wit of that cosmopolitan assemblage. The next morning Boussingault, pioneer investigator of the composition of milk, was elected President for the day. The secretaries had prepared a number of questions for general discussion: Would it be judicious to establish a difference between the terms atom and molecule? Were atoms and molecules entirely different? Would it be judicious to designate by the term molecule the smallest quantity of a body capable of entering into chemical combination? Should the *compound atom* of Dalton be entirely suppressed?

Kekulé of benzene fame spoke first. He was willing to accept the term molecule as distinct from the atom but with reservations. He insisted upon a clear distinction between a *physical* molecule and a *chemical* molecule. Wurtz, Miller and Persoz joined in the discussion. Instead of clarifying the question, an interminable debate ensued in which divergent points of view further confused the animated controversy.

Then the bearded young Italian rose to add his voice to this momentous debate. In the eyes of Stanislao Cannizarro was the glint of a brave crusader. This man had led a colorful life. At fifteen he was studying medicine at the university of his native city of Palermo in Sicily. He then, like Berzelius, turned to chemistry, and went to the University of Pisa and from there to Naples.

Sicily was sizzling with rebellion. The government of the ruling Bourbons was oppressive. A wave of nationalism was sweeping over Italy, led by Mazzini, founder of the revolutionary society of "Young Italy." When Cannizarro heard of a political insurrection at home, he rushed from his laboratory to join the nationalist leaders in Sicily. A warm reception greeted this boy of twenty-one. He became an artillery officer at Messina and was chosen deputy to the newly organized Sicilian parliament. The revolution, however, proved abortive and Cannizarro was obliged to flee to France. His father had been Inspector General of Police in Sicily, and young Cannizarro knew more than the foreign officials thought safe.

From Marseilles he went to Paris and finally found his way to the little chemical laboratory of Michel Chevreul, the grand old man of chemistry who was to live through one hundred and three years of chemical progress. Cannizarro threw himself again into the work of chemistry with the same enthusiasm he had displayed as revolutionist. He never left his beakers and test tubes except to go occasionally to the College of France. Soon he accomplished a research of the first importance—the preparation of cyanamide.

When the political horizon had cleared in his own country, Cannizarro returned to teach. As professor of chemistry at the National College of Alessandria in northern Italy, he brought to this seat of learning the scientific knowledge of France, England and Germany. For the first time his students heard clearly about the odd atoms of the Manchester schoolmaster. Eagerly they listened to this charming teacher lecture about Avogadro's assumption, and of a new method of determining molecular weights. This was all new to them! They were fascinated by the clarity and vivacity of this firebrand. In his enthusiasm he would often forget to dismiss his class on time. When his students reminded him of the lateness of the hour by gently tapping with their feet on the floor, he would smilingly admonish them not "to use the language of beasts." How different he was from their old teachers! The city officials approached him about popular lectures for the general public. Cannizarro was willing, and his special talks on chemistry, citizenship and operas delighted his audiences.

Four years later Cannizarro occupied the chair of chemistry at the University of Genoa, birthplace of Mazzini. While here he married an English lady, Henrietta Withers. Again exciting news reached him. The Sicilians once more had rebelled against their king, Francis II of Naples. On May 11, 1860, one thousand Red Shirts, led by Garibaldi, landed at Marsala to do battle with twenty-four thousand royalist troops. It was an audacious invasion. Garibaldi, hero of Montevideo and candlemaker of New York, forced his way to Palermo and triumphed by sheer courage. Cannizarro was thrilled. He immediately joined a second expedition and landed safely in Sicily. But it was too late. The fighting was over—Sicily was free. He rushed to Palermo to meet his mother and sister whom he had not seen since the stormy year of 1849.

In the midst of these political struggles, Cannizarro received an invitation to attend the Congress of Karlsruhe. As scientist, too, Cannizarro was a champion of reform. He was eager to bring the message of the modest Turin professor who was now dead and forgotten. To him this was as important as the unification of Italy or the *Origin of Species* which Darwin had just published. In an impromptu speech of sparkling eloquence, this young orator of thirty-four combated the ideas of Kekulé and the other chemists who had not seen the light. The world knew little of molecules as conceived by Avogadro. The word had been used since the seventeenth century but it was synonymous with the atoms of the ancients. Scientists had spoken of atoms of hydrogen, an elementary substance, and atoms of water, a compound substance, without distinction. Later they had made some differentiation between the simple atom of oxygen and the compound atom of water. Dalton had been responsible for another blunder. He had used atom and molecule interchangeably.

Amedeo Avogadro had made a bold guess to explain certain facts of gaseous reactions. He had that piercing glance which sees through difficulties that others cannot penetrate. He had caught a glimpse of a truth and followed it through unafraid. A molecule, according to him, consisted not of a *single* atom but of *two* or more atoms chemically combined. Hence a molecule of certain gases was larger than an atom. Cannizarro had studied the work of his countryman. He was convinced of the validity of Avogadro's hypothesis. Perhaps Avogadro's theory had received scant attention because his molecules were not clearly understood. He used the term molecule either alone or with a qualifying adjective. His *integral* molecule meant the molecule of any compound, while his *constituent* molecule referred to the molecule of an elementary substance. Cannizarro clarified the whole situation.

The new molecule of Avogadro promised tremendous improvements in chemistry. Cannizarro knew that Avogadro's molecule would have to fight its way into the society of Dalton's atoms. His voice rang out clear. He made an heroic fight for what seemed a lost cause. He pleaded for the admission of the molecule into its honored place as a worthy peer of its tinier relation. He was certain that Avogadro belonged to the category of illustrious Italian scientists of the breed of Galileo, Torricelli, Volta and Spallanzani.

His audience was lukewarm. Just then Jean Baptiste Dumas of the French Institute appeared in the hall. Tumultuous applause greeted his entrance. Dumas, the dean of French chemists, thirty years back, was one of the very few who had read Avogadro's views, but saw in them only a vague promise, for the world was not ready for them. "We are yet far away," he had declared, "from the time in which molecular chemistry will be founded on sure rules. Let us hope for a near and lasting revolution in this direction." He had come unsuspectingly to witness this revolution. The assembly again applauded wildly as, escorted by the commissioner general, he took a seat near the President's chair. No such loud tribute, not even a single handclap, had been heard at the mention of Avogadro's name.

Another day of debating followed. Again endless discussion, conflicting opinions, and the Congress was ready to adjourn. Dumas thanked the general commissioner for the zeal with which he had organized the meetings, and begged Weltzien to express to the Grand Duke, who was absent from Karlsruhe, the thanks of the members. Then, without accepting any general principle for which purpose they had met, the chemists returned home.

But this conference was not a colossal failure. Two years before Cannizarro had come to Karlsruhe he had written a long letter to his friend, Professor S. de Luca. This letter was later published as an *Outline of a Course in the Philosophy of Chemistry,* based on the theory of Avogadro. Reprints of this *Sunto di un Corso di Filosofia Chimica* were distributed by Cannizarro at the close of the Congress. It had originally met with scarcely any notice, but Cannizarro had hopes that some might see the light. One of the younger chemists who attended the meeting, Lothar Meyer, put a copy of this pamphlet in his pocket. On his way home he read and reread it. "It was as though the scales fell from my eyes," he wrote. "Doubt vanished, and was replaced by a feeling of peaceful clarity." He, at least, had caught the missionary spirit of the Italian. Four years later he incorporated Avogadro's ideas in his *Modern Theories of Chemistry*, and in 1891 Cannizarro was awarded the Copley Medal of the Royal Society for this *Sunto*.

Odling, who had likewise heard the impassioned words of Cannizarro, included his atomic weights in a manual of chemistry but neglected even to mention Avogadro. Hermann Kopp, the leading historian of chemistry of the time, had not even heard of the Italian professor when he wrote his classic *History of Chemistry* fifteen years before the conference. Now, however, in a new edition he made mention of his contribution. Seldom had a man been so entirely neglected as was Amedeo Avogadro.

Here was a tool which unlocked many of the mysteries of atomic weights and formulas, yet its originator had been completely over-shadowed for more than fifty years. What kind of man could the creator of such a speculative theory be? Strangely enough, Amedeo was educated to follow not the imaginative ideas of idle dreamers but, like his father, Philippe, the practical logic of the legal profession. He received his baccalaureate degree in jurisprudence at the age of sixteen and his doctorate in ecclesiastical law at twenty. For three years he busily practiced law. Then suddenly he turned to natural science. This young dreamer with kind, expressive eyes was sick of petty squabbles and the frauds of the local law courts. There were finer ways to pass one's life. Deeply interested in chemistry, physics, mathematics and philosophy, he spent years in their study. He soon attracted attention when, with his brother Félice, he presented to the Academy of Sciences of Turin a memoir on a problem connected with the newly discovered galvanic current. In 1809, at the age of thirty-three, he was appointed professor of physics at the Royal College at Vercelli. When his classic paper on molecules was published in 1811 not a single scientist of the world commented upon it. Even the many-sided Berzelius never heard of Avogadro or his theory. Men were stirred in that year by the accidental discovery and isolation of the element *iodine* by Courtois. The molecules of the Italian professor were not even mentioned.

Avogadro kept teaching and experimenting. He could handle a flask and a balance as well as dream. He measured the increase in the volume of various liquids when they were heated. He studied capillary action, the tendency of liquids to rise in narrow, hair-like tubes. Other problems of physical chemistry occupied his attention outside the classroom. Then in 1820 King Victor Emmanuel I instituted a chair of mathematical physics at the University of Turin, and the versatile scholar Avogadro, who besides Greek and Latin knew German and English, was honored as its first occupant.

He did not hold this position very long. A bloody revolution against the foreign oppressors broke out in Naples. Before it was snuffed out, Piedmont, too, rose in rebellion and demanded a war against Austria. King Victor Emmanuel I abdicated rather than yield to the demands of the revolutionists. His brother Charles Felix succeeded him. The new king was a despot. The uprisings were suppressed and the doors of the University of Turin were closed. Avogadro himself took no part in these stormy affairs. He would not be disturbed. He was given a meager annual pension of about a hundred dollars and the honorary title of Emeritus Professor. But Avogadro could not remain idle. He resumed the practice of law and continued his study of the sciences. Ten years later, Charles Felix died and Charles Albert succeeded him. Mazzini pleaded for greater freedom. The new king was more liberal and was ready to listen to Young Italy. The University of Turin was reopened, the chair of science restored, and Avogadro was reinstated. He continued to teach for another twenty years. At the age of seventy-four he was finally retired and spent the remaining six years of his life in unhampered study and meditation.

Avogadro had married a lady of Biella by whom he had six sons. Luigi rose to the position of general in the Italian army, and another son, Félice, became president of the Court of Appeals. The quiet professor found time to take a fairly active part in the life of his community. Like Lavoisier, he held many offices dealing with public instruction, meteorology, weights and measures, and national statistics. Unlike Lavoisier, however, he lived a very peaceful, uneventful life, until death called him away at eighty. Not a word of eulogy was pronounced at his simple bier. Only brief obituaries appeared in a few scattered scientific journals filled with accounts of the discovery of the first skeleton of a Neanderthal man; of mauve, the first coal tar dye discovered by Perkin in Hofmann's laboratory in London; of a blast furnace for making steel designed by Bessemer. Not a single word about his monumental memoir of the molecules—a glaring example of the neglect of genius.

When a bust of Avogadro was unveiled a year after his death, not a chemist was there to utter a word of homage. Even in his own country he was little known. Only two of his pupils, both physicists, recalled his work. His classic theory of the molecules had appeared originally not in Italian but in French. It was later translated into both German and English and, almost incredibly, was not available in his own language until the opening of the twentieth century. So extremely modest and retiring was this Italian professor that, great as were his contributions in the field of science, when the Scientific Congress met in his own native city, he was not even nominated an officer of that body.

What was the need of the infinitesimal molecule that Avogadro had conjured up out of his meditations? To grasp its real significance we must go back to a memoir which Gay-Lussac had read to the Philomathic Scociety on the last day of the year 1808. Gay-Lussac at twenty-five had been introduced to the eminent naturalist, Alexander von Humboldt. They met at the home of Berthollet in Paris and discussed the composition of water. Together they experimented and found that two volumes of hydrogen gas when sparked with òne measured volume of oxygen gas produced exactly two volumes of water vapor. This was not an original discovery, but Gay-Lussac suspected that "other gases might also combine in simple ratios." Resuming his researches, he discovered that one volume of hydrogen chloride gas when brought in contact with one volume of ammonia gas yielded a white powder, and no residue of either gas was left. The two gases had joined volume for volume. He tried combining carbonic acid gas and ammonia and again found that exactly one part of odorless carbon dioxide had combined with exactly two volumes of suffocating ammonia. His friend Berthollet had shown that a measured volume of nitrogen gas always united with three times its volume of hydrogen to form exactly two measures of ammonia.

Here was a remarkable mathematical simplicity. The combining volumes of gases and the volumes of their gaseous products, could always be expressed in ratios of small whole numbers. No fractions or large numbers were involved. This was the law discovered by Gay-Lussac.

Why this regularity? Was this another example of Nature's striving for simplicity? How could it be explained? Dalton's atoms helped to clarify the question. But here was a fact that even Dalton's little atoms could not explain. Why did one volume of nitrogen unite with one volume of oxygen to yield *two* volumes of nitric oxide gas? Why *two*? Dalton's diagrams showed the formation of just *one* volume of nitric oxide by the union of one atom of nitrogen and one atom of oxygen, thus:

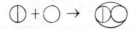

Dalton was asked to explain this inconsistency. He could not fit Gay-Lussac's results into the scheme of his Atomic Theory. He must either alter his theory or reject the conclusions of the Frenchman. He saw no other way out of the dilemma. Dalton, the clumsy experimenter, refused to admit the validity of Gay-Lussac's Law. "The truth is," he maintained, "that gases do not unite in equal or exact measures in any one instance; when they appear to do so, it is owing to the inaccuracy of our experiments." Berzelius knew that Dalton was wrong. He wrote to Dalton: "I think there are parts in your theory which ought to be altered. For instance, that part of it which leads you to assert that the experiments of Gay-Lussac on the volumes of combining gases are inexact. I would rather have supposed that these experiments form the most beautiful proof of the probability of the atomic theory." Dalton could not be convinced and the result was utter confusion.

Another effort was made to bring Gay-Lussac's Law and Dalton's Atomic Theory into harmony. It occurred to some scientists to split the atom in two and thus avoid the difficulty. Thus:

would yield the *two* volumes that Gay-Lussac found and not the single volume that Dalton's theory would compel one to accept. Chemists were desperate. They were ready to agree to any explanation. Dumas himself was willing to break the unbreakable atom of Dalton into quarter atoms. When his friends objected to this scientific sacrilege, Dumas quietly answered, "Chemistry must break apart the atom which physics cannot split. That is all there is to it."

Avogadro alone saw the error that confused the scientific world. He could reconcile the atomic theory of Dalton and Gay-Lussac's law of combining volumes of gases. To him the smallest part of a common elementary gas like hydrogen consisted not of one atom, but of *two* atoms chemically united. This pair of hydrogen atoms he called the *molecule* of hydrogen. Gaseous hydrogen was not H but HH, or H_2 as Berzelius wrote it. In other words, a gas was made up not of single atoms but of groups of atoms called *molecules*. To be sure, Berzelius had spoken of certain elements which seemed always to take part in chemical reactions in pairs of identical atoms, and he had represented them with the symbol of the element with a bar drawn across it, thus: ~~Cl~~. The Congress of Karlsruhe had, in fact, accepted the dashed letters of Berzelius to express "double atoms." But this had by no means cleared up the matter.

Avogadro had reached his conception of pairs of atoms which combined to form molecules by postulating an audacious theory. "*Equal volumes of all gases,*" he declared, "under the same conditions of temperature and pressure contain the *same number of molecules.*" A given volume of any gas, be it simple hydrogen, or compound carbon dioxide, contained the same definite number of molecules. Avogadro's hypothesis was the conclusion of a meditative mind bold enough to fly beyond the limits of the experimental verifications of the balance and the microscope. While he left no positive evidence of the genesis of his hypothesis we may safely believe that he made and rejected numerous assumptions before he finally reached his conclusion.

What evidence had he to back up so bold a statement? He had not tested it in the laboratory. He could not verify it. No balance was sensitive enough to weigh the molecule. For it would take billions upon billions of these tiny particles to turn the scale of the most delicate balance in the world. He surely had not looked into the molecules of matter and detected a twin arrangement of the atoms. For it would take one hundred millions of his molecules placed side by side to make a line one inch long. No microscope had as yet been devised to make such an infinitesimal particle visible. Even had he built a mighty instrument that could magnify the molecule to the limits of human sight, his head would have reeled at the tremendous speed of these tiny molecules flying with a velocity three times as great as that of a bullet leaving the muzzle of a rifle, and colliding five thousand million times per second with neighboring molecules. Stupendous speed of an infinitesimal body!

His theory, when he postulated it, was only a fabrication of the brain. Small wonder that the ears of the world were shut to his announcement. Yet Avogadro did not wrestle with particles millions of times smaller than the head of a pin in the same stupid way that the ancients fought over the number of angels that could dance on the point of a needle. His was the judgment of scientific thinking. One century of advance in physics and chemistry had testified to the truth of his great guess. While at the congress of chemists at Karlsruhe Cannizarro was pleading for his molecules, Clerk-Maxwell, developing the Kinetic Theory of Gases, made the first estimate of the number of particles in one cubic centimeter of a gas. Avogadro's theory was mathematically verified. A mere coincidence, some said. A great work of genius would be the truer and more generous verdict.

Almost one hundred years later, the son of an artillery officer of Lille, France, sat down before a powerful microscope to test by actual experiment the truth of Avogadro's hypothesis. Did equal volumes of all gases really contain the same number of molecules? Other scientists before him had tried to verify it with ingenious schemes. They had beaten gold into sheets so thin that four hundred thousand made a pile hardly an inch thick. They blew soap-bubbles one-billionth of an inch thick. Each film of gold and bubble of soap must be still several molecules thick or it would not hold together. A single pinpoint of indigo colored a ton of water. They were getting nearer to the size of an actual molecule.

Jean Baptiste Perrin, the molecule counter of the University of Paris, followed a different method of attack. He prepared the finest powder in the chemical world from the gum gamboge, and placed a weighed amount of this gum in water. It did not dissolve but the fine particles scattered themselves throughout the water. For three whole weeks this suspension was violently agitated until the heavier particles of gum settled to the bottom and were removed. Only the tiniest grains of the powder remained suspended in the water.

Perrin placed a drop of the water containing the fine particles of gamboge on a slide under the microscope. A powerful beam of light illuminated that drop. Every time a speck of gamboge passed across the field of his microscope he saw a momentary flash of light. Hour after hour and day after day he sat recording these flashes. Weeks passed, and still he kept watching the infinitesimal particles. It was an eye-straining, head-splitting job, but this Frenchman kept on. He counted the number of specks of gamboge—he knew the weight of gum he was using. Then he made his calculations. The number of molecules in a cubic centimeter of a gas was 31,500,000,000,000, 000,000. A staggering number! He checked his results. He tried another suspension—mastic gum in alcohol. Again the same result.

Then Professor Robert Andrews Millikan at the University of Chicago, using an entirely different method, obtained an almost identical figure: 28,500,000,000,000,000,000. What does this number mean? It takes a clock three hundred years to tick one billion. Yet there are, Perrin found thirty billion, billion molecules in a single tiny thimbleful of a gas. This was no idle guess. Perrin and Millikan, both Nobel Prize winners in physics, maintained that we can count the number of molecules in a small volume of a gas with as much accuracy as we can determine the population of a city like New York! Then came another remarkable verification. Dr. Irving Langmuir dissociated the hydrogen molecule at very high temperatures and found it actually to contain two atoms.

What was the immediate importance of Avogadro's theory when Cannizarro proclaimed it to the world? The acceptance of this generalization cleared the scientific air like magic. Inconsistencies, conflicting opinions, impossible conventions and compromises were at once obliterated. Chemistry was again ready to advance.

This conception made possible the true understanding of the atomic weights of the elements. "Instead of taking for your unit of atomic weight the weight of an entire molecule of hydrogen," said Cannizarro, "take rather the half of this weight, that is to say, the quantity of hydrogen contained in the molecule of HCl." When this was done the first accurate table of atomic weights was obtained.

Gay-Lussac, in his memoir, had expressed the hope that "we are perhaps not far removed from the time when we shall be able to submit the bulk of chemical phenomena to calculation." Avogadro's contribution to chemistry showed the way. It led also to a final agreement as to the true formula for water. Dalton had written the

formula for water as   since he believed that *one* atom of

oxygen combined with *one* atom of hydrogen to form *one* atom of water. Thomas Graham, founder of colloid chemistry, as late as 1850 used the formula HO in his textbooks. In our own country in 1863, Edward L. Youmans in his *Class Book of Chemistry* declared there were reasons for believing that the true formula for water was H_2O_2 and not HO. Now for the first time the formula for water was definitely established as H_2O.

Avogadro's hypothesis was of inestimable help in determining the density and molecular weights of a·large number of gaseous products prepared during the rapid development of organic chemistry. This, in turn, resulted in the marvelous strides that were made in the study of the molecular composition of the most complex chemical compounds. His theory showed that the elementary gases, like oxygen, hydrogen, chlorine and nitrogen, all contained two atoms to the molecule, and that, therefore, their molecular weights were twice their atomic weights, ignorance of which had led to innumerable difficulties. Avogadro's theory was found to be almost universally applicable. Later the Dutch scientist van't Hoff extended its application to the problem of solutions.

Cannizarro worked a lifetime to bring Avogadro back to the world. Then at eighty-four, still teaching and editing an Italian chemical journal, the crusader of Palermo passed quietly away in Rome. The following year, on September 24, 1911, an historic meeting took place in Turin to commemorate the hundredth anniversary of the publication of Avogadro's classic memoir. King Victor Emmanuel III was there to pay homage to one of Italy's most illustrious sons. Hundreds of scientists gathered there. Messages and money were pouring in from every scientific laboratory of the world. A monument to the memory of Avogadro was unveiled.

From America, France, Germany, Sweden, England, Spain, Austria and Japan came cablegrams and letters extolling the genius of this Italian professor. Anxious to atone for its grave neglect, the world joined in one of the greatest posthumous tributes to a scientist in history.

"Equal volumes of all gases under the same conditions of temperature and pressure contain the same number of molecules"—this one clear prophetic sentence has immortalized the name of Amedeo Avogadro.

X

WOEHLER

UREA WITHOUT A KIDNEY

ABOUT one hundred and fifty years ago an epoch-making event took place in the laboratory of a young German still in his twenties. He had just returned from the laboratory of Berzelius in Stockholm, to teach in the newly founded municipal trade school in Berlin. A great idea was hatching in Friedrich Woehler's head. He had heard discussions in every scientific circle he had visited of a mysterious vital force, as elusive as phlogiston.

Inside the living body of plants and animals, it was thought, burned a steady invisible flame, and through this flame a mysterious vital force built up the sugars, the starches, the proteins and hundreds of other very complex compounds. This vague creative force existed in the animal and vegetable kingdoms but not in the mineral world. Men believed that the substances which constituted the texture of vegetation differed from mineral substances in that the former could not be built up or synthesized in the laboratory. "Nothing but the texture of living vegetables, nothing but their vegetating organs, could form the matter extracted from them; and no instrument invented by art could imitate the compositions which are found in the organic machines of plants." Man could never imitate the power of this vital force. It was one of those mystic causations of which man was to remain in ignorance all the days of his life. Man's mental machinery and his chemical engines were too puny and simple to reproduce this force of nature. Some even doubted whether these organic compounds obeyed the laws of chemistry. Such was the prevailing opinion of the world in 1828.

Berzelius himself spoke of the impassable gulf which separated organic compounds from inorganic substances. Leopold Gmelin, Friedrich's celebrated teacher at the University of Heidelberg, firmly believed that organic compounds could not be synthesized. Yet Woehler was young and he doubted. He agreed with the

129

eminent French chemist Chevreul that "to regard the distinction as absolute and invariable would be contrary to the spirit of science." If the laws of nature were the thoughts of God, then God would vouchsafe these thoughts to man if only he worked tirelessly to find them. Back in his mind was the suspicion that vital force was another one of those cryptic phrases, a creed which if accepted would destroy the progress of chemistry. Like the young man from China who returned his first watch in America with the plaint that "it died last night," science had endowed those chemical compounds of living matter with the hidden, moving springs of vitalism.

Slowly, carefully, laboriously, Woehler worked away in the sacred temple of his laboratory. If he could only make one of those innumerable substances which until now only the intricate chemical workshop of the living organism had fashioned! What a blow he could strike at this false idea—a blow even more powerful than that which immortal Lavoisier had dealt to the mischievous theory of phlogiston half a century before him! As he dreamed and hoped he kept working, watching his test tubes and flasks, his evaporating dishes and condensers.

Friedrich Woehler had read the recently published work of Chevreul who had shown that many of the fats and other substances occurring in both the animal and vegetable kingdoms were identical. The barrier between animal and vegetable matter had thus been broken down. He was familiar with the work on animal chemistry of Rouelle, magnetic teacher of Lavoisier. These men had taken the first steps.

Woehler's goal was alluring. Experiment after experiment gave negative results but he kept plodding away. Once, in Berzelius' own laboratory in Stockholm, he had made some "peculiar white crystalline substance" which he could not identify. Four years passed. Then one afternoon the miracle happened.

Picture the amazement of this young researcher gazing upon a product which he had made out of lifeless compounds in an inanimate flask. Here under his eyes was a single gram of long, white, needle-like, glistening crystals which Rouelle had first found half a century before in urine and which Fourcroy had later studied and named urea. This white salt had never before been produced outside the living organism.

It was not strange that Woehler recognized at once this crystalline urea. He had started his career in science as a student of medicine and while competing for a prize for the best essay on the waste products found in urine, had come across urea.

Woehler was excited. He was standing upon the threshold of a new era in chemistry, witnessing "the great tragedy of science, the slaying of a beautiful hypothesis by an ugly fact." He had synthesized the first organic compound outside the living body. The restless mind of yound Woehler almost reeled at the thought of the virgin fields rich in mighty harvests which now awaited the creatures of the crucible. He kept his head. He carefully analyzed his product to verify its identity. He must assure himself that this historic crystal was the same as that formed under the influence of the so-called vital force.

When he was sure of his ground, he wrote to Berzelius, "I must tell you that I can prepare urea without requiring a kidney of an animal, either man or dog." The Swede enthusiastically spread the news. The world of science was electrified. Chevreul hailed the achievement with joy. Woehler had actually synthesized urea out of inorganic compounds! What was to prevent others from building up the sugars, the proteins, perhaps even protoplasm, the colloidal basis of life itself? A feeble protest still sounded from the vitalists. Urea was perhaps midway between the organic and inorganic world. For to make urea one must use ammonia which originally was of organic origin. The vital force present in organic substances never disappeared and consequently was capable of giving rise to other organic bodies. So they argued. But even that whisper was soon lost in the great tumult of excitement. It was indeed a brilliant new day for chemistry.

Woehler published his modest memoir on the synthesis of urea in 1828 and a century later Professor Amé Pictet and Hans Vogel of the University of Geneva succeeded in sythesizing cane sugar. Starting with simple hydrogen and carbon dioxide, seventy-year-old Pictet could change them into wood-alcohol, then into formaldehyde which yields glucose, and finally sucrose or cane sugar. For fifty years scientists had worked on this momentous piece of research.

Sir James Colquhoun Irvine, Vice Chancellor of the University of St. Andrews, Scotland, had himself worked on this problem for twenty years. He had almost succeeded in synthesizing this sugar when he read the announcement of Pictet's discovery. Irvine was thrilled and spoke to his students: "Naturally I am keenly disappointed that this synthesis did not occur in our own laboratory, but since it did not I am glad that Pictet gets the credit." No begrudging the work of a fellow researcher. No jealousy. "It is a great achievement," he continued, "It is a milestone in organic and biological chemistry." Such is the essence of science—the spirit of comradeship and good will in the entrancing work of piercing nature's secrets.

What a century of research between Woehler's urea and Pictet's sucrose! Four hundred thousand compounds had been prepared in this branch of synthetic chemistry, while every year more than four thousand new ones were added. No wonder that when Gmelin was preparing his handbook of chemistry he pleaded for chemists to stop discovering to give him a chance to catch up with his work. Woehler, a modest man, would have been the last to claim for himself the distinction of being the forerunner of such tremendous achievements.

Friedrich Woehler was born at the opening of the nineteenth century near Frankfort-on-the-Main. His father, Auguste, a man well educated in philosophy and science, was Master of the Horse to the Crown Prince of Hesse Cassel, who was feared for his violent, impetuous temper. One day during a inspection tour of his stables, something very trifling displeased the Prince who began to abuse his servant. Auguste listened to his vile tongue until the Prince attempted to add a beating to his tongue lashing. Woehler would not put up with such humiliation even at the hands of a royal personage. Seizing a stout riding whip, he struck back fiercely until his master lay bleeding on the ground. Then jumping upon the fleetest horse in the stables and accompanied by a groom who was to return the steed, Auguste fled from Cassel. The Elector, fearing ridicule. did not pursue him.

Thus it came about that Friedrich was born not in the house of his parents but in the home of his uncle, who was clergyman of the village of Escherscheim. He received his early education from his father, who interested him in nature and encouraged him in drawing and in his hobby of mineral collecting. Friedrich carried on a brisk exchange of minerals with his boyhood friends, which he continued even in later life. On one occasion he met the old poet Goethe, who was examining specimens in the shop of a mineral dealer in Frankfort.

Soon this boy added chemistry to his list of hobbies. Through his father he met a friend who had a rich library and a private chemical laboratory where he obtained permission to work. He built voltaic piles out of zinc plates and some old Russian copper coins he had collected. The master of the German mint presented him with an old furnace in which, with the aid of his sister to blow the bellows, he would build a roaring fire. And while he experimented he burned his fingers with phosphorus, and on another occasion was almost killed when a flask containing poisonous chlorine cracked in his hands.

At Marburg University, where his father, too, had been a student, he started to study medicine and won a prize for his investigations on the passage of different waste materials into urine. He performed numerous ingenious experiments upon his dog and even upon himself in preparing for this essay. Some of these experiments were dangerous to his health. He had not avoided them, however, even as twenty years before him, Dr. John Richardson Young, at twenty-two, had given his life at Hagerstown, Maryland, while using himself as a human beaker and test tube to prove that gastric juice and not a mysterious vital spirit was the essential factor in digestion. But chemistry still fascinated him. He built a little laboratory in his private room and prepared cyanogen iodide for the first time. He brought it to his teacher, Professor Wurzer, who reproached him for wasting his time on chemical experiments when he should have been studying his medicine. The sensitive boy was hurt and thereafter never attended the professor's lectures.

Soon the fame of Leopold Gmelin attracted him to Heidelberg. Here he continued his studies, gained the degree of Doctor of Medicine, Surgery, and Midwifery, and made ready to start on his travels to visit the great hospitals of Europe in further preparation for the practice of medicine. But Gmelin had watched this lad work in the chemical laboratory. He had told young Friedrich it would be a waste of time to attend his own lectures. Laboratory work was more important. Gmelin had read with pride his student's paper on the discovery of cyanic acid. He did not, at the time, dream this would in a few years lead to urea, but Gmelin was going to save Woehler for the disciples of Hermes. He spoke to him of the alluring career of a chemist. It was not very difficult to persuade Woehler. Often he had been tempted to turn away from medicine. Gmelin mentioned Berzelius whose fame as chemist had spread throughout Europe. He aroused in Friedrich the hope that perhaps Berzelius would give him permission to work under him in Stockholm.

Woehler wrote to the Swede, and within a few weeks received this answer: "Anyone who has studied chemistry under the direction of Leopold Gmelin has very little to learn from me, but I cannot forgo the pleasure of making your personal acquaintance. You can come whenever it is agreeable to you." Woehler was walking on air. He hurried to Gmelin to tell him the good news. He was to make a pilgrimage to the laboratory of Berzelius.

He started at once. When he reached the town of Lübeck on the Baltic, he learned that he would have to wait six weeks for a small sailing vessel that was to take him to Stockholm. He was too impatient to wait so long in idleness. Through a friend, with whom as a boy he had exchanged minerals, he gained access to a private laboratory where he set to work to find a method of making larger quantities of potassium, that violently active metal which Davy had just isolated.

At last he was on his way to Sweden. When he stepped off the boat the officer of the guard who examined his passport, on learning that he had come from Germany to study under Berzelius, declined to accept the usual fee. "I have too much respect for science and my illustrious countryman," he said, "to take money from one who in the pursuit of knowledge has undertaken so long a journey." Instead of the fee Woehler presented him with a piece of the wonderful potassium he had just prepared.

He reached Stockholm at night and nervously waited for the morning. "With a beating heart I stood before Berzelius' door and rang the bell. It was opened by a well-clad, portly, vigorous looking man. It was Berzelius himself. As he led me into his laboratory I was in a dream." Woehler never forgot his cordial reception by this master.

They wasted no time. Berzelius supplied the young student with a platinum crucible, a wash bottle, a balance and a set of weights, advised him to buy his own blowpipe, and set him to work on the examination of minerals. That was to be his first training in accurate analysis. When Woehler hurried to Berzelius to show him the result of his work his teacher warned him, "Doctor, that was quick but bad." Woehler remembered this valuable advice. Woehler now turned once more to his recently discovered cyanic acid and succeeded in preparing silver cyanate, a compound of this acid.

In the meantime, in the laboratory of Gay-Lussac in Paris worked another young German, Justus Liebig. This handsome, boisterous student, three years younger than Friedrich, was busy with the explosive fulminates. As a lad Liebig, whose father owned a small chemical factory, had seen an itinerant tradesman making fireworks in his native city of Darmstadt. He was eager to learn the secrets of these explosive chemicals. During these researches Liebig prepared a strange compound. This substance was similar in composition to the silver cyanate of Woehler yet vastly different from it in both physical and chemical properties. Here was something very puzzling. How could two compounds made up of the identical elements in exactly the same proportions possess different properties? "Something must be wrong," said Liebig, and straightway he doubted Woehler's results. Perhaps he had misread his paper. He verified the results very carefully. Both Woehler and he were right in their conclusions.

Liebig communicated with his compatriot in Sweden. Woehler could not understand this strange phenomenon. He asked his master Berzelius to help him. The Swedish chemist recognized a tremendous discovery. *Isomers*—this was the term coined by Berzelius to designate chemical compounds having the same composition yet differing in properties—these had been discovered by two young men. This was only the beginning of similar findings in this new field. There were many substances which formed dozens of isomers. The phenonenon of isomerism in the chemistry of the carbon compounds helps to explain the tremendous number of different compounds of organic chemistry.

Later Liebig met Woehler at the latter's home. Woehler told Liebig of his excursion with his famous teacher through northern Norway and Sweden, during which he met Sir Humphry Davy returning from a fishing trip. What an inspiration was the memory of that scene as he stood between Berzelius and Davy, the two foremost chemists of Europe.

At the time of their meeting, Liebig, though twenty-one, was professor of chemistry at the small University of Giessen. He had received this appointment through the influence of Von Humboldt, the celebrated scientist, whom he had met in Gay-Lussac's laboratory in Paris. His salary amounted to only one hundred and twenty dollars a year plus about forty dollars for annual laboratory expenses. It was here that Liebig invented and developed a method of organic analysis still used today.

Woehler was teaching in the city trade school of Berlin and was spending a great deal of time translating into German some of the work of Berzelius from the Swedish, which he had learned while at Stockholm. Liebig admonished him to "throw away this writing to the devil and go back to the laboratory where you belong."

They discussed their mutual researches and their future plans for work to be performed in their respective laboratories. "Liebig expressed joyful assent at once and a research on mellitic acid was selected and carried to a successful conclusion." Fulminic acid was proposed as the next problem, but it was soon abandoned. "Fulminic acid we will allow to remain undisturbed," wrote Liebig. "I have vowed to have nothing to do with the stuff." For Liebig had almost lost his eyesight when some of it exploded under his nose and he was sent away to a hospital to ponder over its dangers. He also reminded Woehler how years before, while still a student at high school, it had exploded in the classroom, and he had been expelled with the verdict that he was "hopelessly useless."

Not that there could not be found men brave enough to wrestle with such obstreperous substances. Nicklés, a Swiss, lost his life in an attempt to isolate fluorine, an element more poisonous than chlorine. Louyet, too, had died of the effects of this gas, while Knox, a Scotchman, ruined his health in its study. Dulong, before them, had lost an eye and three fingers while preparing nitrogen trichloride for the first time, and continued to experiment with this compound even after the accident. On another occasion this same chemical knocked Faraday unconscious. The annals of chemistry contain many such examples of heroism.

In 1832 Woehler lost his young wife whom he had married two years before. It was a sudden shock that threatened to upset him permanently. He went to his friend Liebig for consolation and found it in his laboratory. During this year of bereavement the two young scientists published their joint paper on oil of bitter almonds. They studied a series of new compounds all containing an identical group of atoms which remained unchanged through the most diverse changes which their parent bodies underwent. To this unchanging group of atoms, consisting of carbon, hydrogen and oxygen, they gave the prosaic name of benzoyl. When Berzelius read of this work he saw in it the dawn of a new day in chemistry and suggested for this chemical group or radical the more poetic name of *proin*, the dawn. In Paris the chemical world talked a great deal of these researches.

Their work temporarily completed, Woehler returned to Cassel where he had been called the previous year. "I am back here again in my darkened solitude," he wrote to Liebig. "How happy was I that we could work together face to face. The days which I spend with you slip by like hours and I count them among my happiest."

For five years Woehler remained in Cassel. Here he met and married Julie Pfeiffer, a banker's daughter, by whom he had four daughters; one of them, Emilie, was to act as his secretary and biographer. His work in the field which he had opened had brought him fame. When Professor Strohmeyer, discoverer of the element cadmium, died, Woehler was selected from among a long list of candidates, including Liebig, to fill his chair at the University of Göttingen, a position he held for almost half a century. Liebig never begrudged him this honor.

The two friends continued to work together, and in 1838 they published the results of their experiments on uric acid, another organic compound. It was in this report that these pioneers foresaw the great future of organic chemistry. "The philosophy of chemistry," they wrote, "must draw the conclusion that the synthesis of all organic compounds must be looked upon not merely as probable but as certain of ultimate achievement. Sugar, salicin, morphine, will be artificially prepared." This was indeed prophetic.

The friendship of Woehler and Liebig stands out as a sublime example of scientific fraternity. Liebig spared no words in praise of his friend:

> The achievement of our joint work upon uric acid and oil of bitter almonds was his work. Without envy and without jealousy, hand in hand we plodded our way; when the one needed help the other was ready. Some idea of this relationship will be obtained when I mention that many of our smaller pieces of work which bear our joint names were done by one alone; they were charming little gifts which one presented to the other.

How different this from the too-frequent haggling of scientists over priority of discoveries!

Woehler was a great tonic for the hot-tempered Liebig who, as a student, had been forced to spend three days in jail for taking part in a gang fight. On this occasion he "made scurrilous remarks about those in authority and knocked the hat from the head of not only police officer Schramm but even of Councillor-in-Law Heim." More than once, when Liebig quarreled with a scientific contemporary who opposed his views, Woehler's calm advice smoothed things over. Liebig accused Elihard Mitscherlich, a student of Berzelius, of appropriating the apparatus of others and calling this his own. Woehler pleaded with his friend to stop the quarrel:

> Granted that you are perfectly in the right, that scientifically as well as personally you have cause to complain, by doing this you stoop from the elevated position in which posterity will see you to a vulgar sphere where the luster of your merits is sullied.

Liebig made many enemies. His irascibility had estranged Berzelius whose friendship he valued very highly. He wrote to him wishing for permission to dedicate a book to him. Berzelius thanked him for this honor and incidentally criticized the style of the book. Liebig at once took offense, wrote him an insulting letter, and their friendship was forever at an end.

When Liebig got into trouble with Marchand, again Woehler stepped into the breach.

To contend with Marchand [he counselled] will do you no good whatever or be of little use to science. It only makes you angry and hurts your liver. Imagine that it is the year 1900 when we are both dissolved into carbonic acid, water and ammonia, and our ashes, it may be, are part of the bones of some dog that has despoiled our graves. Who cares then whether we have lived in peace or anger; who thinks then of thy polemics, of thy sacrifice, of thy health and peace of mind for science? Nobody. But thy good ideas, the new facts which thou has discovered—these will be known and remembered to all time. But how comes it that I should advise the lion to eat sugar?

Many of their vacations were spent travelling together. It was difficult to tear Liebig away from his laboratory. Woehler on one occasion tried to persuade him to join him on a trip through Italy. Woehler loved to take these excursions. He would carry his sketch book or easel with him, for he was a fair artist and the beauties of nature enthralled him. Liebig cared more for the smell of the laboratory and the adventures of chemical discovery. "After all what good will it do me to have looked into the crater of Mt. Vesuvius?" he remarked.

In spite of Liebig's shortcomings, Woehler remained his friend to death. Woehler knew his friend and made allowances for his fits of temper. "He who does not know him," said Woehler, "would hardly realize that at bottom he is one of the most good-natured and best fellows in the world."

Woehler, in his youth, had received an excellent education in the fine arts as well as in the sciences. He loved music, was encouraged in his attempts at oil painting by Christian Morgenstern, the landscape painter, and he made a more than superficial study of the German poets. Often his letters and parts of his lectures took on the nature of poetry. In one of his letters to Liebig from Italy we find,

On the highest summit of the Blue Mountain stands the palace of Tiberius, in whose shade I ate splendid grapes and figs while two brown-faced girls, the guides of our horses, danced the Tarantella to the sound of the tambourine.

Woehler built up a famous laboratory at Göttingen. It was among the first of the great teaching laboratories of the world. His fame as chemist and teacher spread over Europe. From every country students flocked to him and his laboratory became a veritable hive, busy day and night. From the United States came Professor James Curtis Booth, his first American student, and also Professor Frank F. Jewett of Oberlin College, who brought back the story of his teacher's discovery and isolation of that extremely light, silvery metal, aluminum. Jewett was fond of talking to his classes of this strange metal which no one had as yet been able to obtain cheaply, in spite of its great abundance in the rocks of the earth. One day as he spoke of the fortune that awaited the man who would solve the problem of a simple method of aluminum extraction, one of his students, nudged the ribs of his young classmate, Charles Martin Hall. "I am going after that metal" said Hall, and on February 23, 1886, he handed Jewett a pellet of the shiny metal. Hall's process was patented that year. This was the beginning of the powerful Aluminum Company of America, which in the years since has been producing millions of tons of pure aluminum.

Woehler's kindly disposition endeared him to another young American student, Professor Edgar Fahs Smith of the University of Pennsylvania. Woehler, black skull cap on his head, would sit for hours on a stool helping a beginner over some difficulty. The Geheimrat once noticed Smith emptying residues of his flasks in the drain outside the laboratory. "Recover your residues so that every thing of value will be saved," Woehler advised him, and together they outlined a method of recovery. When Smith had purified the residues, Woehler sent him to his friend, an apothecary, who bought them, thus saving the American his original expenses.

When Smith was ready for the final examinations for the degree of Doctor of Philosophy in Chemistry, he presented himself appropriately attired in dress suit and white gloves. Near the end of the examination Woehler, who was then old and somewhat feeble, straightened himself in his chair and asked his question. "Herr Candidate, will you tell me how you would separate the platinum metals from each other?" Smith acted somewhat confused, picked at the ends of his white gloves, and then, somewhat haltingly began to repeat the twelve pages of Woehler's treatise dealing with this subject. The Geheimrat, before the American had completed the answer, thanked him profusely and complimented him on his knowledge of the subject. The examination in chemistry was over. The next day, following the usual custom, Smith made a formal call on each of the professors. Woehler complimented him again, saying that his answer at the examination was not only correct but expressed in perfect language. Then Smith confessed that the day before another candidate had tipped him off, and that he had memorized the twelve pages dealing with the separation of the platinum metals as found in his book on Mineral Analysis. "Woehler took it as a great joke and laughed heartily."

In the meantime organic chemistry was making prodigious strides. Marcellin Berthelot, master synthetic chemist of France, went to the ant and learned its secret. He prepared formic acid, the liquid which is responsible for the sting of the insect. Kolbe, crusading student of Woehler, prepared the acid of vinegar without the use of sweet cider or the mother-of-vinegar bacteria. William Perkin, washing bottles in the laboratory of Hofmann in London, mixed at random the contents of two flasks and discovered a method of synthesizing mauve—the first of a long series of coal-tar dyes which rival the colors of nature. Then Kekulé of Darmstadt, falling asleep in front of his fire in Ghent, dreamed of wriggling snakes, and woke up like a flash of lightning with the solution of a knotty problem. He had discovered the structure of benzene— parent substance of thousands of important compounds. Next, his pupil, Adolf von Baeyer, working for fifteen years on indigo, finally discovered its formula and made possible the manufacture of synthetic indigo fifteen years later by the Badische Company which had spent millions of dollars in research on this problem. This achievement rang the death knell of the prosperous indigo-growing industry of India, which soon went the way of its predecessor, the cultivation of woad. Had not Becher complained, "We give our gold to the Dutch for the trumpery color indigo and let the cultivation of woad in Thuringia go to perish."

Strangely enough, both Woehler and Liebig deserted this pregant field of their original triumphs. Liebig turned to the chemistry of agriculture. In 1840 he tested his new theory of soil fertility on a barren piece of land near Giessen. The sceptics laughed but he kept feeding the soil with nothing but mineral fertilizers until he had turned it into as fertile a spot as could be found in all Germany. With one blow he had overturned the firmly rooted belief that plants can thrive only on manure or other organic matter in the soil. He had proved that the vegetable world could construct its organic material from the carbon dioxide and nitrogen of the air and the water of the ground. Others followed this pioneer work. Sir John Lawes at Rothamsted, England, started an experimental station which became one of the most famous of its kind in the world.

Yet Liebig was not happy in the change. "I feel," he wrote, "as though I were a deserter, a renegade who has forsaken his religion. I have left the highway of science and my endeavors to be of some use to physiology and agriculture are like rolling the stones of Sisyphus —it always falls back on my head, and I sometimes despair of being able to make the ground firm."

Woehler, too, had forsaken his first love almost in its infancy. "Organic chemistry nowadays almost drives me mad," he complained. "To me it appears like a primeval tropical forest full of the most remarkable things, a dreadful endless jungle into which one does not dare enter, for there seems no way out." He went to his minerals again and to the study of metals. In Sweden he had watched the master Berzelius at work on his researches of silicon, selenium and zirconium—three new elements. Woehler had learned much during his short stay and, a year before his immortal synthesis of urea had already accomplished a research of the first order—the isolation of the metal, aluminum, for the first time.[11] This same problem had defeated the genius of Davy. By treating a solid salt of aluminum with the intensely active potassium, Woehler was able to tear the metal away from its union and obtain it free as a white powder. But this sample of aluminum was only a laboratory curiosity—it cost a hundred and fifty dollars a pound.

Woehler's span of life covered the troubled days of the Napoleonic and Franco-Prussian Wars. As a lad he had seen the triumphal entry of the hated Napoleon into Frankfort. Sixty years later he heard of the capture of the French flags by the Prussians. Immediately, from Wiesbaden, where as a youth he had searched for urns and lamps in the ancient camps of the Romans, he wrote to Liebig, "The eagles of the captured French flags really consist of gilded aluminum, a metal that was first produced in Berlin in 1827. Such is fate." He modestly refrained from mentioning the part he played in the discovery of this metal.

Woehler isolated two other new elements, beryllium and yttrium, and, because of illness which prevented an accurate analysis, just missed discovering a fourth metal. This metal, vanadium, was soon isolated by N. G. Sefström. Woehler had sent a specimen of a lead ore containing this unknown metal to his friend Berzelius, and marked it with an interrogation point. Berzelius analyzed the mineral and replied with the following story:

> In the remote regions of the north there dwells the Goddess Vanadis, beautiful and lovely. One day there was a knock at her door. The goddess was weary and thought she would wait to see if the knock would be repeated, but there was no repetition. The goddess ran to the window to look at the retreating figure. "Ah" she said to herself, "it is that fellow Woehler." A short time afterward there was another knock, but this time so persistent and energetic that the goddess went herself to open it. It was Sefström, and thus it was that he discovered *vanadium*. Your specimen is, in fact, oxide of vanadium. But [continued Berzelius] the chemist who has invented a way for the artificial production of an organic body can well afford to forgo all claims to the discovery of a new metal, for it would be possible to discover ten unknown elements with the expenditure of so much genius.

The march of organic chemistry still went on after Woehler was dead. He lived long enough to see some of the miracles that succeeded the synthetic production of urea. But mightier developments followed. The story of this advance is like a tale from the Arabian Nights. Emil Fischer, refusing to enter the lumber business to please his father, turns to the chemical laboratory and builds up the most complex organic compounds, link to link and chain upon chain until he synthesizes complex products like $C_{220}H_{142}O_{58}N_4I_2$, and polypeptides which resemble the natural peptones and albumins. No architect could work with greater precision. And when his father dies at the age of ninety-five, Fischer utters the regret "that he did not live to see his impractical son receive the Nobel Prize in Chemistry" in 1902.

In another German laboratory, Paul Ehrlich jabbed mice, rabbits and guinea pigs with injections of strange chemical compounds which he kept changing and discarding by the dozens. He was searching for a differential poison—one which is more poisonous to the microörganism than to its host. Then one glorious morning in 1910, after six hundred and five trials, his synthetic drug—dihydroxy diamino arseno benzene dihydrochloride, that was its chemical name—killed the corkscrew trypanosomes, the deadly microbes that caused syphilis. Six-O-Six, this first real specific against a virulent disease, was a product of synthetic chemistry. In 1910, two synthetic rubber automobile tires were exhibited. They were crude and very expensive. By 1931 the first commercially successful rubber substitute, *neoprene*, was manufactured by Du Pont. Among other rubber substitutes later developed in this country were *Buna-N*, *Perbunan*, *GR-S*, *butyl* and *Buna-S* rubber made both from alcohol and from petroleum. With the entry of the United States into World War II, our manufacture of synthetic rubber was stepped up to a million tons a year. Today the United States alone manufactures all of the synthetic rubber it needs (3 million tons a year) and still has enough to export $175,000,000 of this essential commodity.

Other amazing contributions to chemotherapy followed. Penicillin, streptomycin, sulfanilamide, chloromycetin, aureomycin, terramycin, bacitracin, tetracyclines, chloramphenicol, neomycin, and other antibiotics synthesized since 1935 and used successfully against dangerous streptococci infections, pneumonia, meningitis and other diseases such as trachoma and the rickettsias.

The list of achievements is still incomplete. Chemists have not feared to join battle with any product of the living organism. They have studied the active internal secretions of the ductless glands of the body. These secretions, called *hormones* (from the Greek—to arouse or excite), enter the blood stream in extremely minute amounts as catalysts, and control growth, intelligence and other functions of the nervous system. The first of these hormones to be synthesized (Stolz, 1906) was *epinephrine* (adrenalin), the active ingredient of two tiny capsules found one on top of each kidney. The hormone of these suprarenal tissues was isolated as early as 1900 by the American, John Jacob Abel, and his Japanese co-worker, Takamine. It is the hormone of the he-man and the coward, for the absence or overactivity of the suprarenal bodies has a tremendous influence on human action. During an emotional crisis the adrenals become very active and produce great strength. Their overactivity in the female accounts for the deep-voiced, bearded lady of the circus.

In 1915 Dr. Edward C. Kendall of the Mayo Foundation isolated the hormone of the thyroid gland, *thyroxine*. This needle-shaped crystal containing 65% iodine is found to the extent of less than a quarter of a grain in the whole body, and influences the rate of oxidation in the body. When the thyroid is overactive it produces either a symmetrical giant or a gorilla type of man. When underdeveloped, it results in a misshapen dwarf with the intelligence of an idiot. In 1927, Dr. C. R. Harington of England succeeded in synthesizing this important hormone from coal-tar products. It was a prodigious task. This drug, beta-tetra-iodo-hydroxy-phenoxy-phenyl-alpha-amino-propionic acid, became a blessing to mankind.

The isolation of insulin by Dr. Frederick G. Banting, who was killed in an airplane accident while in the service of Great Britain in 1941, proved a boon to diabetics. This hormone, a polypeptide with a molecular weight of about 6000, surrendered its complex structure in 1955 to Frederick Sanger of Cambridge University, for which he won a Nobel prize. Other hormones were isolated in pure form and them synthesized—*estradiol*, hormone of the female sex gland, *testosterone*, hormone of the male sex gland, *cortin* from the outer layer of the adrenal gland, oxytocin, vasopressin, cortisone, ACTH, and many more.

The discovery, isolation and final synthesis of a whole group of new compounds essential to health in a balanced diet was another triumph of the chemist. These compounds called vitamins A, B_2 or G, C, D, E, K, and several others closely associated with vitamin B_2, such as *niacin, pantothenic acid, inosital, paraamino benzoic acid, choline, pyridoxine(B_6), biotin (H), folic acid and B_{12}* prevent deficiency diseases such as xerophthalmia (an eye disease), beriberi, pellagra, scurvy, rickets, sterility (in rats), excessive bleeding and so forth. Professors

Elmer V. McCollum and Herbert M. Evans, and Dr. Joseph Goldberger were among the early American pioneers in this field of research. Drugs and medicines like procaine, pentothal, butyn, psicaine, tutocaine, aspirin, phenacetin, meprobamate, thorazine, reserpine, urotropin, dramamine, morphine, strychnine, veronal and quinine have been synthesized to alleviate the pains of mankind. The essential oils of the synthetic chemist rival the odors of ancient Arabia and Persia, while his colors outshine the rainbow. Hundreds of new synthetic plastics are eating into the metal, natural fibers, paper, and wood markets. Prostaglandins, a whole family of 20-carbon unsaturated fatted acids found in small quantities in nearly all mammalian tissues and fluids have been isolated and synthesized beginning in 1967. They seem to be involved at the cellular level in regulating gastric secretion, contraction and relaxation of muscles, body temperature, food intake and blood platelet aggregation.

The mind fairly reels at the thought of the possibilities of this new branch of chemistry. Chemistry, once the handmaid, is now the mistress of medicine, for life is largely a matter of chemistry. Our bodies are organic chemical factories. Chemical experiments began controlling the growth of cells, the fundamental biological unit of living organisms, outside the living body. In 1912, Alexis Carrel, a Nobel Prize winner in medicine, took several minute fragments from the heart of a chick embryo and cultivated them. "The bits of tissue went on pulsating and surrounded themselves with connective tissue cells." But in a few days this ceased and "degeneration was imminent." Then Carrel, by carefully regulating the chemical composition of the medium in which the cells were placed, was able to get the heart tissue pulsating again, and many experiments were made with the pure strain descended from this tiny fragment of pulsating tissue. He succeeded in keeping alive minute portions of the original strain for more than thirty years.

Chemists were not ready to lay down their beakers and stirring rods. The nature of the simple cell still defied solution. It was known that the cell structure was based on three kinds of very large polymer-type molecules: nucleic acids, proteins and polysaccharides (such as cellulose). The proteins offered the most stubborn resistance. Their basic building blocks are amino acids. The primary aspect of protein structure is the sequence of these amino acid residues. Genes, protoplasm, many enzymes, viruses, antibodies and hormones such as insulin are proteins. Nucleic acids are polyesters and control the biochemical synthesis of proteins. They are also responsible for the genetic transfer of characteristics during cell reproduction.

The first enzyme or biochemical catalyst to be isolated in pure form was urease about fifty years ago by James B. Sumner of Cornell University. Between 20 and 25 more have since been elucidated. The first virus or disease producing agent was isolated in pure crystalline form by Wendell M. Stanley of the Rockefeller Institute for Medical Research in 1935 from a tobacco plant disease. Sumner and Stanley were honored with the Nobel prize in chemistry.

By the mid 1940's the sequence of the two chains of the protein insulin containing 51 amino acid units was determined by the British chemist Frederick Sanger. He used a paper chromatographic technique set in motion a few years earlier by two British scientists, Archer Martin and Richard Synge. This tool of chromatography, now automated, was first introduced by an Italian-born Russian botanist named Michael Tswett in 1906 and then ignored for a quarter of a century.

In 1947 a protein-like molecule was synthesized by a Harvard University chemist, Robert B. Woodward, who used a chain of amino acid units. Four years later Linus Pauling of the California Institute of Technology showed that some proteins have a secondary structure—a helix structure made possible by intramolecular hydrogen bonding. His research included the use of a complex X-ray diffraction pattern yielded by crystalline proteins. Both of these men later became Nobel laureates in science.

Within the following decade an Oxford University scientist, Dorothy C. Hodgkin, demonstrated that it was possible to improve the method of X-ray diffraction and developed the new three-dimensional viewpoint of molecular structure of extremely complex organic compounds. By the use of this technique two Cambridge scientists, John C. Kendrew and Max F. Perutz, obtained clear, three-dimensional structures of several proteins such as hemoglobin and myoglobin.

In the meantime perhaps the most important single advance in biochemistry in a century was achieved by a team of an American, James D. Watson, Francis D. Crick of Cambridge University, and Maurice H. Wilkins of King's College who deduced a three-dimensional structure of DNA or deoxyribonucleic acid by an analysis of X-ray diffraction patterns and electron microscope photos of DNA fibers. They showed that DNA is a double helix with a twisted ladder structure.

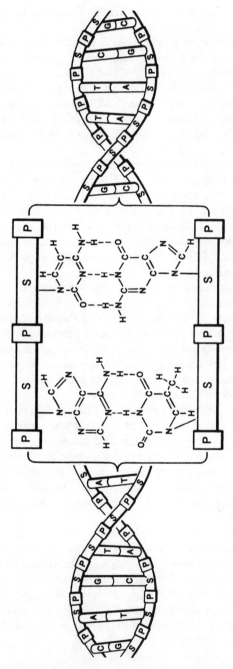

Watson-Crick Model of DNA (Double Helix)

DNA is found in the nuclei of all living cells and in many viruses. It is constructed chiefly from six building blocks. One is a sugar (deoxyribose), another is phosphate (PO_4), and four nucleotide bases named adenine, thymine, guanine and cytosine usually designated A,T,G, and C. It is believed that the sequence of nucleotides in the DNA molecule of the chromosome of cells carries the genetic material which controls heredity.

RNA (ribonucleic acid) is also found in living cells and resembles DNA except that in RNA uracil (U) replaces thymine in one of the nucleotides. RNA translates the genetic material in DNA into the requisite proteins.

The discovery of the double helix turned genes into chemical objects whose structure and function could be analyzed and understood in terms of biochemical machinery. Molecular biology was born.

Giant strides were soon made in the field of genetics on the molecular level. In 1956 Arthur Kornberg, a Brooklyn-born biochemist, set out with a team of researchers to try to produce DNA. Using the necessary amino acids, a chemical to furnish the energy necessary for a chemical change, and an enzyme extracted from certain bacteria to act as catalyst, he added a trace of DNA to act as a primer. He obtained large amounts of new DNA indistinguishable from the bit of DNA he had introduced. However, it turned out not to be the real DNA he was after.

In another related field Stanford Moore and William Stein of Rockefeller University, using column chromatography, were able to determine the full sequence of the 124 amino acids that form ribonuclease (RNase), a small enzyme that came from the bovine pancreas. Christian B. Anfinsen of the National Institute of Arthritis and Metabolic Diseases at Bethesda, Maryland, discovered how this molecule develops its three-dimensional structure. He was interested in the connection between the amino acid sequence and its biologically active conformation. These three men during this research invented the fraction collector and developed a completely automatic amino acid analyzer—an enormous step forward, and a new tool for further protein reseach.

Antibodies which constitute another type of protein presented another chemical bastion to be breached. The fundamental problem of immunology is to determine how the body's own antibody molecule can recognize foreign molecules such as bacteria, viruses and foreign tissues from its own molecules. In the late 1950's Gerald M. Edelman, physician and biochemist at Rockefeller University, and Rodney R. Porter working at Oxford University began tackling this problem independently. They found that gamma globulin (or immunoglobulin) is the specific blood chemical that detects the difference between the unique biological system of one individual from that of any other. By clutching the foreign protein it renders it powerless.

By 1969 Edelman first reported he had deciphered the structure of the entire gamma globulin molecule from material present in a tumor (myelonie). Later he and Porter determined the amino acid sequence of all of its four peptide chains and showed how they were connected. This complete molecule contains 1320 amino acid, 19,996 atoms, and has a molecular weight of about 150,000. They also helped illuminate the relation between chemical structure and the catalytic activity of an enzyme. This was a fundamental advance and confirmed Linus Pauling's belief that it should be possible in time for researchers to make a direct attack on a specific disease on the molecular level. These men, including all of the twenty or more feverishly working in the still-beckoning field of the proteins as described above, all became Nobel Prize laureates. Never before have so many men and women working in the field been so honored in such a short span of time.

Biochemists are also laying siege to another stubborn citadel of proteins, the viruses. These cause, among others, the human diseases of influenza, a virus pneumonia, smallpox, polio, mumps and measles. More spectacular advances in molecular biology continue to be announced. The door has now been opened for the possible synthesis of RNA and DNA. Synthesizing DNA will be creating life since it shapes and directs the machinery of life. It is a chemical agent capable of initiating the living process and building tissues.

Late in 1972 a tentative report was issued from the State University of Ghent in Belgium. Scientists there had made a complete analysis of the chemical structure of a gene. Previous to this, simple artificial genes with known structures had been synthesized. In 1970 Nobel laureate Har Gobind Khorana had manufactured a major part of a real gene. The yeast gene, however, could not direct the production of a protein. In 1975 a mammalian gene, that of a rabbit, was totally synthesized by a team of four Harvard University molecular biologists. It contains 650 nucleotides—as long as a human gene.

The belief in the old vital force which Woehler destroyed is still dead, but in its place there remains another vital power more puzzling than ever. Warburg in Germany, A. V. Hill in England, and L. Henderson and Van Slyke in America worked to unravel the mysterious force which controls the birth, growth and development of living forms. Eugenio Rignano, an Italian philosopher, had hopes that this biotic or vitalistic nervous energy would some day be discovered. Sir Oliver Lodge once told an audience at Oxford University that "it is sometimes said by students of organic chemistry that if we could contrive in the laboratory to continue the manufacture of organic compounds until we had a mass of protoplasm, and were able to subject it to suitable pressure, they would expect that artificial protoplasm to exhibit vitality and manifest one or another form of life." This is no challenge to the conception of God, as some fundamentalists maintain. It is difficult to believe, however, that man will soon be able to produce that entelechy or expanding vital impulse which can breathe the breath of life into the most complex chemical compound he makes. Many still believe that man is probably more than a chemical concatenation of a lump of coal, a whiff of air and a beaker of salt solutions.

Even the achievement of synthetic life would not have frightened the philosopher Emerson. For to him, scientific triumph was not the death but the birth of further mystery.

> I do not know that I should feel threatened or insulted if a chemist should take this protoplasm or mix his hydrogen, oxygen and carbon, and make an animalcule incontestably swimming and jumping before my eyes. I should only feel that it indicated that the day had arrived when the human race might be trusted with a new degree of power and its immense responsibility; for these steps are not solitary or local, but only a hint of an advanced frontier suggested by an advancing race behind it.

However, today, some scientists are apprehensive of the future of this line of research, especially that of so-called genetic engineering. This might, for example, create drug-resistant germs that might escape from the research laboratory and produce havoc. Marshall W. Nirenberg, Nobel Prize winner working at the National Institute of Health, has expressed this concern in this manner: "When man becomes capable of instructing its own cells, he must refrain from doing so until he has sufficient *wisdom* to use his knowledge for the benefit of mankind."

Woehler died in his eighty-third year, following an illness of only three days. After a simple funeral he was buried in Göttingen, the city of his life work. In accordance with his wish only a modest legend was carved on his tombstone—"Friedrich Woehler; Born July 31, 1800; Died Sept. 23, 1882." At Downs, five months before, there passed away another pioneer of science, Charles Darwin, the man who recreated life out of the rocks and fossils of the earth even as Woehler created a new world of compounds out of the same inanimate stones, and with them showed the way to the modern Elixirs of Life.

XI

MENDELÉEFF

SIBERIA BREEDS A PROPHET

OUT of Russia came the patriarchal voice of a prophet of chemistry. "There is an element as yet undiscovered. I have named it eka-aluminum. By properties similar to those of the metal aluminum you shall identify it. Seek it, and it will be found." Startling as was this prophecy, the sage of Russia was not through. He predicted another element resembling the element boron. He was even bold enough to state its atomic weight. And before that voice was stilled, it foretold the discovery of a third element whose physical and chemical properties were thoroughly described. No man, not even the Russian himself, had beheld these unknown substances.

This was the year 1869. The age of miracles was long past. Yet here was a distinguished scientist, holding a chair of chemistry at a famous university, covering himself with the mantle of the prophets of old. Had he gathered this information from inside the crystal glass of some sorcerer? Perhaps, like the seer of ancient times, he had gone to the top of a mountain to bring down the tablets of these new elements. But this oracle disdained the robes of a priest. Rather did he announce his predictions from the stillness of his chemical laboratory, where midst the smoke, not of a burning bush, but of the fire of his furnace, he had seen visions of a great generalization in chemistry.

Chemistry had already been the object of prophecy. When Lavoisier heated some tin in a sealed flask and found it to change in appearance and weight, he saw clearly a new truth, and foretold other changes. Lockyer a year before had looked through a new instrument—the spectroscope devised by Bunsen and Kirchhof. Through this spectroscope he had gazed at the bright colored lines of a new element ninety-three million miles away. Since it was present in the photosphere of the sun he called it *helium* and predicted its existence on our earth. Twenty-one years later,

William Hillebrand of the United States Geological Survey, came across this gas in the rare mineral cleveite.

But the predictions of the Russian were more astounding. He had made no direct experiments. He had come to his conclusions seemingly out of thin air. There had gradually been born in the fertile mind of this man the germ of a great truth. It was a fantastic seed but it germinated with surprising rapidity. When the flower was mature, he ventured to startle the world with its beauty.

In 1884 Sir William Ramsay had come to London to attend a dinner given in honor of William Perkin, the discoverer of the dye mauve.

> I was very early at the dinner [Ramsay recalled] and was putting off time looking at the names of people to be present, when a peculiar foreigner, every hair of whose head acted in independence of every other, came up bowing. I said, "We are to have a good attendance, I think?" He said, "I do not spik English." I said, "Vielleicht sprechen sie Deutsch?" He replied, "Ja ein wenig. Ich bin Mendeléeff." Well, we had twenty minutes or so before anyone else turned up and we talked our mutual subject fairly out. He is a nice sort of fellow but his German is not perfect. He said he was raised in East Siberia and knew no Russian until he was seventeen years old. I suppose he is a Kalmuck or one of those outlandish creatures.

This "outlandish creature" was Mendeléeff, the Russian prophet to whom the world listened. Men went in search of the missing elements he described. In the bowels of the earth, in the flue dust of factories, in the waters of the oceans, and in every conceivable corner they hunted. Summers and winters rolled by while Mendeléeff kept preaching the truth of his visions. Then, in 1875, the first of the new elements he foretold was discovered. In a zinc ore mined in the Pyrenees, Lecoq de Boisbaudran came upon the hidden eka-aluminum. This Frenchman analyzed and reanalyzed the mineral and studied the new element in every possible way to make sure there was no error. Mendeléeff must indeed be a prophet! For here was a metal exactly similar to his eka-aluminum. It yielded its secret of two new lines to the spectroscope, it was easily fusible, it could form alums, its chloride was volatile. Every one of these characteristics had been accurately foretold by the Russian. Lecoq named it *gallium* after the ancient name of his native country.

But there were many who disbelieved. "This is one of those strange guesses which by the law of averages must come true," they argued. Silly to believe that new elements could be predicted with such accuracy! One might as well predict the birth of a new star in the heavens. Had not Lavoisier, the father of chemistry, declared that "all that can be said upon the nature and number of the elements is confined to discussions entirely of a metaphysical nature? The subject only furnishes us with indefinite problems."

But then came the news that Winkler, in Germany, had stumbled over another new element, which matched the eka-silicon of Mendeléeff. The German had followed the clue of the Russian. He was looking for a dirty gray element with an atomic weight of about 72, a density of 5.5, an element which was slightly acted upon by acids. From the silver ore, argyrodite, he isolated a grayish white substance with atomic weight of 72.3 and a density of 5.5. He heated it in air and found its oxide to be exactly as heavy as had been predicted. He synthesized its ethide and found it to boil at exactly the temperature that Mendeléeff had prefigured. There was not a scintilla of doubt about the fulfilment of Mendeléef's second prophecy. The spectroscope added unequivocal testimony. Winkler announced the new element under the name of *germanium* in honor of his fatherland.The sceptics were dumbfounded. Perhaps after all the Russian was no charlatan!

Two years later the world was completely convinced. Out of Scandinavia came the report that Nilson had isolated eka-boron. Picking up the scent of the missing element in the ore of euxenite, Nilson had tracked it down until the naked element, exhibiting every property foreshadowed for it, lay before him in his evaporating dish. The data were conclusive. The whole world of science came knocking at the door of the Russian in St. Petersburg.

Dmitri Ivanovitch Mendeléeff came of a family of heroic pioneers. More than a century before his birth, Peter the Great had started to westernize Russia. Upon a marsh of pestilence he reared a mighty city which was to be Russia's window to the West. For three-quarters of a century Russia's intellectual march eastward continued, until in 1787 in Tobolsk, Siberia, the grandfather of Dmitri opened up the first printing press, and with the spirit of a pioneer published the first newspaper in Siberia, the *Irtysch*. In this desolate spot, settled two centuries before by the Cossacks, Dmitri was born on February 7, 1834. He was the last of a family of seventeen children.

Misfortune overtook his family. His father, director of the local high school, became blind, and soon after died of consumption. His mother, Maria Korniloff, a Tartar beauty, unable to support her large family on a pension of five hundred dollars a year, reopened a glass factory which her family was the first to establish in Siberia. Tobolsk at this time was an administrative center to which Russian political exiles were taken. From one of these prisoners of the revolt of 1825, a "Decembrist" who married his sister, Dmitri learned the rudiments of natural science. When fire destroyed the glass factory, little Dmitri, pet of his ageing mother—she was already fifty-seven —was taken to Moscow in the hope that he might be admitted to the University. Official red tape prevented this. Determined that her son should receive a good scientific education, his mother undertook to move to St. Petersburg, where he finally gained admittance to the Science Department of the Pedagogical Institute, a school for the training of high school teachers. Here he specialized in mathematics, physics and chemistry. The classics were distasteful to this blue-eyed boy. Years later, when he took a hand in the solution of Russia's educational problems, he wrote, "We could live at the present time without a Plato, but a double number of Newtons is required to discover the secrets of nature, and to bring life into harmony with its laws."

Mendeléeff worked diligently at his studies and graduated at the head of his class. Never very robust during these early years, his health gradually weakened, and the news of his mother's death completely unnerved him. He had come to her as she lay on her death bed. She spoke to him of his future: "Refrain from illusions, insist on work and not on words. Patiently search divine and scientific truth." Mendeléeff never forgot those words. Even as he dreamed, he always felt the solid earth beneath his feet.

His physician gave him six months to live. To regain his health, he was ordered to seek a warmer climate. He went to the south of Russia and obtained a position as science master at Simferopol in the Crimea. When the Crimean War broke out he left for Odessa, and at the age of twenty-two he was back in St. Petersburg as a privat-docent. An appointment as privat-docent meant nothing more than permission to teach, and brought no stipend save a part of the fees paid by the students who attended the lectures. Within a few years he asked and was granted permission from the Minister of Public Instruction to study in France and Germany. There was no opportunity in Russia for advanced work in science. At Paris he worked in the laboratory of Henri Regnault and, for another year, at Heidelberg in a small private laboratory built out of his meager means. Here he met Bunsen and Kirchhof from whom he learned the use of the spectroscope, and together with Kopp attended the Congress of Karlsruhe, listening to the great battle over the molecules of Avogadro. Cannizarro's atomic weights were to do valiant service for him in the years to come. Mendeléeff's attendance at this historic meeting ended his *Wanderjahre*.

The next few years were very busy ones. He married, completed in sixty days a five-hundred page textbook on organic chemistry which earned him the Domidoff Prize, and gained his doctorate in chemistry for a thesis on *The Union of Alcohol with Water*. The versatility of this gifted teacher, chemical philosopher and accurate experimenter was soon recognized by the University of St. Petersburg, which appointed him full professor before he was thirty-two.

Then came the epoch-making year of 1869. Mendeléeff had spent twenty years reading, studying and experimenting with the chemical elements. All these years he had been busy collecting a mass of data from every conceivable source. He had arranged and rearranged this data in the hope of unfolding a secret. It was a painstaking task. Thousands of scientists had worked on the elements in hundreds of laboratories scattered over the civilized world. Sometimes he had to spend days searching for missing data to complete his tables. The number of the elements had increased since the ancient artisans fashioned instruments from their gold, silver, copper, iron, mercury, lead, tin, sulfur and carbon. The alchemists had added six new elements in their futile search for the seed of gold and the elixir of life. Basil Valentine, a German physician, in the year when Columbus was discovering America had rather fancifully described antimony. In 1530 Georgius Agricola, another German, talked about bismuth in his *De Re Metallica*, a book on mining which was translated into English for the first time by a (later) President of the United States, Herbert Hoover, and his wife in 1912. Paracelsus was the first to mention the metal zinc to the Western World. Brandt discovered glowing phosphorus in urine, and arsenic and cobalt were soon added to the list of the elements.

Before the end of the eighteenth century, fourteen more elements were discovered. In faraway Choco, Colombia, a Spanish naval officer, Don Antonio de Ulloa, had picked up a heavy nugget while on an astronomical mission, and had almost discarded it as worthless before the valuable properties of the metal platinum were recognized. This was in 1735. Then came lustrous nickel, inflammable hydrogen, inactive nitrogen, life-giving oxygen, death-dealing chlorine, manganese, used among other things for burglar-proof safes, tungsten, for incandescent lamps, chromium, for stainless steel, molybdenum and titanium, so useful in steel alloys, tellurium, zirconium, and uranium, heaviest of all the elements. The nineteenth century had hardly opened when Hatchett, an Englishman, discovered columbium (niobium) in a black mineral that had found its way from the Connecticut Valley to the British Museum. And thus the search went on, until in 1869 sixty-three different elements had been isolated and described in the chemical journals of England, France, Germany and Sweden.

Mendeléeff gathered together all the data on these sixty-three chemical elements. He did not miss a single one. He even included fluorine whose presence was known, but which had not yet been isolated because of its tremendous activity. Here was a list of all the chemical elements, every one of them consisting of different Daltonian atoms. Their atomic weights, ranging from 1 (hydrogen) to 238 (uranium), were all dissimilar. Some, like oxygen, hydrogen, chlorine and nitrogen, were gases. Others, like mercury and bromine, were liquids under normal conditions. The rest were solids. There were some very hard metals like platinum and iridium, and soft metals like sodium and potassium. Lithium was a metal so light that it could float on water. Osmium, on the other hand, was twenty-two and a half times as heavy as water. Here was mercury, a metal which was not a solid at all, but a liquid. Copper was red, gold yellow, iodine steel gray, phosphorus white, and bromine red. Some metals, like nickel and chromium, could take a very high polish; others like lead and aluminum, were duller. Gold, on exposure to the air, never tarnished, iron rusted very easily, iodine sublimed and changed into a vapor. Some elements united with one atom of oxygen, others with two, three or four atoms. A few, like potassium and fluorine, were so active that it was dangerous to handle them with the unprotected fingers. Others could remain unchanged for ages. What a maze of varying, dissimilar, physical characteristics and chemical properties!

Could some order be found in this body of diverse atoms? Was there any connection between these elements? Could some system of evolution or development be traced among them, such as Darwin, ten years before, had found among the multiform varieties of organic life? Mendeléeff wondered. The problem haunted his dreams. Constantly his mind reverted to this puzzling question.

Mendeléeff was a dreamer and a philosopher. He was going to find the key to this heterogeneous collection of data. Perhaps nature had a simple secret to unfold. And while he believed it to be "the glory of God to conceal a thing," he was firmly convinced that it was "the honor of kings to search it out." And what a boon it would prove to his students!

He arranged all the elements in the order of increasing atomic weights, starting with the lightest, hydrogen, and completing his table with uranium, the heaviest. He saw no particular value in arranging the elements in this way; it had been done previously. Unknown to Mendeléeff, an Englishman, John Newlands, had three years previously read, before the English Chemical Society at Burlington House, a paper on the arrangement of the elements. Newlands had noticed that each succeeding eighth element in his list showed properties similar to the first element. This seemed strange. He compared the table of the elements to the keyboard of a piano with its eighty-eight notes divided into periods or octaves of eight. "The members of the same group of elements," he said, "stand to each other in the same relation as the extremities of one or more *octaves* in music." The members of the learned society of London laughed at his Law of Octaves. Professor Foster ironically inquired if he had ever examined the elements according to their initial letters. No wonder—think of comparing the chemical elements to the keyboard of a piano! One might as well compare the sizzling of sodium as it skims over water to the music of the heavenly spheres. "Too fantastic," they agreed, and J. A. R. Newlands almost went down to oblivion.

Mendeléeff was clear-visioned enough not to fall into such a pit. He took sixty-three cards and placed on them the names and properties of the elements. These cards he pinned on the walls of his laboratory. Then he carefully reexamined the data. He sorted out the similar elements and pinned their cards together again on the walls. A striking relationship was thus made clear.

Mendeléeff now arranged the elements into seven groups, starting with lithium (at. wt. 7), and followed by beryllium (at. wt. 9), boron (11), carbon (12), nitrogen (14), oxygen (16) and fluorine (19). The next element in the order of increasing atomic weight was sodium (23). This element resembled lithium very closely in both physical and chemical properties. He therefore placed it below lithium in his table. After placing five more elements he came to chlorine, which had properties very similar to fluorine, under which it miraculously fell in his list. In this way he continued to arrange the remainder of the elements. When his list was completed he noticed a most remarkable order. How beautifully the elements fitted into their places! The very active metals lithium, sodium, potassium, rubidium and caesium fell into one group (No. 1). The extremely active non-metals, fluorine, chlorine, bromine and iodine, all appeared in the seventh group.

Mendeléeff had discovered that the properties of the elements "were periodic functions of their atomic weights," that is, their properties repeated themselves periodically after each seven elements. What a simple law he had discovered! But here was another astonishing fact. All the elements in Group I united with oxygen two atoms to one. All the atoms of the second group united with oxygen atom for atom. The elements in Group III joined with oxygen two atoms to three. Similar uniformities prevailed in the remaining groups of elements. What in the realm of nature could be more simple? To know the properties of one element of a certain group was to know, in a general way, the properties of all the elements in that group. What a saving of time and effort for his chemistry students!

Could his table be nothing but a strange coincidence? Mendeléeff wondered. He studied the properties of even the rarest of the elements. He re-searched the chemical literature lest he had, in the ardor of his work, misplaced an element to fit in with his beautiful edifice. Yes, here was a mistake! He had misplaced iodine, whose atomic weight was recorded as 127, and tellurium, 128, to agree with his scheme of things. Mendeléeff looked at his Periodic Table of the Elements and saw that it was good. With the courage of a prophet he made bold to say that the atomic weight of tellurium was wrong; that it must be between 123 and 126 and not 128, as its discoverer had determined. Here was downright heresy, but Dmitri was not afraid to buck the established order of things. For the present, he placed the element tellurium in its proper position, but with its false atomic weight. Years later his action was upheld, for further chemical discoveries proved his position of tellurium to be correct. This was one of the most magnificent prognostications in chemical history.

Perhaps Mendeléeff's table was now free from flaws. Again he examined it, and once more he detected an apparent contradiction. Here was gold with the accepted atomic weight of 196.2 placed in a space which rightfully belonged to platinum, whose established atomic weight was 196.7. The fault-finders got busy. They pointed out this discrepancy with scorn. Mendeléeff made brave enough to claim that the figures of the analysts, and not his table, were inaccurate. He told them to wait. He would be vindicated. And again the balance of the chemist came to the aid of the philosopher, for the then-accepted weights were wrong and Mendeléeff was again right. Gold had an atomic weight greater than platinum. This table of the queer Russian was almost uncanny in its accuracy!

Mendeléeff was still to strike his greatest bolt. Here were places in his table which were vacant. Were they always to remain empty or had the efforts of man failed as yet to uncover some missing elements which belonged in these spaces? A less intrepid person would have shrunk from the conclusion that this Russian drew. Not this Tartar, who would not cut his hair even to please his Majesty, Czar Alexander III. He was convinced of the truth of his great generalization, and did not fear the blind, chemical sceptics.

Here in Group III was a gap between calcium and titanium. Since it occurred under boron, the missing element must resemble boron. This was his eka-boron which he predicted. There was another gap in the same group under aluminum. This element must resemble aluminum, so he called it eka-aluminum. And finally he found another vacant space between arsenic and eka-aluminum, which appeared in the fourth group. Since its position was below the element silicon, he called it eka-silicon. Thus he predicted three undiscovered elements and left it to his chemical contemporaries to verify his prophecies. Not such remarkable guesses after all—at least not to the genius Mendeléeff!

In 1869 Mendeléeff, before the Russian Chemical Society, presented his paper *On the Relation of the Properties to the Atomic Weights of the Elements*. In a vivid style he told them of his epoch-making conclusions. The whole scientific world was overwhelmed. His great discovery, however, had not sprung forth overnight full grown. The germ of this important law had begun to develop years before. Mendeléeff admitted that "the law was the direct outcome of the stock of generalizations of established facts which had accumulated by the end of the decade 1860—1870." De Chancourtois in France, Strecher in Germany, Newlands in England, and Cooke in America had noticed similarities among the properties of certain elements. But no better example could be cited of how two men, working independently in different countries, can arrive at the same generalization, than the case of Lothar Meyer, who conceived the Periodic Law at almost the same time as Mendeléeff. In 1870 there appeared in *Liebig's Annalen* a table of the elements by Lothar Meyer which was almost identical with that of the Russian. The time was ripe for this great law. Some wanted the boldness or the genius necessary "to place the whole question at such a height that its reflection on the facts could be clearly seen." This was the statement of Mendeléeff himself. Enough elements had been discovered and studied to make possible the arrangement of a table such as Mendeléeff had prepared. Had Dmitri been born a generation before, he could never, in 1840, have enunciated the Periodic Law.

"The Periodic Law has given to chemistry that prophetic power long regarded as the peculiar dignity of the sister science, astronomy." So wrote the American scientist Bolton. Mendeléeff had made places for more than sixty-three elements in his Table. Three more he had predicted. What of the other missing building blocks of the universe? Twenty-five years after the publication of Mendeléeff's Table, two Englishmen, following a clue of Cavendish, came upon a new group of elements of which even the Russian had never dreamed. These elements constituted a queer company—the Zero Group as it was later named. Its members, seven in number, are the most unsociable of all the elements. Even with that ideal mixer, potassium, they will normally not unite. Fluorine, most violent of all the non-metals, cannot shake these hermit elements out of their inertness. Moissan tried sparking them with fluorine but failed to make them combine. (Xenon tetrafluoride and several other "noble" compounds were prepared in 1962. They are no longer regarded as non-reactive.) Besides, they are all gases, invisible and odorless. Small wonder they had remained so long hidden.

True, the first of these noble gases, as they were called, had been observed in the sun's chromosphere during a solar eclipse in August, 1868, but as nothing was known about it except its orange yellow spectral line, Mendeléeff did not even include it in his table. Later, Hillebrand described a gas expelled from cleveite. He knew enough about it to state that it differed from nitrogen but he failed to detect its real nature. Then Ramsay, obtaining a sample of the same mineral, bottled the gas expelled from it in a vacuum tube, sparked it and detected the spectral line of helium. The following year Kayser announced the presence of this gas in very minute amounts, one part in 185,000, in the earth's atmosphere.

The story of the discovery and isolation of these gases from the air is one of the most amazing examples of precise and painstaking researches in the whole history of science. Ramsay had been casually introduced to chemistry while convalescing from an injury received in a football game. He had picked up a textbook in chemistry and turned to the description of the manufacture of gunpowder. This was his first lesson in chemistry. Rayleigh, his co-worker, had been urged to enter either the ministry or politics, and when he claimed that he owed a duty to science, was told his action was a lapse from the straight and narrow path. Such were the initiations of these two Englishmen into the science which brought them undying fame. They worked with gases so small in volume that it is difficult to understand how they could have studied them in their time. Rayleigh, in 1894, wrote to Lady Frances Balfour: "The new gas has been leading me a life. I had only about a quarter of a thimbleful. I now have a more decent quantity but it has cost about a thousand times its weight in gold. It has not yet been christened. One pundit suggested 'aeron,' but when I have tried the effect privately, the answer has usually been, 'When may we expect Moses?' " It was finally christened argon, and if not Moses, there came other close relatives: neon, krypton, xenon and finally radon. These gases were isolated by Ramsay and Travers from one hundred and twenty tons of air which had been liquefied. Sir William Ramsay used a micro-balance which could detect a difference in weight of one fourteen-trillionth of an ounce. He worked with a millionth of a gram of invisible, gaseous radon—the size of a tenth of a pin's head.

Besides these six Zero Group elements, some of which are doing effective work in argon and neon incandescent lamps, in helium-filled dirigibles, in electric signs, and in replacing the nitrogen in compressed air to prevent the "bends" among caisson workers, seventeen other elements were unearthed. So that, a year after Mendeléeff died in 1907, eighty-six elements were listed in the Periodic Table, a fourfold increase since the days of Lavoisier.

Mendeléeff, besides being a natural philosopher in the broadest sense of the term, was also a social reformer. He was aware of the brutality and tyranny of Czarist Russia. He had learned his first lessons from the persecuted exiles in frozen Tobolsk. As he travelled about Russia, he went third class, and engaged in intimate conversation with the peasants and small tradespeople in the trains. They hated the remorseless oppression and espionage of the governments. Mendeléeff was not blind to the abuses of Russian officialdom, nor did he fear to point them out. He was often vehement in his denunciations. This was a dangerous procedure. But the government needed Mendeléeff, and his radical utterances were always mildly tinged with due respect for law and order. Mendeléeff was shrewd enough not to make a frontal attack on the government. He would bide his time and wait for an opportune moment when his complaints could not easily be ignored. On more than one occasion when this scientific genius showed signs of political eruption, he was hastily sent away on some government mission. Far from the centers of unrest he was much safer and of greater value to the officials.

In 1876, Mendeléeff was commissioned by the government of Alexander II to visit the oil fields of Pennsylvania in distant America. These were the early days of the petroleum industry. In 1859, Colonel Edwin L. Drake and his partner "Uncle Billy" Smith had gone to Titusville, Pennsylvania, to drive a well sixty-nine feet deep—the first to produce oil on a commercial scale. Mendeléeff had already been of invaluable service to Russia by making a very careful study of her extensive oil fields of Baku. Here, in the Caucasus, from a gap in the rock, burned the "everlasting flame" which Marco Polo had described centuries back. Baku at this time was the most prolific single oil district in the world and, from earliest times, people had burned its oil which they had dipped from its springs. Mendeléeff developed an ingenious theory to explain the origin of these oil deposits. He refused to accept the prevalent idea that oil was the result of the decomposition of organic material in the earth, and postulated that energy-bearing petroleum was formed by the interaction of water and metallic carbides found in the interior of the earth.

On his return from America, Mendeléeff was again sent to study the naphtha springs in the south of Russia. He did not confine his work to the gathering of statistics and the enunciation of theories. He developed in his own laboratory a new method for the commercial distillation of these products and saved Russia vast sums of money. He studied the coal region on the banks and basin of the Donetz River and opened it to the world. He was an active propagandist for Russia's industrial development and expansion, and was called upon to help frame a protective tariff for his country.

This was a period of intense social and political unrest in Russia. Alexander II had attempted to settle the land question of his twenty-three million serfs. He tried further to ameliorate conditions by reforming the judicial system, relaxing the censorship of the press, and developing educational facilities. The discontented students at the University of St. Petersburg presented a petition for a change in certain educational practices and other grievances. Suddenly an insurrection against the Russian government broke out in Poland. The militant forces again gained control. Russia was in no mood for radical changes; the requests of the students were peremptorily turned down and the more militant ones were arrested. Mendeléeff stepped in and presented another of their petitions to the officials of the government. He was bluntly told to go back to his laboratory and stop meddling in the affairs of the state. Proud and sensitive, Mendeléeff was insulted and resigned from the University. Prince Kropotkin, a Russian anarchist of royal blood, was one of his famous students. "I am not afraid," Mendeléeff had declared, "of the admission of foreign, even of socialistic ideas into Russia, because I have faith in the Russian people who have already got rid of the Tartar domination and the feudal system." He did not change his views even after the Czar, in 1881, was horribly mangled by a bomb thrown into his carriage.

Mendeléeff had made many enemies by his espousal of liberal movements. In 1880, the St. Petersburg Academy of Sciences refused, in spite of very strong recommendations, to elect him member of its chemical section. His liberal tendencies were an abomination. But other and greater honors came to him. The University of Moscow promptly made him one of its honorary members. The Royal Society of England presented him with the Davy Medal which he shared with Lothar Meyer for the Periodic Classification of the Elements.

Years later, as he was being honored by the English Chemical Society with the coveted Faraday Medal, Mendeléeff was handed a small silk purse worked in the Russian national colors and containing the honorarium, according to the custom of the Society. Dramatically he tumbled the sovereigns out on the table, declaring that nothing would induce him to accept money from a Society which had paid him the high compliment of inviting him to do honor to the memory of Faraday in a place made sacred by his labors. He was showered with decorations by the chemical societies of Germany and America, by the Universities of Princeton, Cambridge, Oxford, and Göttingen. Sergius Witte, Minister of Finance under Czar Alexander III, appointed him Director of the Bureau of Weights and Measures.

Mendeléeff broke away from the conventional attitude of Russians towards women, and treated them as equals in their struggle for work and education. While he held them to be mentally inferior to men, he did not hesitate to employ women in his office, and admitted them to his lectures at the university. He was twice married. With his first wife, who bore him two children, he led an unhappy life. She could not understand the occasional fits of temper of this queer intellect. The couple soon separated and were eventually divorced. Then he fell madly in love with a young Cossack beauty of artistic temperament, and, at forty-seven, remarried. Anna Ivanovna Popova understood his sensitive nature, and they lived very happily. She would make allowances for his flights of fancy and occasional selfishness. Extremely temperamental and touchy, he wanted everybody to think well of him. At heart he was kind and loveable. Two sons and two daughters were born to them and Mendeléeff ofttimes expressed the feeling that "of all things I love nothing more in life than to have my children around me." Dressed in the loose garments which his idol, Leo Tolstoy, wore, and which Anna had sewn for him, Dmitri would sit at times for hours smoking. He made an impressive figure. His deep-set blue eyes shone out of a fine expressive face half covered by a long patriarchal beard. He always fascinated his many guests with his deep guttural utterances. He loved books, especially books of adventure. Fenimore Cooper and Lord Byron thrilled him. The theatre did not attract him, but he loved good music and painting. Accompanied by his wife, who herself had made pen pictures of some of the great figures of science, he often visited the picture galleries. His own study was adorned by her sketches of Lavoisier, Newton, Galileo, Faraday, and Dumas.

When the Russo-Japanese War broke out in February, 1904, Mendeléeff turned out to be a strict nationalist. Old as he was, he added his strength in the hope of victory. Made advisor to the Navy, he invented pyrocollodion, a new type of smokeless powder. The destruction of the Russian fleet in the Straits of Tsushima and Russia's defeat hastened his end. His lungs had always bothered him; as a youth his doctor had given him only a few months to live. But his powerfully set frame carried him through more than seventy years of life. Then one day in February, 1907, the old scientist caught cold, pneumonia set in, and as he sat listening to the reading of Verne's *Journey to the North Pole*, he expired. Two days later Menschutkin, Russia's eminent analytical chemist, died, and within one year Russia lost also her greatest organic chemist, Friedrich Konrad Beilstein. Staggering blows to Russian chemistry.

To the end, Mendeléeff clung to scientific speculations. He published an attempt towards a chemical conception of the ether. He tried to solve the mystery of this intangible something which was believed to pervade the whole universe. To him ether was material, belonged to the Zero Group of Elements, and consisted of particles a million times smaller than the atoms of hydrogen.

Two years after he was laid beside the grave of his mother and son, the American Pattison Muir declared that "the future will decide whether the Periodic Law is the long looked for goal, or only a stage in the journey: a resting place while material is gathered for the next advance." Had Mendeléeff lived a few more years, he would have witnessed the beginnings of the final development of his Periodic Table by a young Englishman at Manchester.

The Russian peasant of his day never heard of the Periodic Law, but he remembered Dmitri Mendeléeff for another reason. One day, to photograph a solar eclipse, he shot into the air in a balloon, "flew on a bubble and pierced the sky." But to every boy and girl of the Soviet Union today Mendeléeff is a national hero. A special Mendeléeff stamp in his honor was issued in 1957 on the fiftieth anniversary of his death, and a new transuranium element, Number 101, created in 1955, was named *mendelevium* to commemorate his classic contribution to the science of chemistry.

XII

ARRHENIUS

THREE MUSKETEERS FIGHT FOR IONS

IN THE historic chemical laboratory of the University of Leipzig two men, a German born in Riga, and a Swede, met towards the end of the nineteenth century to plan a great battle against an established theory and the scientific inertia which upheld it. Meanwhile, over in Amsterdam, another scientist, a Dutchman, worked in the same campaign. From this triumvirate came a barrage of scientific experiments which made possible a new era in the field of theoretical and applied chemistry. Here, at Leipzig, the Headquarters of the Ionians, the great struggle was directed.

The three were all young men. Svante Arrhenius was hardly more than a boy. Van't Hoff, the Dutch professor, was thirty-five, and Ostwald, the moving spirit of the revolt, a year younger. The quest for scientific truth had brought these three together, and they vowed to force the venerable authorities of the scientific world to accept the new leaven of the younger generation. The masters, under whom they had cut their scientific eye-teeth, must be shown the folly of ignoring genius among their students.

One of the most difficult problems of that time was a rational understanding of what goes on in a solution when an electric current is sent through it. Even before that memorable day, nearly a century before, when the first experimenter arranged the two poles of his galvanic battery so that an electric current might pass through a solution, this problem had puzzled and perplexed the brainiest of those who followed him. Both Davy and Grothuss had attempted explanations. Faraday, discoverer of electromagnetic induction, had also investigated this subject and had created its terminology. Yet no solution had been found.

The same love of adventure that impelled his countryman Rolf to set sail for the coasts of Normandy prompted Svante Arrhenius to undertake the exploration of a problem that had baffled men grown old in dingy laboratories. An electric current could not be made to traverse distilled water. Neither could solid salt offer free passage to electricity. Yet when salt and water were mixed, their solution became a liquid through which electricity would pass with ease. And, as the electric current passed through this solution, a deep-seated decomposition took place. How could one explain this strange behavior of solutions?

Svante not only wondered but set to work. He was a visionary who soared in the clouds as he watched his test tubes and beakers. He had always been a dreamer, even when as a lad he attended school in his native village of Wijk near Upsala. At seventeen he had graduated, the youngest and ablest student of his class. He had given a brilliant account of himself in mathematics and the sciences. Carried on the shoulders of his friends, he was taken to the nearest hat shop to obtain the white velvet cap—insignia of the university student. At the State University of Upsala, where his father, too, had studied, he chose chemistry as his major subject. He hoped to follow in the footsteps of Berzelius, who, eighty years before, had walked the same halls and listened to the romance of chemistry in the same lecture rooms.

At twenty-two, Svante was ready for his doctorate and went to Stockholm. He had some queer notions of his own about the passage of electricity through solutions. He had done a great deal of thinking and experimenting along this line. Why not choose this problem for his thesis? It did not take him long to decide. He shut himself up in his laboratory. Day after day and often far into the night he filled beaker after beaker with solutions of different salts. One shining glass beaker contained a weak solution of copper sulfate. He labelled it accurately. A second tumbler was filled with a still weaker solution of magnesium sulfate. All over his laboratory table were bottles and flasks neatly marked with formulas and concentrations. Through each of these solutions he passed electric currents. He weighed, measured and recorded all of the results. And, as he watched bubbles of gas issuing from the plates dipped into the various solutions, his hunch, which was to solve the mystery, grew stronger.

Cavendish, a century before, had attempted to compare quantitatively the electrical conductivity of rain-water with various salt solutions. Possessing no galvanometer to register the strength of the currents, he had bravely converted his own nervous system into one. As he discharged Leyden jars through the different liquids he compared the electric shocks which he received. With this crude, heroic method he obtained a number of surprisingly accurate results.

Arrhenius was much better equipped. Great strides had since

been made in the field of electrical measurements. He, too, was an accurate worker and a patient one. For two years he toiled ceaselessly. Tiring, monotonous work, you might say. What joy or fun in sticking shiny electrodes into dozens of glass beakers and watching bubbles of gas or the movements of the dials on galvanometers, ammeters and voltmeters? The sun never shone for Svante during those months in the laboratory. He tried innumerable experiments with more than fifty different salts in all possible degrees of dilution.

"My great luck was that I investigated the conductivities of the most dilute solutions," he wrote later. "In these dilute solutions the laws are simple compared with those for concentrated solutions, which had been examined before." Luck it was, to some extent. But others had observed how the passage of the electric current became easier as more water was added to the concentrated solutions. They, too, had noticed some relation between the strength of an acid and its power to conduct a current. Arrhenius, however, was the first to see clearly the strange relationship between the ease of passage of an electric current through a solution, and the concentration of that solution.

Midst the never-ending washing of beakers and bottles and the perpetual weighings and recordings, Arrhenius stole moments to ponder over the meaning of it all. But first he must finish all of the experimental work. In the spring he had completed it. "I have experimented enough," he said. "Now I must think." He left his laboratory and returned to his home in the country to work out the theoretical part of his research. One night he sat up till very late. In those days the whole world, both of his waking and sleeping existence, was a world of solutions, currents and mathematical data. The rest did not exist. From the sublimated speculations of his experiments, suddenly there crystallized like a flash the answer to the great riddle. "I got the idea in the night of the 17th of May in the year 1883, and I could not sleep that night until I had worked through the whole problem."

Svante had a keen pictorial faculty and a remarkable memory which helped him visualize the whole range of data he had collected during those two years at Upsala. As a boy he would sit beside his father, manager of the university grounds, and help him with the accounts of the estate. He could remember and repeat with ease long rows of figures.

His thesis was now completed. He returned to Upsala with the dissertation in his pocket. He came to Clève, his professor of chemistry, with the new theory formulated in his thesis. "I have a new theory of electrical conductivity," said Svante Arrhenius. Clève, discoverer of the two metals, holmium and thulium, was no doubt a skillful experimenter and investigator of the rare earth elements. But theories to him were abominations to be fought or ignored entirely. In the classroom Arrhenius had listened to him for months. Never once had he heard a single mention of the great Periodic Law of Mendeléeff, even though the Russian's Table of the Elements was now more than ten years old.

Clève turned to this chemical tyro. "You have a new theory? That is very interesting. Good-by." Svante did not lose heart. He knew Clève—he had not expected an enthusiastic response.

As a candidate for the doctor's degree, Arrhenius had to defend his thesis in open debate. This was an event of great interest. The university appointed an opponent. Svante had taken special care in preparing his thesis. His professors at Upsala would be sure to search for the slightest error even of typesetting. He recognized the impossibility of getting them to accept the whole of his heterodox theory. He must not offend existing beliefs too ruthlessly. As a candidate for the doctorate he could not afford to tear down the idols they worshipped and hope to escape damnation. He could not, without danger to the theory he had conceived, make the heretical statements to which his thinking had led him. To save his new theory he was willing to compromise a little. "If I had made such statements in my doctor's thesis it would not have been approved," he later told the scientific world.

Arrhenius feared the enthusiasm of his youth might overstep the bounds of safety. He held himself in check. Carefully he chose the words for his answers. He made sure not to ride roughshod over the established principles of the University of Upsala.

At the end of four hours the questioning was over. Svante, in formal dress, waited breathlessly for the verdict. He expected trouble. The professors appeared to look upon him as a "stupid schoolboy" as Arrhenius remarked years later. They examined his complete record at the university. He had done fairly good work in mathematics, physics and biology.

The final result was announced. In spite of his dissertation, he was grudingly awarded his degree, and as a laurel wreath was placed on his head, a cannon outside boomed the advent of another doctor of philosophy. The award, however, was in reality a veiled condemnation of his theory. His dissertation was awarded a fourth class and his defense a third class.

Svante was almost broken-hearted. "It was difficult to see how the University of Upsala, the University of Bergman and Berzelius, could have condemned a brilliant thesis on the very subject of electrochemistry associated with their names." This was the judgment of Sir James Walker, professor of chemistry at the University of Edinburgh. This discouragement might have ended Svante's career as a chemist. But he was convinced that he had within his thesis a tool which would be a blessing to science. He, the Viking of Truth, was ready to do battle to vindicate his theory. But first he must ally himself with men of power in the field of chemistry. He himself was an unknown—he might look ludicrous in the armor of a chemical crusader.

Upsala was not friendly; he was certain of that. Stockholm, too, was unenthusiastic—had he not submitted his thesis to the Swedish Academy of Sciences only to be met with a cold reception? Sweden, the country of Scheele, Berzelius and Linnaeus, could not see the prophet within its walls.

Svante decided to appeal to the scientific world outside of Sweden. He sent a copy of his thesis to Rudolf Clausius, formulator of the Second Law of Thermodynamics. This German scientist was also the recognized oracle of electrochemistry. More than thirty years before, he had said: "In a solution the atoms composing molecules are constantly exchanging partners and, as a consequence, a certain proportion of the atoms will be uncombined at any instant." This statement seemed then the last word on the subject of Arrhenius' dissertation. "He was a great authority," thought Arrhenius, "therefore it could not be regarded as unwise to share his ideas," at least in part. Arrhenius, therefore, explained that the molecules which are active in solution "*are in the state described by Clausius.*" This expression "did not look so dangerous." But his tactful attempt to win over this German authority also failed. He received no encouragement. Clausius, now old and in feeble health, was not sufficiently interested.

Oliver Lodge also received a copy of his thesis. Lodge was, perhaps, the foremost scientist of England at the time. This was before he had abandoned pure science to grapple with the hidden mysteries of spiritualism. He had not as yet embarked upon his adventures of photographing ghosts and trapping the ectoplasm of departing souls. Still deeply rooted in the soil of scientific experimentation, he recognized in Arrhenius' paper "a distinct step towards a mathematical theory of chemistry." But that was all the active encouragement he gave the young Swedish scientist.

Arrhenius now sent his dissertation to Lothar Meyer. Surely Meyer would have the vision to see and the courage to uphold this new theory! For had he not, independently of Mendeléeff, arrived at the Periodic Law of the Elements? Had not he, Lothar Meyer, back in 1860 at Karlsruhe, listened to the vehement voice of

Cannizarro, and had he not at once championed the great truth of
that Italian crusader? Surely this German would enter the lists in
support of his heterodox theory! But Lothar Meyer, too, was silent.

Wilhelm Ostwald, professor of chemistry at the Polytechnical
School at Riga, also heard from Arrhenius. This champion of
daring chemical causes received Svante's paper on the day his wife
presented him with a new daughter. He was suffering that very day,
from a painful toothache! Ostwald later remarked that it "was too
much for one day. The worst was the dissertation, for the others
developed quite normally."

Arrhenius somehow felt that Ostwald would understand. That
was a lucky hunch. Ostwald read every word of that memoir. He
was tremendously excited. He flew up like a hornet and raged at the
stupidity of the Upsala professors. One could not help recognizing
the genius of this young man. He jumped at the revolutionary idea
that only *ions* took part in chemical reactions. Here was another
momentous cause worth fighting for.

Ostwald lost no time. Dropping all his work, he left at once for
Sweden. He made the long journey from Riga to Stockholm con-
vinced that assistance had to come immediately to the young talent.
The two met in Stockholm in August, 1884.

What was this iconoclastic doctrine of young Svante, which
kindled a blaze and set the chemical world afire? Arrhenius introduced a startling idea. He said that when a solid salt like
common table salt, sodium chloride, was dissolved in water a
tremendous change took place. This change was invisible. Pure
water itself was a non-conductor of electricity. The pure solid salt,
likewise, would not conduct an electric current. But when salt and
water were mixed, an instantaneous change occurred. The molecules of sodium chloride split up, *dissociated* or *ionized* into particles
which, years before, Faraday had labelled *ions* at the suggestion of
William Whewell, an expert in nomenclature. Faraday had
pictured these *ions* as being produced by the electric current.
Arrhenius said they were already present in the solution, even
before the electric current was sent through.

These two parts of the molecule of sodium chloride were absolutely *free*. In solution the ions swam around in all directions. There
were no longer any sodium chloride molecules present. Only sodium
ions and chlorine ions peopled the water. Here was the crash of a
holy idol. Clausius had said that only *some* of the molecules were in
this peculiar condition of dismemberment. Young Svante, the beginner, had dared declare that *all* the molecules in dilute solutions
were disrupted.

If this were true, some asked, then why could not the greenish yellow color of poisonous chlorine be seen? It was a logical and formidable question. Arrhenius answered that the chlorine *ions* differed from the *atoms* of chlorine because the ions were electrically charged. Dissociation had changed the atoms into ions, and the charge of electricity had changed the ion to such an extent that it differed fundamentally from its parent atom.

Arrhenius represented the change as follows:

$$\text{Sodium Chloride } \underline{\text{solution}} \longrightarrow \text{Sodium } \underline{\text{ions}} \; + \; \text{Chlorine } \underline{\text{ions}}$$
$$\underline{\text{NaCl}} \longrightarrow \underline{\text{Na}^+ +} \qquad \underline{\text{Cl}^-}$$

Here was a new chemistry—the chemistry of *ions*—strange, infinitesimal particles of matter bearing infinitely small electric charges which carried an electric current through solutions, and then, as they touched the electrodes, gave up their electric charges and returned once more to the atomic state. This mighty drama took place every time an inorganic acid, alkali or salt dissolved in water. Arrhenius was the first who saw clearly this invisible miracle role of the molecule in solution.

Ostwald grasped the value of this explanation almost at a glance. He was ready to accept the sweeping statement that chemical reactions in solution were reactions between *ions*. What a vast new field of experimentation it opened to science!

Ostwald and Arrhenius spent many pleasant days together in Stockholm. As they walked arm in arm along the shores of beautiful Lake Malar they spoke about ions until they were as real and tangible as so many electrified balls.

> Ostwald of course visited my dear friend and teacher Clève [wrote Arrhenius]. Ostwald spoke to him one day in his laboratory. I came a little later; I was not expected. I heard Clève say: "Do you *believe* solium chloride is dissolved into sodium and chlorine? In this glass I have a solution of sodium chloride. Do you believe there are sodium and chlorine in it? Do they look so?" "Oh, yes," Ostwald said, "there is some truth in that idea." Then I came in and the discussion was at an end. Clève threw a look at Ostwald which clearly showed that he did not think much of his knowledge of chemistry.

But Ostwald would not hurt the old professor. Besides, he was saving his powder for the great battle ahead. "We made plans," wrote Arrhenius, "regarding the development of the whole of chemistry."

Ostwald had been completely won over by the blond, rubicund, blue-eyed Swede. He invited him to come to Riga to continue his investigations in his laboratory. Svante might have gone on the moment. He was weary of the stubbornness of the professors of Upsala and Stockholm. But just then death came to his father, and he was delayed. Later, through the influence of Ostwald, he was given a travelling scholarship and at the close of 1885 his five years of *Wanderjahre* began. He went straight to Riga to work under the inspiration of Ostwald. There were many dubious points to be settled. They wanted to be absolutely certain. They needed reams of experimental ammunition to meet the terrific onslaught of the sceptical world of science. Ostwald, in the meantime, was working on the conductivity of acids.

The winter of 1886–7 was approaching. Arrhenius had spent almost a year with Ostwald. Friedrich Kohlrausch at Wurtzburg had been busy experimenting on the conductivity of solutions and had discovered that all the ions of the same element, regardless of the compound from which they were formed, behaved in exactly the same way. Arrhenius heard of his valiant work in this new field, and determined to leave Ostwald for a while and study with Kohlrausch. Surely he could learn something from this skillful German. At Wurtzburg, too, Arrhenius was to meet Emil Fischer. This young man, walking in the footsteps of Woehler, was busy with the synthesis of complex organic compounds. Arrhenius stopped to talk to him about his ions. Fischer was fascinated by the new conception, but he warned the Swede that most chemists would not readily accept his theory. It was too visionary, too revolutionary.

Arrhenius must present a foolproof theory or be damned by the chemical world as the parent of a monstrosity. In Kohlrausch's laboratory he jumped into the work again with the fervor of a fanatic. He must bring to the unbelieving world of science inexorable facts and invulnerable data. He read voraciously every piece of research that touched upon his subject. His star was bound to rise; soon he came across a memoir by Jacobus Hendrik van't Hoff.

Van't Hoff was a dreamer with a mind that leaped above the commonplace facts of chemistry and dared postulate new ideas. At twenty-two he had founded a new branch of chemistry, "stereochemistry," or the chemistry of atoms in space. He, too, had met with stubborn opposition. The world was up in arms against the "space chemistry" of this upstart. Kolbe, a distinguished German chemist, likened his stereochemistry to the belief in witchcraft. It was pernicious and dangerous. He raved against this fledgeling.

A certain Dr. van't Hoff, an official of the Veterinary School at Utrecht [Kolbe wrote] has no taste for exact chemical investigations. He has thought it more convenient to bestride Pegasus, evidently hired at the veterinary stables, and to proclaim in his *Chemistry in*

Space how, during his bold flight to the top of the chemical Parnassus, the atoms appeared to him to have grouped themselves throughout universal space.

Van't Hoff was not perturbed. He photographed the most decrepit horse to be found in the veterinary stables, labeled it Pegasus, and hung it on the walls of the University of Utrecht.

Van't Hoff had fought his way to recognition, championed by the same Wilhelm Ostwald. His "distorted theory" grew into a robust idea which did much to develop the field of organic chemistry. Now van't Hoff, thirteen years older, wrote about a theory of solution which suggested that dissolved substances obeyed the same laws as gases. Arrhenius read the paper very carefully. In it he found experimental data which was to help him fashion his own theories into a wonderfully consistent whole. He recognized in the Dutchman's memoir a great argument for his own theory of ionization.

Arrhenius was eager to work with van't Hoff. Time was passing rapidly and there was still much to be done. It was now the summer of 1887. But first he must meet Ludwig Boltzmann at Gratz with whom he worked until the following spring. Then Arrhenius set out for Amsterdam. On his way he stopped at Kiel to talk with Professor Max Planck, who became keenly interested in his theory and spent some time investigating it. This man Planck was another visionary who at the opening of the twentieth century, was to enunciate the "quantum" principle, a law of nature that shook the whole scientific world.

The friendship of Arrhenius and van't Hoff began when they met for the first time in Amsterdam. As van't Hoff worked side by side with Arrhenius for months, their devotion grew. Few men worked with more unselfishness. They talked about each other's theories. They discussed solutions, ions, gas laws and osmotic pressure. They pledged themselves to do battle for a common cause.

But Arrhenius was beginning to miss the fire of Ostwald, the human dynamo, whose essential characteristic was energy. He was almost ready for his final memoir on the chemical theory of electrolytes. He needed the effective aid and the cheering encouragement of his commander. Ostwald had written telling him of his new appointment as professor at the University of Leipzig. Arrhenius went there immediately. In the presence of Ostwald he could not help but gain renewed confidence in his theory. They brought together all the puzzling facts of electrolysis of solutions. Here they sat and planned the great Battle of the Ions. Ostwald had the foresight and shrewdness of the modern campaigner and public relations man. He was ready to launch a drive that was to end in a wave of enthusiasm for the ideas of Arrhenius. He first used the weapon of his newly founded scientific journal—the *Zeitschrift für Physikalische Chemie*—to broadcast the new theory of dissociation. He knew that the great notoriety which would be given to the theory even by opposition would suffice to launch a tremendous amount not only of discussion but of experimentation.

Ostwald's campaign was effective. Europe began to hear about Arrhenius and his strange ions. The young students in Ostwald's laboratory had been the first to hear the odd name of Svante Arrhenius. In the halls outside the laboratory where they gathered to smoke—they were forbidden this luxury in the laboratory—they spoke in whispers about this man whom their master had taken under his wing. Sir James Walker recalled how one day he "peered out of the laboratory and saw a stoutish, fair young man talking to Ostwald near the entrance hall. It was Arrhenius. We were made acquainted by Ostwald. He was the simplest and least assuming of men. He gave himself no airs."

When, in 1887, Arrhenius' classical paper "On the Dissociation of Substances in Aqueous Solutions" appeared in the first volume of the *Zeitschrift*, there was printed beside it van't Hoff's memoir on the analogy between the gaseous and dissolved state. As was anticipated, great opposition was aroused. The Battle of the Ions was raging in earnest. Ostwald led his small but valiant army of Ionians like a true warrior of old. His two solitary lieutenants were Arrhenius and van't Hoff. The host of the opposition was a formidable one. There were many in the workshops of science who would not swallow these ions. Even Mendeléeff opposed them, because he did not consider the theory in accordance with facts. His opposition, however, was not so severe. He believed that "the conception of electrical dissociation, although retarding the progress of the theory of solutions, was useful in giving the motive for collecting a store of experimental data to be embraced by a truer explanation in the future." Others, more severe, brought argument after argument to bear against these ions.

Ostwald, the great chemical crusader, leader of forlorn and victorious hopes, was impatient. "Let us attack them," he boomed, "that is the best method." He opened the pages of his chemical journal to the champions of the great cause. He invaded the enemy's territory. He worked heroically in his own laboratory. He instituted the first laboratory for instruction in physical chemistry in history. Students came to him from all over the world. From England came Ramsay. From America came Harry Clary Jones of

Johns Hopkins, Wilder Bancroft of Cornell, Arthur Amos Noyes and William David Coolidge of the Massachusetts Institute of Technology, and Theodore W. Richards of Harvard. Ostwald had difficulty in speaking English; he filled his mouth with zwieback to get the correct sound of "the." His students were amazed at his energy and enthusiasm. The young Americans, especially, looked up to him with reverence, for he had been the sole chemist in all Europe who, more than ten years before, had recognized the work of the modest retiring American, Josiah Willard Gibbs of Yale, one of the greatest scientific products of his generation.

In 1890, the three musketeers of physical chemistry met in England, where they were invited to discuss the theory of solution with a committee of the British Association. Opinion was now divided as to the merits of the new theory. Many frankly admitted they were not competent to pass judgment. Professor Percival Pickering maintained that "the theory of dissociation is altogether unintelligible to the majority of chemists." They wanted to ask more questions of these wild Ionians. Ramsay, who had studied under Ostwald, tried to clear the way for the acceptance by English scientists of the views of Arrhenius. Lodge, too, was present, and was not antagonistic to the new theory. But Lord Kelvin of Glasgow was not convinced. Sir William Tilden also was hostile, nor would the French chemists accept the theory of Ionization. It was a bitter uphill battle. Ostwald, Arrhenius and van't Hoff parted with renewed declarations to see the fight through.

Ostwald wrote to Arrhenius to come and settle down in Leipzig as a professor at the university, but he chose to stay in Sweden, and accepted a minor position as lecturer and teacher at the Technical High School of Stockholm. Here he remained for four years and found time, between his ion-chasing and bottle washing, to marry Sofia, the daughter of Lieutenant-Colonel Carl Rudback. A son was born to them, Olav Vilhelm, who, as a young man, joined the ranks of the workers on soil science and agricultural botany.

His post at the Technical High School was now to be converted into a professorship at the University of Stockholm. The news of this impending change spread. The enemies of the Ionians gathered to prevent the appointment of Arrhenius. He could not be ousted without some semblance of trial. It was agreed to subject him to an examination. What humiliation! Arrhenius, laughing inwardly at this farce, presented himself before the trio of learned scientists. Lord Kelvin, the eminent British scientist, was one of the examining committee. Dr. Hasselberg, a Swede, and Christiansen, a Dane, completed the group of inquisitors.

Far away at Leipzig Ostwald heard of this and roared, "It is preposterous to question the scientific standing of such a giant as Arrhenius." He wrote to Stockholm and fought hard for his friend. The examination, however, came off as scheduled. Arrhenius, not at all disconcerted, answered the volleys of questions quietly and confidently. This time he was not going to distort the truth of ionization even for a Kelvin.

When the examination was over and the report submitted, a new tumult was raised. Kelvin opposed the theory in general. He could understand nothing, he said, which could not be translated into a mechanical model. For this reason he had likewise rejected Maxwell's electromagnetic theory of light. Only the Dane submitted an enthusiastic judgment of the competence of Arrhenius. His own countryman, Hasselberg, declared that his answers "were not physical enough" to make him fitted for a professorship. What a comedy! The university authorities kept searching, in the meantime, for a foreign professor to fill the newly created position. Had they succeeded in obtaining one eminent enough to accept the chair of chemistry, they would have sidetracked Arrhenius altogether. But Ostwald kept fighting tooth and nail for his friend. The University of Stockholm feared a scientific scandal. And just as the Bunsen Society in Germany was electing him an honorary member for his Theory of Dissociation, Arrhenius was finally made professor at the University of Stockholm.

The struggle for recognition was still going on. The theory had opponents aplenty. Professor H. E. Armstrong of England likened it to phlogiston. From Oswald's own laboratory, at Leipzig, Louis Kahlenberg had graduated with a Ph.D., *summa cum laude,* the highest honor attainable. He had dug into the theory of Arrhenius but was not convinced of its truth. The theory had entirely neglected the existence of chemical reactions in solutions other than water. Arrhenius had declared that chemical reactions took place only between *ions* in solution. But Kahlenberg had undeniable proof that some reactions took place in solutions which could not conduct an electric current and hence, according to Arrhenius, contained no ions. And here was another objection. Silver nitrate dissolved in benzonitrile allowed the electric current to traverse it, yet this solution contained no ions. Curious exceptions to the theory of Arrhenius.

Kahlenberg went back to the University of Wisconsin and worked ten years as professor of chemistry to disprove the truth of the conception of Arrhenius. He tried the queerest experiments, which seemed miles away from his subject. But Wisconsin was a place for freak experiments anyway. Here Stephen Moulton Babcock and his young assistants performed the most outlandish experiments on men and beasts and finally discovered the "hidden hunger" of the vitamins. For weeks Kahlenberg gathered together an experimental group of fifteen people in his office. There were twelve young men between the ages of twenty and thirty, three young women of the same ages, one woman of sixty, and a man of sixty-three. He had them taste all sorts of beverages and carefully record their reactions. When he finally disbanded this council of taste he had come to the conclusion that here, at least, Arrhenius was right. *Hydrogen ions* were responsible for the sour taste of acids. A strong acid was one which contained a large number of these hydrogen ions, and a weak acid contained only a few of these ions.

But what of the numerous cases which would not fit into Arrhenius' scheme of things? Kahlenberg was just as emphatic in opposing the ionic theory as his teacher Ostwald was in defending it. "The difficulties," Kahlenberg declared, "which the theory of electrolytic dissociation encounters are really insurmountable." He challenged the Ionians to "try to recall any real marked improvements or discoveries in the realm of electrolysis which are directly traceable to the influence of the dissociation theory. These polemical discussions," he asserted "are doing considerable good in that they emphasize how inadequate the dissociation theory really is; they represent the begining of the end of that theory." To Kahlenberg the theory was but a web of naked fancies.

As late as 1900 Kahlenberg fought against the theory and prophesied its doom. But he lived long enough to witness the triumph of ionization.[12] "The chemistry of atoms and molecules gave place to the chemistry of ions," declared Jones. Ostwald used the completed theory of Arrhenius with such skill and understanding that he laid the basis of a new analytical chemistry upon the bedrock of ions. Electrolysis, electroplating and other applications of electrochemistry have their foundations deeply rooted in this new theory. Physiology, medicine, and bacteriology, too, found it very helpful.

In the meantime, beginning in 1893 and continuing until his death in 1919, a new and comprehensive theory of complex ion formation was developed by a Swiss chemist named Alfred Werner. This theory born of a dream confirmed and expanded the theory of Arrhenius. Werner showed that metals combine chemically in two ways. The first was through the formation of ionic or primary linkage. The second was through the formation of *nonionic* or secondary linkages. For example, in the compound potassium ferrocyanide, $K_4Fe(CN)_6$, the primary bond is between K^{+1} and

$Fe(CN)_6^{-4}$. The secondary bond is between Fe^{+2} and $(CN)^{-6}$. This type of chemical linkage has specific chemical orientations about the central metal ion which influence the behavior of this complex ion.

Together with Kekulé and van't Hoff, Werner is recognized as one of the founders of modern atomic spacial and structural theory. Arrhenius lived long enough to see the Nobel Prize for 1913 awarded to Werner for this great contribution.

The authorities had made no mistake in promoting Arrhenius. Within two years after his appointment as professor, he was elected President of the University. His great battle was being won. His fame began to spread. Five years later the Royal Society of England honored him with the Davy Medal. In the following year came the crowning recognition. He received the Nobel Prize, the highest honor in science.

In June, 1904, Arrhenius spoke before the Royal Institution, and the following week sailed for America, on his first visit to the United States. At the St. Louis Exposition to which he had been invited, he again saw Ostwald and van't Hoff. The three musketeers were still riding. They met again to take stock of the new theory. It had fared well. Two of the musketeers were Nobel Prize winners, and Ostwald was soon to be similarly honored.

On the way home Arrhenius was offered a professorship of chemistry at the Berlin Academy of Sciences, the same honor which van't Hoff had previously accepted. King Oscar II of Sweden planned a more tempting offer to keep him at home. The King founded the Nobel Institute for Physical Research at Stockholm, and Arrhenius was made director. Oxford and Cambridge honored him with degrees.

On the twenty-fifth anniversary of the appearance of Arrhenius' thesis, Ostwald dedicated a whole volume of his *Zeitschrift* to it. Bancroft of Cornell, LeBlanc of Leipzig, Le Chatelier of Paris, Ciamiciau of Bologna, Van Deventer of Amsterdam, H. C. Jones of Johns Hopkins, Wegscheider of Vienna, and other distinguished scientists filled the pages of that journal and crowded out many who begged to be allowed to contribute. The battle had been won with glory.

In 1911 Arrhenius again visited this country to deliver a series of lectures at our principal universities. He spoke at Johns Hopkins, Yale, Ohio State University, the College of the City of New York and Columbia. He was invited by the Chemists' Club of New York to talk to its members on May 17, because on that night, twenty-eight years before, he had received the inspirational flash of the true meaning of electrolysis. The Willard Gibbs Medal was presented to him by the American Chemical Society. Arrhenius made many friends here. He spoke a clear, grammatical English, and his American audiences paid unstinted tribute to the genius of this Swede who could speak with as much lucidity and interest about his tiny, electrified, invisible ions, as he could about the vast universes building up in the celestial furnaces.

When the battle was over and the victorious Ionians had put away their armor, Ostwald, the picturesque standard bearer of radical theories, purchased a country estate in Gross Bothen, appropriately named it *Energie*, and settled down to further work in chemistry. Van't Hoff had died in 1911. Arrhenius, still as vigorous and acute as he had been a generation back turned from his original triumph to other fields of speculation. His fertile mind became active in the field of cosmology. His meditations led him to a new theory—the birth of the solar system by the collision of great stars. Cosmology was not the only branch of contemplative science he cultivated. He speculated as to the nature of comets, the aurora borealis, the temperature of celestial bodies and the causes of the glacial periods. He observed a strange periodicity of certain natural phenomena. He reflected upon the world's supply of energy, and studied the conservation of natural resources. Like Becher and Ostwald, he dreamed of a universal language, suggesting a modified English. He was a true polyhistor. There was hardly a field of science which he left unnoticed, and in all he presented original if not altogether universally accepted ideas.

He did more than speculate. He hurried to Frankfort to study the treatment of disease with serums. He was one of those who watched Paul Ehrlich shoot injections of fluids into the blood stream of animals suffering from malignant diseases. Arrhenius marvelled at his dexterity and almost superhuman perseverance. He made a careful study of the work, and was the first to attempt to explain the chemistry of this serum therapy.

Arrhenius also spent three weeks at Manchester, in the laboratory of Rutherford whose new discovery was convulsing the scientific world. He wanted to learn more about it at first hand. The young New Zealander fascinated the Swede. Later, when he came to America, Arrhenius made a trip to the marine biological laboratory of Jacques Loeb. Arrhenius had met this experimental biologist while a student at Strassburg. He had come now to watch him demonstrate how the unfertilized egg of a sea-urchin could be made to develop by chemical means. It was one of the most thrilling experiences he had witnessed. A carefully prepared chemical solution had performed the function of wriggling sperm. Loeb had seen the importance of Arrhenius' theory of ionization and had made use of it in his study of the physiology of the lower animals.

Arrhenius pondered over the problem which Woehler had evoked when he synthesized urea. Could life on this earth have originated from the inanimate without the intervention of some vital force? Arrhenius could not believe this. Rather he felt that life on this planet had started from a living spore carried or pushed from some other planet by sunbeams or starbeams until it finally fell upon the earth. Giordano Bruno, philosopher and poet, had been burnt at the stake in 1600, in the presence of Pope Clement VIII, for daring to say that other worlds might be blessed with life. This was no longer a dangerous idea but still revolutionary. Waves of light, Arrhenius maintained, actually pushed small particles of matter away from a star and brought them to the earth trillions of miles away. Arrhenius pictured these spores swept through the ether like corks carried by the waves of the ocean. He calculated the size of particles that could be moved by this light pressure, and found it to be within the limits of the size and weight of bacteria. He estimated the speed of this interstellar movement, and found it would take only three weeks for spores to be propelled from Mars to the earth, and nine thousand years from the nearest star. This theory of panspermia was challenged by the contention that any life-bearing seed would have perished in the frigid temperatures of interstellar space. But the theory was still safe, at least from this attack. Bacteria, subjected to temperatures very close to those reached between celestial bodies, lived after removal from liquid helium.

Such was the rich versatility of Arrhenius. He helped to popularize science by writing *Worlds in the Making, Life of the Universe, Destiny of the Stars,* and it is difficult to believe that this imaginative man, who possessed the literary ability of a poet, was not particularly interested in literature or the fine arts. His chief, perhaps his only delight, was in natural truth and natural beauty. He mixed very little in the political life of his country. Only on rare occasions did he talk about matters of government. He was opposed to the dissolution of the union of Norway and Sweden in 1905, but later his feelings in the matter changed, and he expressed the hope that Britain might give Ireland similar freedom. During the first World War he openly sympathized with the Allies, much as he owed to Germany during his early years of struggle.

In the early part of 1927, when Arrhenius was past sixty-eight, his failing health compelled him to retire from the Directorship of the Nobel Institute. Sweden honored him without stint. He was granted a full pension for the remainder of his life. But scarcely had he left the Institute when news reached the world that this great figure had joined the eternal caravan of those who had watched the crucible. After a public funeral at Stockholm, his body was taken to Upsala and buried near the University of Berzelius and Linnæus. His life adds further testimony to the native genius of Sweden.

XIII

CURIE

THE STORY OF MARIE AND PIERRE

INTO a desolate region in Southern Colorado, in the latter part of 1920, came a small army of men to dig for ore. Almost every acre of America had been searched for such a mineral. Twenty years before it could have been imported from Austria, but conditions had changed. The Austrian Government had placed an embargo upon its exportation. So Joseph M. Flannery—he was the leader of this band of men—had to be satisfied with the sand in barren Colorado. There was nothing left to do but dig it out of this God-forsaken place.

Flannery's gang, three hundred strong, worked feverishly to collect tons of this sand called carnotite. They dug, sweated and often swore at the insanity of a boss who took them so far away from civilization. Into wagons they threw the canary yellow ore, and sure-footed burros hauled it over eighteen miles of roadless land half a mile above sea level. At the end of that mean trail Flannery had set up a concentration mill, the nearest water supply to the ore mines. In this mill five hundred tons of carnotite were chemically treated until only one hundred tons were left. This dirt was crushed into powder, packed into hundred-pound sacks and shipped sixty-five miles to Placerville. At this railway center the bags were loaded into freight cars destined for Canonsburg, Pennsylvania, twenty-five hundred miles away.

Here two hundred men were waiting to reduce this mass of powder to but a few hundred pounds. Workers skilled in the handling of chemicals used tons of acids, water and coal to extract the invaluable treasure from the ore. Not a grain of the precious stuff hidden in this mound of powder was lost in innumerable boilings, filterings and crystallizations. Months passed, and at last all that remained of the Colorado sand was sent, under special guard, to the research laboratories of the Standard Chemical Company in Pittsburgh. And now began the final task—a careful and painstaking procedure of separation. A year's work to extract from these five hundred tons of dust just a few crystals of a salt!

For this thimbleful of glistening salt five hundred men had struggled with a mountain of ore. It was the most precious substance in all the world—a hundred thousand times more valuable than gold. For this gram of salt one hundred thousand dollars had been spent. A fabulous price for a magic stone!

Into a steel box lined with thick walls of lead, enclosed in a casket of polished mahogany, were placed these tiny crystals in ten small tubes. The precious casket, weighing fifty pounds, was locked and guarded in the company's safe to await the arrival of a visitor from France.

On May 20, 1921, in the reception room of the White House stood the President of the United States. Around him sat the French Ambassador, the Polish Minister, scientists, Cabinet members, judges and other men and women well known in the life of America. Before the President stood a frail, delicate figure dressed in black with a black lace scarf thrown over her shoulders. The room was fragrant with the scent of flowers—she loved flowers. This woman, who had been honored by kings and queens, stood here before the spokesman of a hundred thousand women. The President began to speak: "It has been your fortune to accomplish an immortal work for humanity. I have been commissioned to present to you this little phial of radium. To you we owe knowledge and possession of it, and so to you we give it, confident that in your possession it will be the means to increase the field of useful knowledge to alleviate suffering among the children of man."

Radium—that was the magic element which had brought Flannery and his gang of men into desolate Colorado to dig for carnotite. Almost twenty-five years before, this woman, with but one assistant, her beloved Pierre, had accomplished the miracle of Flannery's five hundred men backed by a great modern financial organization with every scientific invention at its disposal. She had accomplished this wonderful work in an abandoned old shed in Paris. She had solved a problem and blazed a trail that Flannery and others have since travelled with less travail.

For many years, in the chief laboratory of the Radium Institute of the University of Paris, this woman, until she was sixty-six, worked silently with her test tubes and flasks while all the world waited for another miracle. Even to the end the years had not completely broken this immortal bottle-washer. She remained broad-shouldered and above average height. Her splendidly arched brow was crowned with a mass of wavy gray hair, once blond. Her soft, expressive, light blue eyes were full of sadness.

Prophetic Mendeléeff had met this woman when she was a young girl mixing chemicals in her cousin's laboratory in her native city of Warsaw. He knew her father, professor of mathematics and physics in the high school. Mendeléeff predicted a great future for Marie if she stuck to her chemistry. Marie looked up at her father, smiled, and said nothing. This modest and retiring girl, who had lost her mother when still an infant, loved her father passionately. Every Saturday evening he would sit before the lamp and read masterpieces of Polish prose and poetry. She would learn long passages by heart and recite them to him. Her father was to her one of the three great minds of history—Karl Gauss, mathematician and astronomer, and Sir Isaac Newton were the other two. "My child," remarked the professor when she confided this to him, "you have forgotten the other great mind—Aristotle." And little Marie accepted his amendment in all seriousness.

Poland in those days was not a free Poland. It was part of Russia. Since 1831 the czarist government from St. Petersburg persecuted its refractory subjects who had unsuccessfully revolted in the hope of gaining complete independence. Tyrannical Russia imposed many restrictions. The Polish language was forbidden in the newspapers, churches and schools. The old University of Warsaw, whose professors were compelled to teach in the Russian language, was only a ghost of what it had once been. And the Russian secret service was omnipresent.

When Marie was seventeen, conditions at home compelled her to become governess in the family of a Russian nobleman. She kept in constant touch with the political affairs of her native country. Poland under Russian rule was suffering. Secretly there had sprung up groups of young men and women who vowed to overthrow the foreign oppressor. Among the most fervid of these plotters were some of her father's students. They assembled clandestinely to teach in the Polish language those subjects they knew best, and Marie joined one of these groups. She had heard how, four years before her birth, Russian cannon had been fired upon women kneeling in the snow. She hated the Cossacks with their twisted hide whips. She even wrote for a revolutionary sheet—a dangerous practice, but she was as fearless as she was bitter.

The Russian police rounded up some of the young rebels. Marie escaped the net, but to avoid bearing witness against one of her unfortunate friends, she left Warsaw and the hated Russians. In the winter of 1891, at the age of twenty-four, she arrived in Paris. Paris, the city of her scientific triumphs, was a place of bitter suffering during her first years. She rented a small room in a garret; she could afford no better quarters. It was bitter cold in winter time, and stifling hot in the summer. Up five flights of steps she was forced to carry water and the coal for the little stove that gave her some warmth. She had to stint, for her daily expenses, carefully figured, dare not exceed half a franc. Her meals were often reduced to nothing more than bread and chocolate. On the rare occasions when she allowed herself the luxury of a meal of meat and wine she had to acquire a new taste for these foods.

Marie did not mind these privations. She had come to Paris to study and teach. Europe was agog over the strange ions of a young teacher at Stockholm. Pasteur, old and broken in health, was the idol of France. Marie began to dream of a career in science. Strange that she should have such fancies at a time when science was a closed field for women. But she was dreamer enough to believe herself to be the woman whom destiny had selected to play a tremendous role in science. Had not Mendeléeff told her so? Quick as a flash, she made up her mind. She went to the Sorbonne and matriculated. It meant washing bottles and taking care of the furnace in the laboratory to meet expenses. But Faraday had done it—why could not Marie?

In the laboratory of Paul Schutzenberger, founder-director of the Municipal School of Physics and Chemistry of Paris, worked Pierre Curie, "a tall young man with auburn hair and limpid eyes." He had graduated from the Sorbonne, and was now doing research work with his brother Jacques on electrical condensers and the magnetic property of iron. In 1894, at the home of a mutual friend, Marie met Pierre. "I noticed," she wrote later, "the grave and gentle expression of his face, as well as a certain abandon in his attitude suggesting the dreamer absorbed in his reflections."

They began a conversation which naturally concerned scientific matters. How else could Marie have approached this silent man? Then they discussed "certain social and humanitarian subjects." Marie was happy for "there was between his conceptions and mine, despite the difference between our native countries, a surprising kinship." Pierre, too, was joyful. He was amazed at the learning of this girl, and when he frankly admitted his astonishment, Marie twitted him with, "I wonder, Monsieur, where you can have imbibed your strange notions of a woman's limitations."

At twenty-two, Pierre had written, "Women of genius are rare, and the average woman is a positive hindrance to a serious-minded scientist." He was thirty-five now, and his contact with life had not changed his ideas much. Yet Pierre was captivated. He could not hide it, undemonstrative as he usually appeared. He expressed a desire to see this magnetic woman again. Marie walked on air. She wanted to know this dreamer. The sadness of his face drew her to him. Marie came to Professor Paul Schutzenberger and begged for permission to work beside Pierre. Her request was granted, for Schutzenberger was fond of Pierre. The shy, bashful, sixty-five-year-old scientist had devoted his life to the pursuit of science. Pierre, his young, idealistic disciple, was a kindred spirit. So here in the laboratory of the Ecole Municipale, Pierre and Marie met day after day as teacher and pupil, suitor and admirer.

Pierre was beginning to experience a radical change of opinion about women. Before long Pierre, who might have been a man of letters, wrote to Marie: "It would be a lovely thing to pass through life together hypnotized in our dreams: your dream for your country, our dream for science. Together we can serve humanity."

Marie was ready to go through life working at his side in the citadel of science. Their courtship was a short and happy one, and in July, 1895, they were married. Pierre, although brought up in a Catholic home, believed in no cult, and Marie at the time was not practicing any religion. Marie's father and sister came from Poland to greet them. It was a civil ceremony. Only a few friends were present. Marie wore the same dress as usual. It was a simple wedding. They had neither time nor money for elaborate ceremonies. They were both intensely happy.

The problem of furnishing a home was not a very serious one for two beings who cared nothing for convention. They rented three rooms overlooking a garden and bought a little furniture—just the barest necessities. Pierre was made professor of physics at the Ecole Municipale. He was earning now six thousand francs a year, and Marie continued with her studies. They allowed themselves no luxuries except the purchase of two bicycles for short week-end trips to the country, when they went picnicking alone among the chickens and flowers which Marie loved.

They were both back in the laboratory when, in Würtzburg, Wilhelm Konrad Roentgen discovered a ray of great penetrating power. On January 4, 1896, he described these X-rays, as he called them, to the members of the Berlin Physical Society. And hardly had the news of the discovery of these X-rays, which could penetrate solid objects and reveal the bony framework of a man, reached the world when an accident of great importance happened in the darkroom of the modest laboratory of Professor Henri Antoine Becquerel. It was known that phosphorescent substances after exposure to sunlight became luminous in the dark. He was trying to find out whether such phosphorescent substances gave off Roentgen's rays.

It was not the sort of accident to reach the front pages of newspapers, although its result was world-shaking. From this accidental observation came a train of events which culminated in the triumphal work of Mme. Curie. Quite by accident, Becquerel had placed a piece of uranium ore upon a sensitized photographic plate lying on a table in his darkroom. Uranium salts had been known since 1789; they had been used to color glass. There was nothing very remarkable about this substance.

But one morning Becquerel found more than he expected. He noticed that in this completely darkened room the plate covered with black paper had been changed under the very spot on which the ore was placed. He could not understand this! Perhaps someone had been playing a prank. Now he deliberately tried the experiment to satisfy himself. The same effect was noticed. The photographic plate had been affected without any visible light and only under the uranium ore. How could he explain this strange phenomenon? He repeated the experiment with other ores containing the element uranium. In every case a spot was left on the plate. He analyzed the ores to determine the amounts of actural uranium they contained, and saw at once that the intensity of effect was directly proportional to the amount of uranium present in each ore.

Becquerel, famous scion of a family eminent for its researches on fluorescent light, was ready to draw a definite conclusion. He announced that it was the uranium salt present in each ore which was *alone* responsible for the strange effect produced on the photographic plate. But he did not cling very long to this belief. He tested the chief ore of uranium, pitchblende, a mineral which came from northern Bohemia. It was a strange rock; it puzzled him. Instead of giving a photographic effect directly proportional to the amount of uranium present, this ore was much more powerful than its uranium content could account for. Becquerel now made the simplest inference. "There must be," he said, "another element with power to affect a photographic plate many times greater than uranium itself."

Marie's lucky day had dawned. Becquerel recognized in this Polish girl at the Sorbonne a scientist of the first order. He had watched her at work in the laboratory. Even as she weighed chemicals and adjusted apparatus he observed the dexterity of a trained and gifted experimenter. Yes, she had heard the startling news. He presented the problem to her. Would she undertake this piece of research?

She talked it over with Pierre. Her enthusiasm captivated him. She told her husband that, in her opinion, the increased activity of the ore from Bohemia was due to a hitherto unknown element more powerful than uranium. "This substance," she told Pierre, "cannot be one of the known elements, because those have already been examined; it must be a new element." Pierre was working on crystals, and she on the magnetic properties of metals in solution. Both dropped all their work to join in the great adventure of tracking down the unknown cause of the great power of pitchblende. Mendeléeff, hearing of this, consulted his Periodic Table. There was room for such an element. Marie was bound to find it.

The Curies had no money to undertake the search—they borrowed some. Neither had they any idea how much time it would take. They wrote to the Austrian Government which owned the pitchblende mines. The Austrian officials were willing to help. Soon, from the mines of Joachimsthal, there arrived in Paris one ton of pitchblende. Marie was sure that in this hill of sand the undiscovered metal lay hidden.

Those were hectic days for the Curies. They worked incessantly. Not a moment was wasted; the search was too alluring. They boiled and cooked the great mound of dirt, filtered and separated impurity after impurity. When the poison gases threatened to stifle them under the leaky roof of their improvised laboratory, Marie herself lifted and moved large vats of liquid to the adjoining yard. It was the work of men, protested Pierre, but Marie told him she was strong. She could do superhuman work. For hours at a time she stood beside the boiling pots stirring the thick liquids with a great iron rod almost as large as herself. The stifling fumes made that shed a hell, but to Marie beside her Pierre it was heaven. There stood Pierre lifting great batches of heavy chemicals and dreaming of scientific conquests.

"We lived in a preoccupation as complete as that of a dream," remarked Marie years later. When the cold was so intense that they could not continue their work, she would brew some tea and draw closer to the cast-iron stove. The bitter winter of 1896 came and found that mad couple still laboring in their hangar. Marie was bound to break under this terrific strain. Soon pneumonia made her take to bed, and it was three months before she was strong enough to return to her boiling cauldrons. Pierre, too, at the end of each day's work was broken with fatigue. But the search went on.

In the month of September, 1897, a daughter was born to the Curies, Pierre's boyhood friends came to congratulate them. Debierne, discoverer of actinium, Perrin, the molecule counter, and Georges Urbain were among the visitors. The mother, as she lay helpless, kept thinking of her job under the shed. When the child was but a week old, Marie walked into that workshop again to test out something that had occurred to her as she lay in bed. However, she cared for baby Irene with the same devotion she gave to science. Pierre, of course, helped her, and in the evenings when he returned from the shack to assist Marie, they spoke now of three things— baby Irene, science and Poland.

It became a serious difficulty for Marie to take care of Irene and continue her scientific work. But a way out was soon found. Pierre's mother had just died, and his father, a retired physician with a taste for research, came to live with them. Grandpa watched and cared for his little girl, while her parents grappled with a mound of sand.

In the meantime, the pile of pitchblende had dwindled down to a hundred pounds. They made their separations by a method of electrical measurement which exposed the more powerful fractions of their material from the inactive parts. Often in the midst of some chemical operation which could not be suspended, Pierre would work for hours at a stretch, while Marie prepared hasty meals which they ate as they continued their task. Another year of heroic work. Again Marie was ill. Pierre was ready to give up, but Marie was courageous. In spite of all their sufferings, Marie confessed that "it was in that miserable shed that we passed the best and happiest years of our life."

They were fighting a lone battle. No one came to help. When almost two years of constant work were behind them, the news of the great experiment leaked out, though they had tried to keep it secret. Pierre was invited to accept a chair of physics at the University of Geneva. It was a tempting offer. He made the trip to Switzerland, but was back before long. The great work would be in danger if he were to accept. Marie was happy again.

By now they had extracted a small amount of bismuth salts which showed the presence of a very active element. This element appeared to be about three hundred times as potent as uranium. Marie set to work and isolated from this bismuth salt a substance which resembled nickel. Perhaps it was a new element. She subjected it to every known test, and in July, 1898, she announced the discovery of a hitherto unknown element, which she named "polonium" in honor of her beloved country. The reality of this new element was at first questioned. It was suspected to be a mixture of bismuth and some other element. But its existence was soon confirmed.

Others might have been satisfied with this discovery of an element hundreds of times more active than uranium. But not the

Curies. They kept working with portions of that ton of pitchblende, now boiled down to amounts small enough to fit into a flask or test tube. This fraction of chemicals appeared to possess properties much stronger than even polonium. Could it be possible? Marie never doubted it. She looked at this bit of material, the residue of two years of tedious extractions by repeated crystallizations. It was a very tiny amount; she must be more than careful now. She examined every drop of solution that came trickling through the filter. She tested every grain of solid that clung to the filter paper in her funnel. Not an iota of the precious stuff must escape her. Marie and Pierre plodded on. One night they walked to the shed. It had been a dissecting room years ago; it was now a spookier place. Instead of "stiffs" laid out for dissection, they "saw on all sides the feebly luminous silhouettes of the bottles and capsules containing their product. They were like earthly stars—these glowing tubes in that poor rough shack." They knew that they were near their goal.

Bémont, in charge of the laboratory at the Sorbonne, was called in to help in the final separations. Bottle after bottle, crystallizing dish after crystallizing dish, was cleaned until not a speck of dust was left to contaminate the last product of their extractions. Marie did the cleaning. She was the bottle washer who was first to gaze upon a few crystals of salt of another new element—the element *radium*, destined to cause greater overturning of chemical theories than any other element that had ever been isolated. This was the end of that long trail under the abandoned old shed in Paris.

Pierre was given the position of professor of physics at the Sorbonne, and Marie was put in charge of the physics lectures at the Higher Normal School for Girls at Sèvres, near Paris. She taught, studied, worked in her laboratory and helped take care of Irene. Baby Irene was growing up. In her spare moments Marie found time to make little white dresses. She knitted a muffler for her, and washed and ironed the more delicate garments. Even now she had to watch her pennies. Pierre was superb. He helped her at every turn.

Marie was ready to study every property of the queer new element. She intended to include this work in her thesis for the degree of doctor of science; as a teacher she needed this title. After five more years of research, she presented her thesis. The examining committee of professors was made up of Henri Moissan, inventor of the electric arc, Gabriel Lippmann, developer of color photography, and Bonty. Marie presented her complete work on radioactivity, as she named the effects produced by polonium, radium, uranium, and similar elements. She described radium, an element millions of times more active than uranium. Unbelievable, yet true! The professors were astounded by the mass of original information brought out by this woman. They hardly knew what to ask. Before her, these eminent scientists seemed mere schoolboys. It was unanimously admitted that this thesis was the greatest single contribution of any doctor's thesis in the history of science.

The news was made public. A strange element had been discovered by a woman. Its salts were self-luminous; they shone in the dark like tiny electric bulbs. They were continuously emitting heat in appreciable quantities. This heat given off was two hundred and fifty thousand times as much as that produced by the burning of an equal weight of coal. It was calculated that a ton of radium would boil one thousand tons of water for a whole year. This new element was the most potent poison known to mankind—even acting from a distance. A tube containing a grain the size of a pinhead and placed over the spinal column of a mouse paralyzed it in three hours; in seven hours the animal was in convulsions and in fifteen hours it was dead. Radium next to the skin produced painful sores. Pierre knew this. He had voluntarily exposed his arm to the action of this element. Besides, his fingers were sore and almost paralyzed from its effects. Becquerel had complained about it to Marie. "I love it," he had told her, "but I owe it a grudge." He had received a nasty burn on his stomach from carrying a minute amount of radium in a tube in his vest pocket when he went to London to exhibit the peculiar element to the Royal Society. Its presence sterilized seeds, healed surface cancer and killed microbes. It colored diamonds and the glass tubes in which it was kept. It electrified the air around it, and penetrated solids.

The world marveled at the news. Here was another one of nature's surprises. Chemists were bewildered. A woman had not only pushed back the frontiers of chemical knowledge—she had discovered a new world waiting to be explored. From every laboratory on the face of the earth came inquiries about this magic stone. The imagination of the world was kindled as by no other discovery within the memory of man. Overnight the Curies became world famous.

Then began the tramp of feet to the hiding place of the Curies. The world was making a beaten path to the door of these pioneers. Tourists invaded Marie's lecture rooms. Journalists and photographers pursued then relentlessly. All sorts of stories came back of this strange couple—Pierre the reticent, dreamy, publicity-hating philosopher, and Marie the sad-faced mother who sewed and cooked and told stories to her dark little girl. Newsmongers invaded the privacy of her home and went so far as to report the conversation between Irene and her little friend, and to describe the black and white cat that lived with them. They described Mme. Curie's study; "a writing table, two rather hard armchairs, two others with straw bottoms, a couple of bookcases with glass doors through which you see volumes, papers, and vials thrown together pell mell, an iron stove in the middle of the room. Curtains, rugs, and hangings absent, letters and telegrams piled high on the table."

Marie and Pierre complained. "These are days when we scarcely have time to breathe, and to think that we dreamed of living in a world quite removed from human beings!" They wanted to be left alone, but it was of no avail. Letters, invitations, telegrams, visitors bothered and distracted them. The world clamored for the Curies. They must come out of their laboratory for a few hours at least. Lord Kelvin, England's greatest scientist, personally invited them to come to London to receive the Davy Medal of the Royal Society.

This was only the beginning of still greater honors, many of which they refused. They would rather have laboratories than decorations, was Pierre's reply, on being offered the ribbon of the Legion of Honor. Within a few months the Nobel Prize was awarded them, to be shared with the man who had started Marie on her triumphant research—Becquerel of Paris. The money from this prize was soon gone, to pay the debts incurred to keep their experiments going. They could easily have capitalized their discoveries, but they had not labored for profit. Their work was one of pure science, their sole object to serve humanity, and they refused emphatically to patent their discoveries. Almost a century before, Sir Humphry Davy, too, had been urged to patent his newly invented miner's safety lamp, which could have brought him an annual income of ten thousand dollars. He had refused. "I have enough," he had said, "for all my views and purposes. More wealth would not increase either my fame or my happiness."

The case of the Curies was so different. Theirs was still a severe struggle. And yet they refused fabulous profits. Every crystal of radium salt which they wrenched from mountains of rock they turned over to hospitals without charge. When, in February, 1905, they succeeded in isolating a few grains of the new salt, they sent it to the Vienna Hospital in recognition of the help of the Austrian Government on providing them with the first load of pitchblende. Even that gram of radium salt, gift of American womanhood in 1921, was willed at once to the Institute of Radium of Paris for exclusive use in the *Laboratoire Curie*.

Marie's joy had now reached the skies. Irene was now a lovely little child of seven. Pierre had lost some of his sadness. Things were becoming a little easier for them. Then another baby daughter came—Eve Denise. Their cup of happiness was filled to the brim. But death was soon to stalk in the house of the Curies. In the afternoon of the 19th of April, 1906, a messenger knocked at the door of their home at 108 Boulevard Kellermann. One of the loveliest unions in all the history of science had come to a tragic end. A few minutes before, Pierre had been speaking to Professor Perrin at a reunion of the Faculty of Sciences. They had talked about atoms and molecules and the disintegration of matter. Pierre was on his way home. As he was crossing Rue Dauphine a cab knocked him down, and as he fell, the wheels of a heavy van coming from the opposite direction passed over his head. He died instantly.

Marie listened to the story. There was no tearing of hair or wringing of hands. Not even tears. She kept repeating in a daze, "Pierre is dead, Pierre is dead." This blow almost struck her down. She mourned silently. Messages of condolences came pouring in. Rulers of nations and the most eminent scientists of the world shared her great grief. For a time it seemed she would never be able to resume her work. Within a few weeks, however, she was back in her laboratory, more silent than ever. She was to consecrate the rest of her life in the laboratory to the memory of Pierre.

Then France made a wonderful gesture. Marie was asked to occupy the chair of physics vacated by the death of her husband. This was indeed contrary to all precedent. No woman had ever held a professorship at the Sorbonne. Tradition was smashed. There was muffled whispering in the halls of the University of Paris. Men with long beards shook their gray heads against such a blunder. Some believed that whatever inspiration there had been in her work on polonium and radium was due to the fact that she had been working under the guidance and stimulation of a profoundly imaginative man, whom, furthermore, she loved very dearly. That, they whispered behind closed doors, was the only reason for her creative work in the past. "Wait," they said, "a few years more, and Marie will have disappeared from the stage like a shadow." They dare not be heard lest they wound more deeply the broken heart of Mme. Curie. There was no open opposition. The magic word radium stilled the voices of those who might have cried out.

Then it was announced that Mme. Curie was to lecture in the great amphitheatre of the Sorbonne. This was to be her first lecture. Men and women from all walks of life came to Paris to hear her, members of the Academy, the faculty of science, statesmen, titled ladies and great celebrities. Lord Kelvin, Ramsay and Lodge, were among the audience. President and Mme. Fallières of France had come, and King Carlos and Queen Amelia of Portugal were also present to do honor to this woman. "On the stroke of three an insignificant little black-robed woman stepped in through a side door, and the brilliant throng rose with a thrill of homage and respect. The next moment a roar of applause burst forth. The timid little figure was visibly distressed and raised a trembling hand in mute appeal. Then you could have heard a pin drop."

She began her lecture in a low, clear, almost musical voice. There was no sign of hesitation now. She spoke French with but a slight Polish accent. There was no oratorical burst of enthusiasm; she was like a passionless spirit, the very personification of the search for scientific truth. Her audience expected to hear her extol the work of her predecessor. "When we consider," she began, "the progress made by the theories of electricity—." Her listeners were spellbound. Not a word of her great tragedy. She continued Pierre's last lecture on polonium almost at the exact point where he had left off. When she finished, there was a burst of applause that rang even in the ears of the hundreds that remained outside unable to gain admittance. None waited for the report of this historic lecture with more eagerness than her sister Dr. Dlushka at Zakopane in the Carpathian Mountains, and her brother Dr. Sklodowski in the hospital of her native Warsaw. And old Mendeléeff, dying in St. Petersburg of infected lungs, smiled again as he received the news. Andrew Carnegie, hearing of it in America, provided a fund to help her research students.

There were a few who still whispered about tradition, inspiration, women and science. They still doubted the individual greatness of Marie. She heard those faint rumors, but said nothing. She was as silent as a sarcophagus.

The element radium must be isolated—free and uncombined with any other element. That was the task she set herself. Debierne, boyhood friend of Pierre, was to aid her. Radium was a stubborn element. It was difficult to pry it loose from its chloride. And there was so little of the salt to work with! Numerous methods of separation were tried unsuccessfully. Marie lived in the laboratory. She never took time for the theatre or the opera; she refused all social engagements. France hardly saw her. Finally, in 1910, Mme. Curie passed an electric current through molten radium chloride. At the negative mercury electrode she began to notice a chemical change. An amalgam was being formed. She skillfully gathered up this alloy and heated it in a silica tube filled with nitrogen under reduced pressure. The mercury boiled off as a vapor, and before her eyes lay at last the elusive radium—brilliant white globules that tarnished in the air. This was her crowning achievement. It was fitting that she who had first isolated its salts should be the first to gaze on the free element itself.

Here was a piece of brilliant work performed by Marie without Pierre beside her. The whispers were stilled forever. For this epochal work Marie became the recipient of the Nobel Prize for the second time, the only scientist up to that time ever so signally honored.

Mme. Curie was persuaded to become a candidate for membership in the Academy of Sciences of Paris, which Pierre had joined in 1905. The taboo of sex was again raised in that circle of distinguished scientists. No woman had ever been elected to that body. There was "an immutable tradition against the election of women, which it seemed eminently wise to respect." Level-headed scientists suddenly became excited. There was much heated discussion. Marie, of course, remained in the background. When, on January 23, 1911, the vote was taken, Mme. Curie failed of election by but two votes, and Professor Edouard Branley, inventor of the coherer used in the detection of wireless waves, was selected instead. France never lived down this episode of bigotry.

In the summer of 1913 Mme. Curie went to Warsaw to found a radium institute, returning to the University of Paris in the fall. Then, in 1914, while the hordes of German soldiers were advancing almost within sight of the Sorbonne, this brave woman made a secret and hurried trip to Bordeaux, with a little package safely tucked away in a handbag. While great guns roared the opening of the Battle of the Marne, and Paris taxicabs filled with light-blue uniformed men dashed madly out of the city on their way to the front, this woman fled from Paris for the South. She ran away, not for fear of German bayonets, but in dread lest the little tube she carried in her bag might fall into the hands of the enemy. When the tube of radium was safely hidden in Bordeaux, Marie made haste to return to Paris to do her bit for the country of her adoption. Air raids did not disturb her now, nor the dangers of a ruthless invasion.

Mme. Curie planned a great undertaking. She collected all the available radiological apparatus in Paris; there was very little outside of the capital. She issued a call for young girls to be trained in the use of this wonderful new tool of medicine. One hundred and fifty girls were selected and for eight weeks she lectured and trained them to be radiological operators. Irene, now seventeen, who had refused to leave Paris under bombardment, was among the volunteers.

Mme. Curie learned to drive a car and transported instruments to be installed in the army hospitals. And while this woman, then fifty, loaded heavy pieces of apparatus, Irene did ambulance service near Amiens, where the old cathedral shook under incessant cannonading. Irene even went into Ypres where chlorine choked the lives out of helpless soldiers. Mother and daughter worked like Amazons.

When the invading German army had been driven back, Mme. Curie returned to Bordeaux, packed the precious tube of radium salt in her bag, and brought it again to Paris. The first year of the war saw the the completion of the Radium Institute of the University of Paris. Curie was made Director. In a little room in the Institute on rue Pierre Curie, devoted to X-rays and the extraction of radium, she worked feverishly all through the war. While the slaughter of thousands went on, Marie worked heroically to save a few battered, shattered hulks. She loved freedom more than she hated war, and when the peace was signed, she declared: "A great joy came to me as a consequence of the victory obtained by the sacrifice of so many human lives. I have lived to see the reparation of more than a century of injustice that has been done to Poland." Her native land was now an independent country. Professor Ignace Moscicki, who also worked with beaker and test tube in the chemical laboratory, became President of this Republic.

In 1921 she was asked what she preferred to have most and promptly replied: "A gram of radium under my own control." This woman who had given radium to mankind owned none of the metal herself, though the world possessed one hundred and fifty grams of it. Within a few months, however, a gram of radium, gift of the women of America, was hers.

Eight years passed and again America showed its profound interest in Mme. Curie. With the radium which she received in 1921 she was also given a small annuity. This she immediately used to rent some radium for a hospital in Warsaw. While in the hospitals of New York there were fourteen grams of the salt of this curative element, in all of Poland with its twenty-five million inhabitants there was not a gram of this substance. Mme. Curie felt this keenly but was powerless to help. Her friends invited her to come to New York to receive another gift which would enable her to give Poland a gram of radium.

Her doctors were opposed to another trans-Atlantic trip. She was anemic and weak. Her heroic sacrifices for science had played havoc with her strength. Yet she insisted on undertaking this journey, and risked her life once more. Her visit, however, was made as confidential as possible. On October 15, 1929, she arrived in New York. All red tape was cut. She was given the freedom of the port. A distinguished delegation quietly met her at the pier. She was spared the American ordeal of handshaking which had so distressed her on her previous visit.

President and Mrs. Hoover met this pale-faced woman at the front door of the White House and after an informal family dinner she was escorted to the National Academy of Sciences. Here the President of the United States presented her with a silver-encased draft of fifty thousand dollars, with which to purchase a gram of radium in Belgium. Since the discovery, in 1921, of rich radium ore deposits in upper Katanga of the Belgian Congo, Belgium had cut the price of radium in half. Otherwise she would have again received American-produced radium.

During this second visit she remained in seclusion most of the time except when she attended a few public functions. In New York she was the guest of honor at a dinner of the American Society for the Control of Cancer. In Detroit she took part in the celebration of the Golden Jubilee of Edison's perfection of the incandescent electric lamp. She also attended the ceremonies in connection with the dedication of the Hepburn Hall of Chemistry of St. Lawrence University at Canton, New York, where a bas-relief of her was unveiled. Here the honorary degree of Doctor of Science was added to the other degrees which Yale, Columbia, Wellesley, Smith and the Universities of Chicago and Pennsylvania had already conferred upon her. Owen D. Young invited her to visit the Research Laboratories of the General Electric Company through which she was conducted by Whitney, Langmuir and W. D. Coolidge—as eminent triumvirate of scientists as ever graced any sanctum of science.

On November 8, she embarked for France to return once more to the laboratory of the Curie Institute. France could not see America outdo her in veneration for this great woman. Before she returned, the French Government voted a million and a half francs for the construction of a huge factory-laboratory for the study of radioactive elements. The plans for this unique laboratory had been outlined by Mme. Curie and Professor Urbain, Director of the Chemical Institute of the University of Paris.

More than half a century has passed since presidents and kings first came to the Sorbonne to honor this woman. Her slow, noiseless step is no longer heard there. On July 4, 1934, this indomitable spirit passed away, her death hastened by the effects of the potent salt of her creation which her long supple fingers had fondly handled for so many years. And the world still wonders which was greater, her epoch-making scientific conquests, or the nobility of her self-effacing life absorbed in the adventure of science.

XIV

THOMSON AND RUTHERFORD

THE ATOM HAS A NUCLEUS

WHILE the Curies in Paris toiled in a workshop that closely resembled the laboratories of the ancient gold cooks, in a cloistered cell at Cambridge a group of young Englishmen were battering down the walls that held the tiny atom intact and indivisible. The Curies had given them the tool of power with which to lay siege to the citadel of the atomic world.

In 1897, when the search for radium was leading the Curies to glory, the bubble of the atom as the ultimate reality of matter was pricked by a great Master who stood at the fountainhead of a brilliant group of disciples gathered in the Cavendish Laboratory of Experimental Physics. Chemistry had borrowed lavishly from the storehouse of physics. Now the great advancing problems in chemistry were questions which the physicists were better equipped to solve, but the chemist worked hand in hand with the physicist—here was a great scientific entente. The borderland between physics and chemistry was obliterated.

The Master was a man familiarly known to his students as "J. J." His rise in the ranks of pure science had been phenomenal. J. J. Thomson was born near Manchester towards the close of the year which witnessed the death of another dreamer in pure science—Amedeo Avogadro. While originally wishing to become a practical engineer, his career in pure science was due, strangely enough, to its being impossible for him to make the necessary arrangements for engineering. He attended Owens College, where a scholarship for research in chemistry had recently been made possible by a fund of twenty thousand dollars raised by the citizens of Manchester in memory of John Dalton, architect of the atoms. From Owens College he went to Cambridge, there to become the third of that trinity of discoverers of the ultimate particles of matter—Atoms, Molecules, Electrons.

At Cambridge, Lord Rayleigh was in charge of the Cavendish Laboratory, established hardly a decade before by a descendant of the family of Cavendish. Rayleigh was the successor to the first occupant of the chair of experimental physics, James Clerk-Maxwell, that great genius who laid the foundations of the electromagnetic theory of light. Five years later Lord Rayleigh decided to resign. Asked to name his successor, he pointed without hesitation to his most gifted pupil, Joseph John Thomson. This news created an uproar. A lad of only tweny-eight mentioned as successor to Clerk-Maxwell and Rayleigh! What if Thomson had shown unmistakable signs of genius when, at twenty-five, he had won the Adams Prize for an essay which attacked as unscientific the theory that atoms were vortices of whirlpools in the ether? This essay was unquestionably an admirable presentation of the fallacies of the Vortex Theory. But he had done very little experimentation. Most of his work was in mathematics, and even in this field his record of honors so far had not been the highest. In the traditional Tripos at Cambridge, an examination for honors in mathematics, he had come out not at the head of his group, but only as Second Wrangler. But Maxwell had been beaten for Senior Wrangler honors.

Three eminent scientists constituted the Board of Electors which was to make the final choice—Lord Kelvin, the Scotchman who in Glasgow worked out the intricate problems of the first Atlantic Cable; Sir George Gabriel Stokes, investigator of fluorescence; and Professor George Howard Darwin, second son of Charles Darwin. They saw inside that massive head of Thomson an imaginative yet crystal-clear mind with powerful penetrating power. The lad from Manchester was chosen. "Shades of Clerk-Maxwell," declared one well-known professor, "things have come to a pretty pass in the University of Newton when mere boys are made professors." Michael Pupin, the eminent American scientist, coming from a cracker factory in New York to study physics under Clerk-Maxwell at Cambridge, was frightened away when he learned that a young lad, only two years his senior, had been put at the head of the famous Cavendish Laboratory. "I thought he was too young to be my teacher of physics," he complained.

And so it came about that a mere boy filled the chair of two illustrious predecessors, and under his leadership the Cavendish Laboratory became the dominant center of scientific research in the world. Here was carried on more important research per square foot than in any other part of the earth. Here men's minds soared to heights never dreamed of before. The spirit of the boy Thomson was to pervade that sanctum of science for nearly half a century.

In the lightning flash which splits the heavens Thomson saw a force in which lay the key to the mystery of the material world. He chose as his field of research the realm of electricity. A year before he entered Cambridge, Thomson had heard of a peculiar glass tube or globe constructed by his countryman, William Crookes. By means of a vacuum pump, Crookes drew almost all of the air out of this tube so that only an infinitesimal fraction of the original molecules of air remained in his sealed glass container. With the aid of an induction coil he discharged a high voltage current of electricity through his highly evacuated globe.

Then Crookes observed a ghostly fluorescence issuing from the negative plate, or cathode, of the glass tube. What could account for this spooky light? The molecules of the thin air in his tube were illuminated by a pale, dim light, and a greenish yellow fluorescence formed on the glass walls of the instrument. Crookes was not the first to look upon these strange rays of light. William Watson, English apothecary and physician, almost a century and a half before had passed the electric energy of his improved Leyden jar through a glass tube three feet long, partly exhausted of air. "It was," he recorded,"a most delightful spectacle when the room was darkened to see the electricity in its passage. The coruscations were of the whole length of the tube between the plates."

But was it really light he beheld? Light, as every responsible professor had taught, was neither ponderable nor material. Yet these *cathode rays* could be made to bend under the influence of a strong electromagnet brought near the tube. Crookes was flabbergasted. Light, and yet unmistakably matter! How to reconcile the two irreconcilables? He could not.

For want of a better name, he termed these cathode rays a *fourth state of matter*—for it was neither gas, liquid, nor solid. He ventured another name—*radiant matter*. That was the best he could do. But the mystery still remained. Crookes, as he gazed upon those cathode rays and saw the flight of myriads of disembodied atoms of electricity, just missed discovering the Electron.

However, Crookes, son of a tailor, had done valiant service. He had given mankind a new instrument of discovery. With it Roentgen discovered X-rays, and with it Thomson was to accomplish other wonders. This Crookes' tube was the forerunner of the giant five-foot cathode tube through which later at the General Electric

Company at Schenectady, a stooped, near-sighted, curly-haired scientist, Dr. William D. Coolidge, shot a million volts of direct current electricity and produced effects which were staggering.

Thomson was to learn more about this "borderland where Matter and Force seemed to merge into one another, that shadowy realm between Known and Unknown." He began to construct these strange big-bellied cathode tubes, and worked his vacuum pumps until the air inside the tubes was twenty million times thinner than the air he breathed. There were only seven students working in his laboratory at the time. He called Richard Threlfall to his side to work with him on the passage of electricity through these tubes. He sent currents of high potential through them and watched the ghastly glow of the cathode rays in the blackness of his research den. They were spooky streams of light. Thomson was like a conjurer trying to evoke the spirit of the great secret of matter.

He wondered at the cause of the undeniable bending of that beam of light by a magnet. The stream of light was deflected as if it were made up of so many iron filings attracted by a magnet. He began to understand why Crookes, pulling at his long curled mustache, had been puzzled almost to madness.

Thomson varied the conditions of his experiments. He changed the degree of evacuation of his tubes. He used different cathodes, altered the intensity of electricity which was sent through the tubes. Years passed. His data kept piling up, and as the facts and figures mounted, Thomson's mind, too, soared high.

In 1890, in the midst of his researches, he married Rose Elizabeth, daughter of Sir George E. Paget, and two years later, George Paget Thomson was born to follow in the footsteps of his father. In 1894 he was elected President of the Cambridge Philosophical Society, and then made a trip to America to lecture at Princeton University on "Electrical Discharges Through Matter." He was gradually evolving a new theory. It was not to be a creed; to him any theory was only a plan or guide to work by.

Faraday's study of electrolysis had led him to suspect atoms of electricity and his laws of electrolysis strongly hinted at discrete particles of electricity. Helmholtz of Potsdam, in 1881, before the Royal Institution, was actually bold enough to declare that "electricity is divided into definite elementary portions which behave like atoms of electricity." That same year Thomson, at twenty-five, had measured the mass of a small pith ball before and after electrification to determine whether electricity possessed mass. He examined the phenomenon of a moving electric discharge and found that more work was required to give a definite speed to an electrically charged sphere than to the same sphere uncharged. This astonishing result indicated to him that an electric charge possessed inertia—the distinguishing characteristic of all *matter*.

He was back at Cambridge now, as busy as ever. Then one Friday evening, on the 30th of April, 1897, Joseph John Thomson announced to the Royal Society his epoch-making conclusion of twenty years of work. "Cathode rays," he declared, "are particles of negative electricity." He denied the ultimate reality of the atom! Since 1800 the Daltonian atom had been regarded as the primordial substance from which every material of the universe had been built. It had been generally accepted as the indivisible brick of the universe. Another sacred cow of chemistry had been slaughtered!

More than two centuries before, Robert Boyle, revered by Englishmen as the father of chemistry, had declared the elements to be "the practical limits of chemical analysis." He believed them to be substances "incapable of decomposition by any means with which we are at present acquainted." But, he added, "there may be some agent so subtle and so powerful as to be able to resolve the compounded corpuscles into those more simple ones, whereof they consist." Robert Boyle, of course, never dreamed of the new chemistry and the new physics. But Thomson did. He had an abiding faith in the simplicity of nature. "There must be something simpler than ninety-two separate and distinct atoms of matter," he whispered to himself. And now he had found that something!

It was the *electron* or corpuscle, as he had first called it. The stream of cathode rays which the magnet had deflected was made up of electrons, torn away from the atoms of the gas in the tube. These electrons were part of the atom, and were alike no matter where they originated. They were negative particles of electricity and were ponderable. The electron was also the smallest particle of matter which moved with a velocity of 160,000 miles per second. Every one of the ninety-two atoms of the chemist contained these same electrons.

That is what Thomson told the world. Would reputable scientists believe him? Thomson was not a Becher, creator of phlogiston. He was going to establish definitely the existence of his chemico-physical monstrosity—a disembodied atom of electricity. He was going to prove its reality by calculating its mass. No man ever set himself a more difficult task. And no man, without the dexterity and imagination of Thomson, could have ever hoped to succeed.

He measured the amount of bending which the cathode stream of electrons suffered in the presence of magnets of known strengths. Through a maze of experimental details, figures and calculations,

Thomson arrived at a number. He had determined the ratio of the electric charge of the electron to its mass—the "e/m," as it is called. This mathematical genius and experimental wizard was as infallible as any human experimenter could be. He announced the calculated mass of the electron as two thousand times less than that of the atom of hydrogen, the lightest substance hitherto known.

The world was not altogether convinced. True, the latter part of the nineteenth century was bewildering in its great scientific discoveries. Men had seen such vast miracles and revolutions in science, that they were afraid to deny the validity of Thomson's work. But still they doubted. After all, they said, it was only a "calculation." Thomson himself was not satisfied.

He called in his research students. Their number had doubled since he had been put in charge of the laboratory. Nearly every afternoon they met in his room for tea. "J. J." was at his best at these informal gatherings. He was wonderfully human. Science was not the only subject discussed. "Thomson's vigorous radical utterances were very warmly discussed and often among the cosmopolitan collection of students political discussions became very animated." The conversation would often turn to less serious matters. "The gossip of the laboratory went round and a story had to be a pretty tall one if he did not manage to cap it." John Zeleny, professor at Yale University, vividly remembered those days when he worked with Thomson on the mobilities of gaseous ions. "We lived," he recalled, "in an atmosphere sparkling with new thought, and enjoyed a free and happy comradeship."

The Master talked over his own researches with his students. The whole subject of the reality of the electron was discussed. There were two Wilsons in his laboratory at the time. Suddenly he turned to C.T.R.—that was the way he addressed Charles Thomson Rees Wilson. This boy, too, had originally come from Owens College. Thomson had been watching him at work with his "dust counter." Wilson had noticed that particles of dust acted as nuclei around which moisture condensed as tiny droplets of water when the air was suddenly cooled by expansion. These dust particles were too small to be photographed, but when they were surrounded by droplets of water they became easily visible and could be photographed. He thus devised an ingenious method of counting dust particles of the air.

Thomson spoke to him. He had that extraordinary gift of stimulating originality in his students. His whole laboratory smacked of dexterous schemes and subtle, ingenious devices for cornering nature in its most inaccessible places. In such an atmosphere C.T.R.

had worked. Thomson asked him this question: "Can you photograph the elusive electron?" There was nothing left to do but attempt it, even though it came perilously near the work of a magician.

That dust counting had given Wilson some wonderful training. Perhaps an electrical particle would act in the same way as tiny dust specks. He tried the experiment, and after innumerable trials he triumphed. He saw through his powerful microscope water vapor condensing into tiny droplets around Thomson's negatively charged particles or electrons.

And now to prove the objective reality of electrons to every Tom, Dick and Harry of a chemist or physicist. If he could only capture these moving particles long enough to imprint them on a photographic plate! It savored of the miraculous. One atom, two thousand times heavier than an electron—even a million uncharged atoms could not be photographed, yet C.T.R. felt that he could trap a single electron. For nothing was impossible to a disciple of the Master.

Wilson began to work on a super-camera which would photograph an electron. It was a tremendous job. Months passed. The Curies had discovered radium, Marie had read her immortal thesis on radioactivity, and still he experimented. In 1903 Thomson left for the United States for the second time, to lecture at Yale and Johns Hopkins Universities. He returned with another bundle of degrees to find C.T.R. still constructing his camera. And while the Master was being honored with the Nobel Prize and knighthood, C.T.R. still worked. Then, in 1911, the work was completed. "The photographic plate is placed in a light frame supported by a silk thread. The plate is kept at the top of a light-tight metal case and as soon as the desired conditions are obtained, the plate is lowered by means of a winch, the axle of which works in an air-tight ground joint." The device was a masterpiece of workmanship. Would it work?

The whole camera was sealed in a glass chamber in which electrons could be produced at will. Charcoal tubes were attached to the camera and immersed in liquid air, for charcoal at this very low temperature absorbed gases greedily thus keeping the chamber highly evacuated. When everything was in readiness the plate was lowered into the field of the electrons, and a photograph was taken. The vacuum in the apparatus was destroyed, the film removed and developed. Wilson had won again. He had arrested the flight of electrons and had drawn their pictures. A tangled skein of threads representing the path of single electrons after expulsion from their atoms appeared on the plate. These fog tracks of electrons were faint, to be sure, but they were undeniably there. Wilson had imprisoned a single electron, surrounded by a droplet of water, moving dizzily through space. Here was incontestable proof of the reality of the electron. In 1927 Wilson received the Nobel Prize.

In the meantime, Thomson and another of his English students, Harold A. Wilson, later professor at Rice Institute, Houston, Texas, were attempting with the aid of C.T.R.'s "cloud-chamber method" to isolate and determine the mass of a single electron. They did it in a fashion, but the achievement of this remarkable work belongs to one who, here in America, after reading about Thomson and his school, had set out to trap a single electron and actually measure it. One might as well attempt to capture the mote in a sunbeam and weigh it on a grocer's scale.

In the science laboratory of the University of Chicago worked Robert Andrews Millikan, a man about C.T.R.'s age. He had carefully read accounts of the work already done in the Cavendish Laboratory at Cambridge. Then he set to work to construct a new piece of apparatus. It consisted of two brass plates about one-third of an inch apart. In the center of the upper plate he bored a hole the diameter of a very thin needle, and illuminated the space between the plates by a powerful beam of light. He connected the brass plates to a battery which supplied ten thousand volts.

By means of an ordinary commercial atomizer he sprayed oil into the air above the upper plate. These drops of oil were one ten-thousandth of an inch in diameter. Millikan was certain that eventually one single drop of this oil spray would find its way through the tiny hole to the space between the plates. For hours at a time he watched this space through the eyepiece of a powerful microscope. Suddenly he noticed, against the black background of his field of vision, a single neutral droplet of oil, like a glowing four-pointed star, fall gently through that space. Millikan repeated the experiment, and observed the similar behavior of each drop of oil. It took half a minute to make the fall of a fraction of an inch. Reversing the polarity of the plates did not affect its motion.

Now he had to act quickly. He was going to strip an electron from an atom of this neutral oil droplet. Radium could do this. He held a small tube of radium so that its rays would strike the oil drop. Something happened. The neutral droplet slowed down in its fall. "When this occurred," Millikan knew, "the droplet was no longer neutral; it had lost some of its electrons and become positively charged." By observing the change in speed with which it travelled he could determine how many electrons it had lost. He noticed that the droplet always travelled at definite rates of speed. There was a certain *minimum* speed. The speed would be suddenly doubled, then tripled. "It was easy to see," wrote Millikan, "that the slowest speed was the *result of the loss of one electron.* This proved conclusively that the smallest invisible load which I was able to remove from the droplet was actually one electron and that all electrons consist of exactly the same quantity of negative electricity."

Millikan worked very accurately. His method was foolproof. By controlling the current he was able to keep his droplets, stripped of electrons, floating between the plate for hours while he left his laboratory to dine or lecture. With the same apparatus he tried another series of experiments, using droplets of mercury, and even droplets of glycerine. These specks of matter were much heavier than the oil, but the same incontrovertible results were obtained.

By means of this electrical balance, thousands of times more sensitive than the most delicate mechanical scale, Millikan had isolated and determined the mass of an electron which agreed closely with the value obtained by Thomson, i.e., eighteen hundred and fifty times less than the mass of a single atom of hydrogen. Some criticized this value. Ehrenhaft of Vienna, for example, worked with tinier drops and reported smaller charges than Millikan had found. But the Austrian's figures were later proven erroneous.

Thomson heard of this remarkable achievement. He did not wonder that it had taken three years of patient labor to accomplish. It was not at all strange that the electron had eluded man so long. "The population of the earth is a billion and a half," Thomson said. "The smallest number of molecules we can identify with ordinary means is about seven thousand times the population of the earth. In other words, if we had no better test for the existence of a man than we have for that of an electrified molecule we should come to the conclusion that the earth is uninhabited." A clear-cut practical analogy from a fanciful dreamer.

What did all this mean? Just one thing. Matter and electrical energy were one. The electron, a negative particle of electricity, entered into the composition of every atom. But it was only part of each atom. What else composed the structure of the atom? This question was even more difficult to answer.

We must go back once more to the Cavendish Laboratory at the time when Thomson counted among his associates William H. Bragg of crystal structure fame, Richard T. Glazebrook and William Napier Shaw, who later became assistant directors, Harry

F. Reid, American geologist of Johns Hopkins, and Arthur Schuster of Owens College who was to give the first correct explanation of Röntgen's X-rays. It was a select group, which had to be kept small because of the lack of greater facilities. Now Thomson opened the doors of his laboratory to other research students. New ground was about to be broken again in science. In October, 1895, within a few hours of each other came two recruits—John Sealy Townsend from Dublin, and a twenty-four-year-old boy fresh from the University of New Zealand.

Ernest Rutherford had come from a small homestead near the village of Nelson, New Zealand, where his father was an odd-job man and simple farmer. The boy had heard of this ancient college whose very breath was reverence for pure science. Here honor students from all over the world fought valiantly for the mastery of nature. Scions of distinguished families came from luxurious palaces to vie with peasant boys from rolling plains and stuffy garrets. Nowhere else in the world could one breathe this sacred atmosphere.

Rutherford, who had received honors in mathematics and science, had been enabled to come to England by the help of a scholarship from home. As he caught the first glimpse of the sacred pile of Trinity College his heart leaped. This temple was the shrine of Newton and Maxwell. Standing before the stained glass windows of the Chapel he vowed to make himself worthy of these masters. Michael Pupin, a decade before, had made that pilgrimage from America. In the forenoon of the day of his arrival, he had seen "a monastic looking procession of serious and thoughtful men in black caps and gowns suddenly change into gay groups of lively youths." The afternoon was reserved for play. But Rutherford was not to be found in the afternoons in white flannel trousers and gay colored blazer. He was to work every minute of the day.

The New Zealander at once caught the spirit of the Master. He found him human, helpful, inspiring. What a wonderful mind he had! He loved, too, Thomson's characteristic smile. And he soon began to feel ecstatic pleasure on hearing the heavy footsteps which every Cavendish man recognized as solely J.J.'s. Among a cluster of brilliant men Rutherford worked here for four years. His research was mainly in the field of the newly discovered X-rays.

Then Thomson was asked to name one of his students to fill the chair of physics at McGill University. J.J. had a splendid group of twenty-five research workers in his laboratory. Blindfolded he could have picked a man among that band without danger of making a mistake. But to him Ernest Rutherford was the brightest jewel. How this man could work! Tirelessly, dexterously—with the skillful fingers of a pianist and the imaginative mind of a visionary. Thomson hated to lose this dynamic being, but he realized that at Montreal, in his own laboratory, Rutherford was bound to accomplish wonders. Rutherford, too, was reluctant to leave—there was only one J.J. But he was destined to make the trip to Canada to shed luster on McGill University for almost ten years. He had written to his fiancée, Mary Newton, in New Zealand, that he was expected "to form a research school in order to knock the shine out of the Yankees."

Before Rutherford left the Cavendish Laboratory he had taken an active part in the many discussions over the work of Becquerel, Roentgen, and the Curies. Here was a virgin field full of possibilities. He chose it at McGill, and began working with uranium and thorium, a kindred element. By 1900, he had already noticed a peculiar phenomenon in connection with the latter substance. It gave off a minute amount of a gas very rich in radioactivity. He carried out precise experiments to determine the nature of the gas and found, to his astonishment, that it was a hitherto unknown substance. He named this gas thorium "emanation."

Early that summer he went back to New Zealand, married Mary Newton and returned with his bride that September. The following March a daughter, Eileen Mary, was born to them—their only child. It turned out to be a very happy marriage.

Rutherford, like Thomson, surrounded himself with research students. He had already encountered Frederick Soddy, originally from Oxford, but who had been appointed Demonstrator in Chemistry at McGill University. Soddy was only twenty-three, but he had a mind as keen as Rutherford's. These kindred spirits worked together for two years, and towards the end of 1902 they published jointly, in the *Philosophical Magazine,* a new theory of radioactivity.

Atoms of radioactive elements, they declared, were not stable entities. They were constantly changing and withering away. During this breaking down process, beta particles (electrons), gamma rays (more penetrating than X-rays), and positive particles were thrown off by the radioactive elements. Rutherford called these particles *alpha rays.* Atoms of radium, spontaneously and utterly beyond his control, were slowly flying to pieces propelled by an internal explosion which nature alone could govern. Neither the extreme cold of liquid air nor the intense heat of an electric furnace influenced this disintegration. Heraclitus, the Greek, was right, "Change was everywhere, nothing was stable."

Rutherford knew the weight of the radium salt he was using. From this and other data he had collected he was able to calculate the speed of the disintegration of radium. In a gram of radium thirty-five billion atoms of radium were disintegrating every second. This meant that radium was losing its activity at the rate of one per cent every twenty-five years. At the end of seventeen hundred years, he calculated, radium would have lost half of its strength. A slow process, yet a definite one. Soddy, back in Europe, was in the meantime collecting alpha particles from disintegrating radium and weighing them. His experiments led to results corroborating Rutherford's results, and afforded convincing evidence of the essential correctness of their data. And incidentally this data enabled him to deduce values for the weight of an individual atom.

The process of disintegration and ejection of alpha particles took place in several others of the heaviest elements. Uranium, for example, took six billion years for half of it to disappear. Amazing facts backed by careful experiments, and capped by as daring a theory as had ever been expounded. And all this by a man scarcely out of his twenties, working with a young man of twenty-five. The whole accepted structure of chemistry seemed to be standing upon shifting sands! Another established belief—the immutability of atoms—had been dealt a death blow.

Rutherford's fame was spreading. In 1903, at the age of thirty-two he was elected a Fellow of the Royal Society. More students came to Montreal. The Old World was making a pilgrimage to the New to sit at the feet of a man who had come out of the Antipodes. Young Otto Hahn arrived in 1905 and worked with Rutherford for a year before returning to Germany. Thirty-three years later this man discovered atomic fission which changed the world, and in 1944 he received the Nobel Prize for this achievement. The United States, too, was soon represented in the laboratory of Professor Rutherford. Bertram Boltwood came there from Yale University to measure the weight and speed of formation of alpha particles from the disintegration of radium. There was a world of work still left undone. Thomson had discovered that the negative rays given off by radioactive elements were identical with his negative particles of electricity or electrons. Rutherford wondered what the positively charged alpha particles might be. Why did all radioactive substances eject these particles? He knew that alpha particles moved with tremendous speeds and could penetrate thin paper. They could even pass through very thin glass, although the walls of an ordinary tube stopped their flight. He was going to trap these alpha particles, and examine them by means of a spectroscope which could detect one-tenth of a millionth of a gram of a metal.

Rutherford was another Thomson. One who knew him well thus described him. "He is a man resembling the alpha particle in his local concentration of energy. He is inimical to leisure. He can arouse enthusiasm in anything short of a cow or a cabinet minister. Frank and genial, he can discuss almost any subject and smoke almost any tobacco."

In mid-1906 Rutherford received an invitation to occupy the chair of physics at the University of Manchester. He had turned down a similar offer from Yale University. Columbia University and Leland Stanford also offered him a professorship. But he wanted to return to England, and in May, 1907, he came to Manchester. Within three weeks he set up the first emanation electroscope for measuring the electric charge on products of radioactive disintegration. The following year he was awarded the Nobel Prize in Chemistry for his researches and discovery of the meaning of radioactive disintegration.

During that same year Hans Geiger, a young German scientist who had come to Manchester a year before Rutherford appeared, invented a new and very useful tool for detecting the products of radioactive disintegration. It became known as the *Geiger counter* and was destined to be almost a household word in the years to come.

Rutherford was already at work trying to determine the nature of the alpha particle. As a source of alpha particles, Rutherford took some radium emanation. It was not a simple task to construct the apparatus he planned. He broke hundreds of tubes and tried different kinds of glass, until finally he made a double tube, one sealed inside the other. Rutherford filled the inner tube, an extremely thin one, with emanation before sealing it to the outer tube. After two days he examined the space between the tubes, which had been carefully exhausted of all gas. Only alpha particles could penetrate the thin walls of the inner tube and get into this void. Yet what was his astonishment when the spectroscope showed unmistakable evidence of the presence of *helium* gas between the tubes. He tried the experiments a number of times. Yes, it was true! The alpha particles had passed through the thin walls and were identified as atoms of helium: In 1909, five years after he had first dimly suspected it, he announced the identity of alpha particles as positively charged atoms of helium. Here was a significant revelation, and it was accepted. The world had learned to believe this man. When Thomson at Cambridge heard of this masterful proof, he shook his massive head, and thought with pride of this human powerhouse. Rutherford's contributions to science were later recognized by King George V and he was knighted in 1914 as his Master had been ten years before.

Ever since Thomson had discovered the electron as part of every atom, Rutherford had pondered over the nature of the rest of the atom. His study of radioactivity had revealed a little. Surely, he thought, there must be in the neutral atom of all elements some positive electricity to counteract the negative electron. Thomson had postulated this theory. Arrhenius, fighting for his ions, had spoken of positively electrified atoms in solution. Even Berzelius, a century before, had introduced the idea of electrically polarized atoms. Was this positive electricity distributed together with the negative electrons throughout the whole atom or, as J. J. Thomson thought, was it concentrated in the tiny center or nucleus of the atom? To find the answer to this problem the imaginative mind of Rutherford soon hit upon an ingenious method of attack.

If he was to conquer the inner citadel or nucleus of the atom he must use projectiles small enough to enter it. Yet his projectiles must be powerful enough to disrupt the most stable thing in the universe. The mightiest battering ram ever used by man must be puny in comparison to the energy of the bullet which he must use. He knew all about the alpha particle. He had identified and christened it. He understood its colossal powers. It possessed the greatest individual energy of any particle known to science at that time—seven million electron volts. The mass of this tiny positive particle of helium was eight thousand times as much as that of an electron. It was ejected from radium with the stupendous velocity of twelve thousand miles per second, a speed which would bring us to the sun, ninety three million miles away, in a little more than two hours. It moved three hundred times faster than a meteor.

A simple tool was at hand. Sir William Crookes had reported flashes produced by alpha particles striking the chemical zinc sulfide. He invented a small simple instrument to help him observe these flashes and called it a "spinthariscope" (derived from the Greek word *spintharis,* a spark). In his book on radioactivity published in 1904, a year after the spinthariscope became available to him, Rutherford wrote, "In the scintillations of zinc sulfide we are actually witnessing the effect produced by the impact of single atoms of matter, that is, the alpha particle."

With the aid of a microscope in a darkroom Rutherford was able to observe the sudden appearance of sparks of light. It looked promising and he called on young Geiger to try it out. As a source of alpha particles he used radium emanation. Then for about half an hour he would just sit, waiting for his eyes to become sensitive to very small light effects. Using a telescope for viewing, he would count the scintillations for hours at a time, never emerging from the darkroom until the day's work was done. "Geiger worked like a slave on this project," wrote Rutherford. He had tried this counting job himself but gave it up—he had not enough patience for this kind of work especially when he had a man like Geiger to do it.

[Then] one day [Rutherford recalled many years later] Geiger came to me and said "Don't you think that young Marsden (only twenty years old and still with no degree) whom I am training in radioactivity methods ought to do a small piece of research? Why not let him see if any alpha particles can be reflected from a solid surface and perhaps scattered through a large angle?

Geiger had already reported to Rutherford that he had noticed that the alpha particles sometimes emerged not at the end of a straight line path but of one somewhat bent (some slight scattering) by very thin pieces of heavy metal, of the order of about one degree. Most of the projectiles passed right through the targets unbent. Some, however, after plowing, through thousands of atoms, were suddenly thrown off their straight course as they struck or were repelled by something extremely heavy and stable enough to turn aside these tiny bombs.

Rutherford continued:

Now I may tell you in confidence that I did not believe they would be, since we knew that the alpha particle was a very fast massive particle with a great deal of energy, and you could show that if the scattering was due to the accumulated effect of a number of small scatterings the chance of an alpha particle's being scattered backwards was very small. Then I remember Geiger two or three days later coming to me in great excitement saying, "We have been able to get some of the alpha particles coming backwards." It was quite the most incredible event that has ever happened to me in my life. It was almost as incredible as if you fired a 15-inch shell at a piece of tissue paper and it came back and hit you.

Rutherford made some preliminary calculations and was convinced that one could not get any scattering of that order of magnitude unless one was dealing with a system in which the greater part of the mass of the atom was concentrated in a very tiny, hard central core. Rutherford, Geiger and Marsden then discussed the nature of the huge magnetic or electric forces which could turn and scatter an alpha particle while it was passing through a thin film of the metal gold. It was decided to continue the experiments and undertake a thorough study of all the angles at which particles were swerved as they passed through a gold film and also through films of such other metals as lead, tin, copper, platinum, silver, iron and aluminum.

When sufficient data had been collected after a long and tedious undertaking, Rutherford told Marsden to polish up the report with Geiger in a form suitable for publication.

By the summer of 1909 this paper was read before the Royal Society. Six months later Geiger sent in his own paper on the scattering of alpha particles. It was then that Rutherford reported that "I had the idea of an atom with a minute massive center carrying a charge." Rutherford soon formulated the laws of the scattering effect. Finally on the evening of March 7, 1911, Rutherford made his first public announcement of his theory after a dinner to which Charles Galton Darwin, mathematical physicist, grandson of the great Charles Darwin who enunciated the theory of evolution, as well as several other young scientists trained in his laboratory including Geiger, Marsden, Kasimir Fajans and Henry J.G. Moseley. A brief extract of this announcement was published in the *Journal of the Literary and Philosophical Society of Manchester* in which a little more than a century before John Dalton had first announced his atomic theory.

Rutherford "for convenience" first called his discovery the "central positive charge" but soon renamed it the "nucleus." Later research found the diameter of the nucleus to be only 10^{-12} cm. wheras the diameter of the atom was 10^{-8} cm. that is, 10,000 times greater. In the Rutherford atom, the electrons were distributed in space outside the nucleus. The determination of their true number and exact location was still in the future. The composition of the nucleus was also an enigma.

Just prior to the outbreak of World War I, on July 28, 1914, a score of different researches were going on in Rutherford's laboratory. Nearly all of these experiments dealt with radioactivity and the problems of the structure of the atom. When a week later England declared war against Germany, Rutherford's laboratory ceased to be the busy hive of research students fighting a battle against the atomic world. Almost overnight the men scattered to join the military or to do other pressing government projects. Rutherford himself had to devote most of his time to the problem of submarine detection.

But Sir Ernest Rutherford managed to steal time to continue some of his work on the structure of the atom. With the aid of C. T. R. Wilson's newly invented camera he photographed the paths of alpha particles he shot through the gas nitrogen. Due to the enormous difference in weight, "an electron could have little more effect on an alpha particle than a fly on a rifle bullet." Thousands of the fog tracks did turn out to be straight lines as he expected. But here in the picture was one which seemed suddenly to have been thrown off its course. The alpha particle, submicroscopic projectile, must have struck something heavy and stable enough to turn the mighty bullet off its direct path. Or, perhaps, the positively charged alpha particle had approached close enough to some massive nucleus, similarly charged, which repelled and deflected it through an angle close to 180° at times. To be sure, the alpha particle had ploughed its way a distance of more than two inches through tens of thousands of nitrogen atoms before it was deflected. There might be, he thought, something very solid in the center of the atom to twist the flight of a projectile with an energy four hundred million times that of a rifle bullet.

Of what was the heavy central core of the atom composed? Rutherford dimly suspected it might be a group of positively charged hydrogen atoms, for he had found them after the bombardment. He was uncertain about any way to explain their presence. It was difficult enough to isolate, photograph and determine the mass of an electron. The positive part of the atom was even more resistant to investigation. Rutherford and some of his assistants continued to bombard the atoms of other elements. They used three metals, sodium, gold, and aluminum, and then a non-metal, phosphorus. In every case positively charged hydrogen atoms were ejected from the nucleus of the atom. The spectroscope had positively revealed hydrogen to them. There was no other conclusion to draw. This charged atom of hydrogen must be present in the nucleus.

Here was the counterpart of the negative electron. This positive charge of electricity was like the electron; it could be deflected by powerful magnets, it obeyed the same laws. The great difference between them lay in their different masses—the positive particles in the nucleus were each almost two thousand times as heavy as the electron. In 1920 at the Cardiff meeting of the British Association in Wales, Rutherford christened the new arrival *proton*, just as twenty-two years before Thomson has announced the discovery of the electron.

Here was Rutherford's greatest single contribution to science. He gave us a new picture of the structure of the atom — an atom resembling the solar system with its massive nucleus of positive electricity around which, at a relatively great distance from this sun, revolved small planetary electrons. The atom was a tiny universe made up of nothing but negative electrons and positive protons.

With the end of World War I hostilities in November, 1918, Thomson at sixty-two retired as head of the Cavendish Laboratory to become Master of Trinity College, Cambridge. It was no surprise that Rutherford, his most distinguished pupil, was selected to succeed him. Somewhat reluctantly, Rutherford left Manchester to take up his new duties.

To celebrate his appointment, A. A. Robb, F. R. S., at one time a worker in Rutherford's laboratory, composed the following verses called *"Induced Activity."* At the annual dinner of the Cavendish Laboratory it was lustily sung to the tune of "I Love A Lassie."

We've a professor,
A Jolly Smart professor,
Who's a director of the lab in Free School Lane.
He's quite an acquisition
To the cause of erudition.

He's the successor
To his great predecessor,
And their wondrous deeds can never be ignored.
Since they're birds of a feather,
We link them both together,
J. J. and Rutherford.

What's in an atom,
The innermost substratum?
That's the problem he is working at today.
He lately did discover
How to shoot them down like plover,
And the poor little things can't get away.
He uses as munitions
On his hunting expeditions
Alpha particles which out of Radium spring.
It's really most surprising,
And it needed some devising,
How to shoot down an atom on the wing.

Soon after his arrival at Cambridge Rutherford was ready to perform a piece of research which stands as his last crowning achievement. It was another classic contribution to chemistry—the first artificial transmutation in history. In 1919 he changed nitrogen into oxygen, and made one of the dreams of ancient alchemy come true. During his bombardment of nitrogen with swiftly moving alpha particles (helium ions) the nucleus of the nitrogen atom was penetrated and a hydrogen ion or proton was ejected changing nitrogen into oxygen. This may be represented thus:

$$\text{Nitrogen} \quad + \quad \text{Helium} \rightarrow \text{Hydrogen} \quad + \quad \text{Oxygen}$$
$$14 \quad\quad + \quad\quad 4 \quad \rightarrow \quad 1 \quad\quad + \quad\quad 17$$
$$\text{(atomic weights)}$$

In 1926, a dinner was given at Cambridge in honor of Thomson. Disciples who had passed through his laboratory during the last half century were now spread over the world in many schools and laboratories. Some of them had forgotten the cherished years of his comradeship and inspiration. A paper was delivered to the Master of Trinity College: "We, the past and present workers in the Cavendish Laboratory, wish to congratulate you on the completion of your seventieth year. We remember with pride your contributions to science, and especially your pioneer work in the structure of the atom." It was signed by two hundred and thirty men, among whom could be counted the greatest scientific geniuses of our time, including five Nobel Prize winners in science.

Ernest Rutherford, was of course, present as toastmaster. He read a few of the telegrams of congratulation which poured in. They made a simple, inspiring tribute. From Copenhagen came the voice of Niels Bohr, another Nobel laureate, extolling J. J. for "having opened the gates to a new land." George E. Hale of the Mt. Wilson Observatory in California wrote a message of good cheer as he scanned the heavens for new stars. Millikan, measurer of the electron, sent a message expressing the conviction that Thomson's electron would "probably exert a larger influence upon the destinies of the race than any other idea which has appeared since Galileo's time." It was no inflated compliment to link the names of Galileo Galilei and Joseph Thomson. Thomson had given to the world its knowledge of the smallest entity in the whole universe. This tiniest of all things is omnipresent.

There sat its discoverer, still mentally alert, with the same characteristic smile—the same human yet almost god-like J. J. Through the great gathering of eminent scientists and philosophers the spirit of that young, old man was dominant. Suddenly the voice of the assembly burst forth in the song of the "Jolly Electron":

There was a jolly electron—alternately bound and free—
Who toiled and spun from morn to night, no snark so lithe as he,
And this the burden of his song forever used to be:—
I care for nobody, no, not I, since nobody cares for me.

Though Crookes at first suspected my presence on this earth
'Twas J. J. that found me—in spite of my tiny girth.
He measured first the "e by m" of my electric worth:
I love J. J. in a filial way, for he it was gave me birth.

Then Wilson known as C. T. R. his camera brought to bear,
And snapped me (and the Alphas too) by fog tracks in the air.
We like that chap! For a camera snap is a proof beyond compare:
A regular star is C. T. R.—we'd follow him anywhere.

'Twas Johnstone Stoney[13] invented my new electric name.
Then Rutherford, and Bohr too, and Moseley brought me fame.
They guessed (within the atom) my inner and outer game.
You'll all agree what they did for me
I'll do it for them, the same.[14]

And as the strains of those verses sung to the tune of the "Jolly Roger," echoed through that room, the Master still dreamed of other scientific conquests. He was developing a new method of chemical analysis by means of "positive rays." Again he was doing the impossible—imprisoning on an impalpable balance molecules which have an existence of but the ten-millionth of a second. Upon his photographic plates he located tracks "corresponding to molecules found neither in the heavens above nor the earth beneath— grotesque monsters of the world of molecules as N_3 and H_3," which live and die in a fraction of a second.

Rutherford lectured widely around the world including Germany several times to attend scientific conferences. During the emergence of Hitlerism in the early 1930's he saw this menace encroaching on the freedoms of men both in and out of the scientific laboratories. He played an active role in helping the victims of Hitler's race theory. Through his influence many of Germany's finest scientists and other intellectuals found refuge in England and the United States.

Thomson, too, continued his scientific work almost to the day of his death, at the age of eighty-four, which the world heard about on August 30, 1940. A vast amount of new information had come out of countless laboratories since that day when the electron was born. The giant new field of electronics has since been developed and expanded upon our ever growing knowledge of this powerful and versatile entity. Hundreds of new tools and machines such as the cathode ray oscillograph, the betatron, and the electron microscope, radio, television, and radar equipment were all born from the electron and the discoveries that followed its arrival.

Three years before Thomson died, Rutherford, too, passed away. It was the result of a freak accident—a fall from a tree he was pruning in his garden. The news stunned the world. J. J. called him "one of the greatest in the whole history of science."

In 1913 Rutherford had been knighted for his scientific contributions and in 1930 had been elevated to Baron Rutherford of Nelson and thereafter was called Lord Rutherford. He was buried in Westminster Abbey alongside of Newton and the ashes of Thomson.

Now that Thomson and Rutherford are no longer here and the time has come to erect monuments to them, "perhaps the noblest symbol that could be placed thereon for Thomson would be e/m," and for Rutherford the name for the newest element number 104 announced by Ghiorso and his team at Berkeley, might be accepted as *rutherfordium* just as their discoverers had proposed.

XV

MOSELEY

THE WORLD IS MADE OF NINETY-TWO

It has been the fate of some men to accomplish in their youth a work of surpassing importance, and then to have their career suddenly cut short by a great catastrophe. Such is the story of Henry Moseley, whose life work was done in less than four years. Before the world had heard of him, he was gone.

In the summer of 1914, while the English school of scientists was hot on the trail of the mystery of the chemical elements, one of Professor Townsend's students at Oxford stopped to say good-by to him. The boy was to board a steamer that morning for Australia to take part in the forthcoming meetings of the British Association for the Advancement of Science. With him was his mother Amabel, later the wife of William Johnson Sollas, professor of geology at Oxford.

The arrival of mother and son in Australia coincided with England's declaration of war against Germany. Moseley would have enlisted at once, but he had certain engagements to fill. He had just completed an astounding piece of research which threw a flood of light on the inner structure of the atom. Two weeks after Britain's entry into the war, he took part in the great scientific discussion on the structure of atoms at Melbourne led by Ernest Rutherford, who emphasized the importance of Moseley's work. A week later, at Sydney, Moseley read his paper on the nature of the elements. He explained in a simple way his classification of the elements.

As soon as the Australian meeting of the British Association was over. Moseley hastened home to offer his services to the government. He could work at home, Rutherford pleaded with him, in one of the war research laboratories, but he refused. He wanted active service at the front where some of his colleagues were already fighting. So during the madness of those early war days he was granted a commission in the Royal Engineers. Rutherford made another move to get Moseley transferred from Brookwood, Surrey, where he had been sent for training.

On June 13, 1915, Moseley, however, was already on his way to the front at Gallipoli as signalling officer in the 38th Brigade of the First Army. He had a charm of manner, a frankness and fearlessness, that endeared him to his men and fellow officers in trench and billet. His letters to his mother from the East were full of cheer. He wrote nothing of the hardships and terrors of the campaign in the Dardanelles. Rather he sent her messages of his observations of nature as he rambled over the hills near the trenches. Like his father, he was a keen and observant naturalist. As a boy he had known most every bird and bird's nest in the neighborhood of his home. He had also been greatly interested in prehistoric implements, and on holidays, on the Isle of Wight, he used to search the blue clay deposits with his mother or sister, and found some excellent specimens which are now in the Pitt-Rivers Museum in Oxford. During one vacation he picked up a very beautiful arrowhead in a small cairn in the Shetland Isles. Moseley had been very proud of this specimen and showed it to his friend, Julian Huxley, when he came down from Balliol College to spend his vacation with him. Charles Galton Darwin joined them at times, and the three boys, all of the same age, grandchildren of three famous scientists who had made these very rocks and stones tell weird stories of the birth of the world, were now immersed in a great world struggle.

In less than two months Harry's letters to his mother ceased. From one of his fellow officers came the dreadful news:

> Let it suffice to say that your son died the death of a hero, sticking to his post to the last. He was shot through the head, and death must have been instantaneous. In him the brigade has lost a remarkably capable signalling officer and a good friend; to him his work always came first, and he never let the smallest detail pass unnoticed.

Little did this officer realize the greater tragedy that occurred when young Moseley was stricken at Suvla Bay as he lay telephoning to his division that the Turks, two hundred yards away, were beginning to attack. But there were many who realized the colossal loss. Said Millikan:

> In a research which is destined to rank as one of the dozen most brilliant in conception, skillful in execution, and illuminating in results in the history of science, a young man twenty-six years old

threw open the windows through which we can glimpse the subatomic world with a definiteness and certainty never dreamed of
before. Had the European War had no other result than the snuffing
out of this young life, that alone would make it one of the most
hideous and most irreparable crimes in history.

As Moseley lived so he died, bequeathing in his soldier's will made
on the battlefield all his scientific apparatus and private wealth to
the Royal Society of London for the furtherance of experimental
scientific research in "pathology, physics, physiology, chemistry or
other branches of science, but not in pure mathematics, astronomy
or any branch of science which aims merely at describing, catalogu-
ing, or systematizing."

When Harry was four years old his father, Henry Nottidge
Moseley, professor of comparative anatomy at Oxford, died. He was
a very strong man, never fatigued by either physical or mental
exertion, but he had lately overworked and began to suffer from
cerebral sclerosis. When his end came in 1891, the upbringing and
education of the boy was left entirely in the hands of his wonderful
mother. So well did she prepare him that at the age of thirteen he
entered Eton with a King's scholarship.

His life and experiences at school were those of an ordinary
healthy English boy. He early showed his liking for mathematics,
and when he went to a boarding school at the age of nine, it was
found that he knew the rudiments of algebra, although he had
never been taught them. In the home school room he had sat,
presumably writing copies which did not interest him, and instead
listening to his two elder sisters being taught the beginnings of
algebra, which he found very entertaining. This genius for mathematics was later to aid him in his great research.

After five years at Eton he entered Trinity College, Oxford, with
a Millard scholarship in natural science. He had also done
brilliantly in the classics; his mind was not lopsided. Harry exhibited the gifts of his distinguished family. His father had such
keen intellectual powers that it was said he had only to be put down
on a hillside with a piece of string and an old nail and in an hour or
two he would have discovered some natural object of surpassing
importance. His grandfather, Henry Moseley, had been a celebrated mathematician, physicist and astronomer at Kings College,
London. On his mother's side, his grandfather, John Gwyn Jeffreys,
had been an eminent oceanographer and authority on shells and
mollusks. His elder sister, Mrs. Ludlow Hewitt, distinguished herself
at Oxford in biology and contributed a valuable paper on a new
subject in science, the rudimentary gill of the crayfish.

Before Harry graduated in 1910 with a B. Sc. degree and honors in natural science, he was dreaming of a career in pure science. He made a visit to Ernest Rutherford at Manchester who welcomed him warmly. This famous teacher saw in him one of those rare examples of a born investigator. In July the formal acceptance came and a few months later, upon his graduation from Oxford, he proceeded at once to Rutherford's laboratory.

He was soon working very hard at Manchester. He used to come down to his mother's home on a rare weekend. His mother had bought a piece of land close to the New Forest in southwestern England, and had a little home built, the plans of which were made by Harry when he was eighteen, and accepted as workable by the builder. He took great delight in the new garden around the house, which was simply a piece of heather land, and planned and arranged this garden entirely himself. He planted it with many rare and unusual trees and shrubs. Nothing gave him more enjoyment than to see his garden growing up and doing well. "The only trees which did not succeed," wrote his mother from Banbary Road, "were a row of Sophoras, made in Germany, as we said, for the English nurserymen having none in stock had imported them for us. He (Harry) always left me with many garden tasks to carry out in view of improvements."

Moseley had the good fortune to be trained under the guidance of a master experimenter. When Moseley came to him, Rutherford's world was still that of the baffling phenomenon of radioactivity. It was standard practice for every newcomer to his laboratory to go through a period of intensive training and instruction, consisting of a rigorous course of study and experimentation in electricity, magnetism, optics, and, of course, radioactivity. It included some practical laboratory exercises devised by Rutherford and his young assistant Hans Geiger, inventor of the Geiger counter.

Moseley's first piece of research assigned to him dealt with a problem concerning the life of an emanation of actinium, one of the radioactive elements. This period was so short that special, delicate devices had to be constructed to detect it. Together with the young Polish scientist, Kasimir Fajans, who had just arrived from Heidelberg, Germany, he soon found that during the radioactive disintegration of actinium, the first product formed, actinium emanation, in turn produced a radioactive product actinium A which emitted alpha particles with a half-life of about $1/500$ of a second. Actinium A turned out to have the shortest half-life ever recorded up to that time. Moseley and Fajans published their findings, "Radioactive Products of Short Life," in *Philosophical Magazine* in 1911. This was Moseley's first published paper.

The following year, he was busy on another ticklish bit of research. He was trying to determine whether any limit could be set to the strength of the electric charge of an insulated body containing radium. As radium continued to lose negative electrons, it became more and more strongly charged positively. What could be the limit of this positive charge? The difficulties to be overcome were tremendous, but Harry went serenely along as if he were playing some simple game. The radium by losing electrons kept building up a difference of potential in the vacuum tube until it had reached one hundred and fifty thousand volts. This charge he was able to increase until the radium emanation withered away and disappeared.

Meantime, in June 1912, news had reached the scientific world that Max von Laue of the University of Zürich had discovered a peculiar property of crystals then exposed to X-rays. X-rays, consisting of extremely short waves in the ether (ten thousand times smaller than those of ordinary light), are produced when a stream of electrons falls upon the metal reflector of a Crookes tube. Max von Laue found that pure crystals of salt split up X-rays like light—the minute spaces between the atoms of the crystals acting like a diffraction grating and producing an X-ray spectrum. Professor William Henry Bragg at Leeds University and his son, William Lawrence at Trinity College, using this discovery, developed a method which enabled them to determine the inner structure of pure salts. X-rays were allowed to pass through very thin sections of crystals and then photographed. They found that crystals were made up of regularly spaced rows of atoms, (not molecules), about one twenty-millionth of an inch apart. From mathematical calculations, the Braggs made a real pattern of the crystal in three dimensions. Here was the foundation of the modern concept of crystal structure by X-ray analysis.

Moseley followed closely these experiments of the father and son at Leeds. Then he and Darwin, a mathematician and boyhood chum, photographed the X-rays produced by electrons striking the positively charged platinum plate of a Crookes tube and then passing through a crystal grating. Here was the germ of the classic research which was to bring Moseley, the modern crystal gazer, imperishable fame.

Shortly before Laue's discovery, Rutherford had already been led to propound a theory of the nucleus of the atom. He had a vague hunch that the main mass of the atom was concentrated in a tiny nucleus of positive electricity surrounded by enough electrons to make the atom electrically neutral. In 1911, with the aid of his students, Geiger and Marsden, he had actually determined what appeared to suggest the number of positive charges in the atoms of gold and other elements and found them to be equal to approximately one-half their atomic weights. He had also found that the higher the atomic weight of the element, the greater was the positive charge in the nucleus.

Rutherford ventured a prophetic hypothesis. "*The charge in the nucleus of every element*," he said, "ought to be *proportional to the atomic weight* of the element." Could this guess stand the critical test of experiment?

This was a problem for the most brilliant of his students. He called Moseley into conference. Rutherford was like his old Master at Cambridge. They discussed this research thoroughly, and before Moseley left him a decision had been reached. X-rays were known to be of two kinds. The first was due merely to the stoppage of electrons. The second was sent out from the anticathode of a Crookes tube and depended upon the metal or metals of which the anticathode was composed. Charles G. Barkla, then of the University of London, had discovered this atomic phenomenon and had determined the length and penetrating power of these rays by absorbing them in very thin sheets of metallic aluminum—a research which earned him the Nobel Prize in 1917. Moseley, using Bragg's X-ray spectrometer method, was to compare the photographs of the X-ray spectra of different elements, and thus hopefully help determine the nature of the electric charge in their nuclei.

Here in Rutherford's laboratory Moseley also had the good forturne to meet young Niels Bohr who had come in the spring of 1912 to "get some experience." He had been immensely impressed and stimulated by the Dane as he followed his explorations into the structure of the atom. In the presence of his friend Darwin, Moseley questioned Bohr at length again when the latter returned to Manchester in 1913 to confer with Rutherford and to discuss his second paper on the structure of the atom.

Bohr spoke especially of the need to know more about the number of electrons in each ring which he had pictured in the atom, the nature of the X-ray spectra of the elements, and the proper sequence of the arrangement of the elements. At this time Moseley told Bohr "there is every reason to suppose that the integer which controls the X-ray spectrum is the same as the number of electrical units in the nucleus." A study of these X-rays, now that their wavelengths could be experimentally determined, might provide more information about his electron rings and the number of positive charges in the nucleus.

Moseley had also gotten the idea that perhaps the way to settle the problem of the number and arrangement of the elements was the systematic determination by experiment of their reflected X-rays. It was then that Bohr "learned for the first time about Moseley's plan to settle this problem."

He started these experiments and by November, 1913, had written to Bohr that he felt that the results he was getting seemed "to lend great weight to the general principles which you use, and I believe that when we really know what an atom is as we must within a few years your theory even if wrong in detail will deserve much of the credit." He wrote to Bohr on several occasions for advice and criticism.

From the very beginning Moseley was fully aware of the tremendous importance of this piece of research. He told this to his mother. He never wrote to her about it, but would discuss it with her at home. He did not see his mother so often now. He worked in his laboratory day and night. If genius is an infinite capacity for taking pains, as Carlyle believed, then Harry Moseley possessed genius.

> His powers of continuous work were extraordinary, and he showed a predilection for turning night into day. It was not unusual for an early arrival at the laboratory to meet Moseley leaving after about fifteen hours of continuous and solitary work through the night. This trait he inherited from his father no doubt.

Here was an outstanding example of a man working with religious fervor, thinking of no recompense save the joy of wholehearted work for the love of pure science.

Moseley fixed a metal plate at the anticathode of a Crookes tube. This metal acted as target for a stream of electrons sent out from the cathode, or negative pole. When the metal was thus excited it emitted its characteristic X-rays. These rays were then allowed to fall as a narrow beam on a crystal mounted on the table of a spectroscope. The reflected rays were then photographed. Moseley soon discarded Barkla's method for determining the nature of the X-rays and perfected this new method of photographing X-ray spectra.

Now he was ready to repeat this procedure with as many elements as could be treated in this manner. Above aluminum in the Periodic Table were twelve elements which could not be adapted to this method of attack. He started with the thirteenth element, the metal aluminum. He invented an ingenious device to speed up his experimental observations. He arranged a series of plates of the different elements on a movable platform in the Crookes tube so that every one of these elements could easily be made the anti-cathode. This piece of apparatus, still on view in the Oxford Science Museum in Oxford, England, delighted him. He was like a boy amusing himself with some mechanical contraption of his own invention.

Formidable problems presented themselves at every turn. In some cases the rays were so easily absorbed by the glass that Moseley had to devise a window in the glass tube thin enough for the rays to pass through. It was not an easy task. He built a window in the glass wall of the tube and covered it with goldbeater's skin, a very thin membrane of the large intestine of an ox, but the extremely high vacuum in the tube resulted in oft-repeated ruptures of the window. It was a delicate and tedious job to keep renewing the window and re-exhausting the tube. But Moseley had the patience of his ancestors, and went quietly about making the many annoying changes.

When he imagined he had overcome the most difficult part of his experiment another problem presented itself. To avoid absorption of the X-rays, the whole photographic apparatus, including crystal and spectroscope, had to be enclosed in a glass vessel exhausted of air. Again, with characteristic energy, he accomplished an almost impossible job.

He worked with such breathless activity that within six months he had examined the X-ray spectra of thirty-eight elements, from aluminum to gold. Moseley studied the results of his measurements. Different elements gave rise to X-rays of different wave lengths. He confirmed a definite relationship—the heavier the element the shorter and more penetrating the X-rays produced. He arranged all his figures on graph paper. He plotted the numbers of the elements, representing their position in Mendeléeff's table, against the inverse square roots of the vibration *frequencies* of a characteristic line of their X-rays. The elements actually arranged themselves on a straight line in the exact order of their atomic weights.

Moseley had left Manchester and gone back to Oxford to live nearer his mother. Professor Townsend gave him a private room in his laboratory where he could work quietly and independently. Here he completed his last research in science. What could these figures and graphs mean? Moseley heard the weak whisper of Nature yielding another of her secrets. The whisper gradually became louder. A strange story was told to young Moseley: "There is in the atom a fundamental quantity which increases by regular steps as we pass from each element to the next. This quantity can

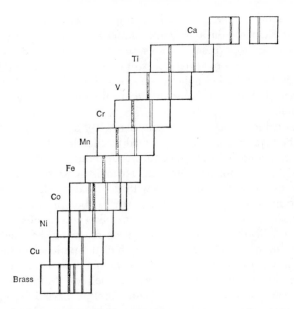

Moseley's "staircase." The line pattern is shifted to the shorter wave length as the charge of the nucleus increases. (After drawing in Moseley's paper "The High-Frequency Spectra of the Elements, Part 1," *Philosophical Magazine,* 1913.)

only be the charge on the central positive nucleus." This was the essence of his fundamental discovery.

In 1912, at the age of twenty-six, Moseley published his results— he had discovered the *Law of Atomic Numbers.* He prepared a new Table of the Elements more fundamental than that of his Russian predecessor. He gave the world an infallible road map of all the elements of the universe—a chart based, not on atomic weights, but on *atomic numbers.* Mendeléeff's romantic blueprints had served science for fifty years. Now a new and more enduring structure was reared, fashioned by the cunning brain of a youth.

The first element in his Table was hydrogen, with an atomic number of one; uranium, with an atomic number of 92, was the last element. For the first time, a scientific limit to the number of building blocks of the universe was set. There could be no other elements besides these ninety-two, said Moseley. It was an astounding declaration. More than seventy elements had been announced during the last generation to fill sixteen gaps in Mendeléeff's Table. Moseley's work showed that almost all of them were spurious discoveries. It definitely settled the claims of men who had "discovered" new elements. Ogawa, a Japanese, had announced *nipponium* as the missing element eka-manganese of the Russian. Ramsay had disproved its reality—it had no room in Moseley's list of elements. Coronium, nebullium, cassiopeium and asterium, "found" in the spectra of the heavenly bodies and placed in Mendeléeff's Table between hydrogen and lithium, were definitely thrown out as nonexistent. There were no places for these so-called elements in Moseley's List.

His Table of Atomic Numbers brought a new harmony into the classification of the elements. It helped determine the proper placing of a number of elements which Mendeléeff's Table could not explain. He found the atomic number of potassium to be 19, while that of argon was 18, even though their accepted atomic weights called for the reverse order. The positions of cobalt and nickel, and iodine and tellurium, were similarly corrected. The discrepancies of Mendeléeff's Table had been ignored for the sake of harmony. Atomic numbers were immensely more fundamental. They were absolutely trustworthy; not an error could be detected.

When the news of Moseley's discovery reached France, Professor Georges Urbain of the University of Paris rushed to Oxford to meet this man. Urbain, sculptor, musician and eminent authority on rare elements, was baffled by a number of elements found in certain Scandinavian ores, in the sands of North Carolina, and the igneous granite of the Ural Mountains. Between the elements barium and tantalum were fifteen others so closely allied in properties that it was extremely difficult to separate them completely. These fifteen elements were the "rare earths." Mendeléeff had been confronted by them when he arranged his Periodic Table. He admitted the "position of the rare earths to be one of the most difficult problems offered to the Periodic Law." He could find nò place for them in his list of the elements.

No one had found a way to clarify this forbidding group of mysteries — lanthanum, cerium, praseodymium, neodymium, samarium, europium, gadolinium, terbium, dysprosium, holmium, erbium, thulium, ytterbium, and lutecium. Crookes had expressed the situation confronting chemists rather pessimistically: "The rare earths perplex us in our researches, baffle us in our speculations, and haunt us in our very dreams. They stretch like an unknown sea before us, mocking, mystifying, and murmuring strange revelations and possibilities."

Moseley's Table of Atomic Numbers had places for all of these fifteen elements. They fitted beautifully into spaces 57 through 71. His work on the X-ray spectra of the elements had settled once and for all time the position and number of the rare earths. This in itself was a remarkable achievement.

Perhaps Moseley could help unravel the mess of the rare earths with his new method of analysis. Urbain handed the young Englishman an ore containing an unknown number of the rare earths mixed together in minute amounts. "Tell me," said Urbain, "what elements are present." Moseley did not keep him waiting long. His mystic crystal was at his side. He went through some strange, dexterous movements with his spectroscope, followed by short rapid calculations on paper. Turning to the French savant, Moseley told him the complete story of the rare earths which had taken Urbain months of laborious analytical operations to find out for himself. Erbium, thulium, ytterbium and lutecium, of atomic numbers 68, 69, 70. and 71, were present, but the element corresponding to 61 was absent.

Urbain was astounded! He put another question to this miracle boy. "Can you tell me the relative *amounts* of the rare earths which are present in this sample?" Urbain felt that perhaps here Moseley might fail. But the Oxford scientist was equal to the task. Urbain was dumbfounded. He returned to France, marvelling at the brilliancy of such a lad. When, a year later, Professor Urbain received Rutherford's letter notifying him of the death of Moseley, the French scientist recalled his memorable visit. "I had been very much surprised when I visited him at Oxford to find such a very young man capable of accomplishing such a remarkable piece of work. The law of Moseley confirmed in a few days the conclusions of my efforts of twenty years of patient work."

In Moseley's Table were gaps for seven missing elements with atomic numbers of 43, 61, 72, 75, 85, 87, and 91. Since Mendeléeff's death not a single one of these elements had been discovered. After the appearance of Moseley's work, however, all of these gaps were filled. Moseley had worked out the X-ray spectra of these hidden elements and prophesied that "They should not be difficult to find." His predictions were fulfilled, for others followed his ingenious line of attack. In 1917 the existence of element No. 91, the eka-tantalum of Mendeléeff's Table, was established. Otto Hahn, who later was to discover the process of nuclear fission, and Lise Meitner in Berlin who later explained this process, discovered the new element which was named protactinium. It was isolated as a pure metal in 1934 by Aristid V. Grosse in Germany. In 1923 Professor Georg von Hevesy and Dr. Dirk Coster, working in the laboratory of Dr. Bohr in Copenhagen, discovered *hafnium*, element No. 72, in an ore of

zirconium, which very closely resembled it. One of the first specimens of this new element to be isolated was deposited in the
American Museum of Natural History. It is not an extremely rare
element; cyrtolite, an ore found near New York, contains as much
as five per cent of hafnium. It constitutes one part in 100,000 of the
crust of the earth, yet it had remained hidden so long because of its
close resemblance to other elements of the rare earths. Then on
June 15, 1925, Walter Noddack and Ida Tacke of Berlin announced
rhenium, missing element No. 75, brought to light by the X-ray
spectrum analysis of Moseley.

In 1937 Carlo Perrier, Emilio Segrè, and B.N. Cacciapuoti prepared an isotope of element No. 43 in an atomic reactor at
Berkeley, California. This radioactive element was later named
technetium. Two years went by when from the Radium Institute of
Paris came the report of the discovery of a radioactive isotope of
element No. 87. Its discoverer, Mlle. Marguerite Perey, named it
francium after her native land. In 1940 another of the 92 natural
elements was obtained by Segrè and associates at the University of
California. Element No. 85 was named *astatine.* Finally in 1945, the
elusive rare-earth element No. 61, which had given Moseley such a
merry chase, was announced in atomic reactor experiments by a
team of scientists—J.A. Marinsky and L.E. Glendenin working with
C. D. Coryell. It was named *promethium.*

When Mendeléeff announced the Periodic Table of the Elements,
he frankly stated, "It has been evolved independently of any conception as to the nature of the elements. It does not in the least
originate in the idea of a unique matter and it has no historical
connection with that relic of the torments of classical thought." He
was alluding to the ancient idea of the unity of all matter. Plato had
said "Matter is one." Sporadically, this idea of a primordial substance from which everything else originated had been enunciated
by philosophers and pseudo-scientists. The Monists, a band of
mystics far away from laboratories, talked of the unity of matter,
and of the evolution of the sex-linked elements from some queer
ethereal substance. The world paid little attention to their abstract
conclusions.

Then, in 1815, there was printed in the *Annals of Philosophy* a
paper in which the writer suggested that the *protyle* of the ancients
was really hydrogen. The author had calculated the atomic weights
of a number of elements and had found them to be whole numbers,
multiples of the atomic weight of hydrogen. Thus he listed the
atomic weights of zinc, chlorine and potassium as 32, 36 and 40.

When he was confronted with a number of elements whose atomic weights were far from integers, he considered the accepted weights erroneous, and declared that the future, with improved methods of analysis, would prove the atomic weights of these elements also to be whole numbers.

Had the author of this idea been Berzelius, he might have created more than a slight ripple. But the anonymous writer proved to be a young English physician, William Prout. His theory that all the elements were formed by various unknown condensations of hydrogen was not taken seriously. The great mass of facts of chemical analysis was against Prout. He himself had done very little accurate determining of atomic weights, but used the data of others, especially those results which fitted his theory.

Prout's theory acted temporarily as a ferment. Berzelius, and later the eminent Belgian chemist, Jean Servais Stas, carried out some extremely accurate atomic weight measurements. This search of the fourth decimal place of atomic weights brought to light many cases of atomic weights which were unmistakably far away from whole numbers. "I have arrived at the absolute conviction," declared Stas, "that the law of Prout is nothing but an illusion, a mere speculation definitely contradicted by experience." The world of chemistry settled back again and forgot Prout and his protyle. Prout returned to his practice of medicine in London. He made another bid for fame a few years later when he announced the discovery and importance of hydrochloric acid in the gastric juice, and then for nearly a century the name of Prout remained forgotten.

Moseley's epochal work on atomic numbers suddenly brought Prout's theory back from the limbo of the past. Perhaps, after all, the idea of the oneness of the elements was not all twaddle. Had not J. J. Thomson shown the electron to be common to all elements? And Rutherford had proven beyond the shadow of a doubt that electric particles were present in the nuclei of all the elements. The great harvest of experimentation of the last fifteen years made it appear almost certain that all the elements were closely related. And now Moseley peeped into the kernel of the atoms and confirmed Rutherford's assumption of the number of positively charged electrical units inside these different atoms.

Prout's conclusions seemed more plausible now. "If the views we have ventured to advance be correct," Prout had written, "we may consider the protyle of the ancients to be realized in hydrogen." Surely evidence seemed to point to the presence of positively charged hydrogen units in the nuclei of all atoms. But there was still a great barrier to the acceptance of this belief. If all the atoms were composed of condensations or multiples of hydrogen, then every element should have an atomic weight equal to some perfect integer, since the atomic weight of hydrogen was one. There could be no place for fractional atomic weights. How could one explain away the atomic weight of chlorine, known to be 35.46, or of lead, fixed at 207.2? Surely these fractional atomic weights could not be the result of experimental errors.

What a powerful weapon could be forged by a clear scientific explanation of the apparent inconsistencies of Prout's theory. Pregnant doubts and questions had already been raised. Crookes, in 1886, addressing the British Association at Birmingham, had made a bold and original statement. "I conceive that when we say the atomic weight of calcium is 40, we really explain the fact, while the *majority* of calcium atoms have an actual atomic weight of 40, there are not a few which are represented by 39 or 41, a less number by 38 or 42, and so on."

It was an audacious speculation and coming from one of the most eminent scientists of England, it had to be seriously considered. Could it really be possible that Dalton was wrong—that all the atoms of the same element were not alike in weight, although similar in properties? Was it really true that what chemists for a hundred years had considered the unchangeable atomic weights of the elements were only the *average* relative weights of different atoms? Lavoisier had said, "An element is a body in which no changes cause a diminution in weight." Was he really in error?

Paul Schutzenberger's study of the rare earths during the close of his life led him to recognize the possibility of different atoms of the same element. Curie's radium and radioactivity provoked more doubts and misgivings. The discovery of *ionium,* identical in chemical properties with thorium and almost similar to it in weight, had for a long time defied the labors of chemists. The following year *mesothorium I* was isolated and was thought to be chemically the same as radium, but differing from it slightly in weight. Emanations and other radioactive elements seemed to lend proof to the speculations of Crookes. Perhaps atomic weights were really averages of atoms whose weights were actually whole numbers. Ramsay declared that the existence of such a large number of elements with atomic weights very nearly whole numbers, was not an accident. The chances were a billion to one that this was fortuitous, he said. By 1910, many level-headed, serious-minded researchers began to whisper the thoughts of Crookes. Frederick Soddy, co-author with Rutherford of the revolutionary theory of radium disintegration, spoke out boldly in favor of Crookes' idea of mixtures of atoms.

At the British Association meeting in Birmingham the year before the opening of World War I, a paper was presented on the homogeneity of neon—there were some doubts as to whether its atomic weight was constant. Soddy, too, standing in the forefront of the new battle, started a great discussion. He had found two samples of a radioactive element with identical physical and chemical properties yet differing in atomic weights. Theodore W. Richards, the first American to receive the Nobel Prize in chemistry, had also investigated the subject of changing atomic weights and found ordinary lead to have an atomic weight of 207.20, while that of lead from a radioactive uranium ore from Norway was 206.05. No one could doubt these figures; Richards was the most accurate investigator of atomic weights of his generation.

Soddy came out firmly for his belief in the existence of the same elements having different atomic weights. He had the boldness to give a name to such elements. *Isotopes*—elements in equal places—was the word he coined. What an upheaval this created. What was left of chemistry and all its pretty theories? Was it all a house of sand? In 1897, on the discovery of radium, Professor Runge of Göttingen had cried out, "Nature is getting more and more disorderly every day." What would he have said now? Every time a scientist dug into the foundations of chemistry another rotten, unsafe timber was discovered!

Would not scientists leave some things alone for a while and rest satisfied with the existing structure? It did not appear so. Men scratched their heads and vexed the elements once more. Chemists were afraid to accept these disclosures. Had not the whole scientific world been taught for more than a century that elements had immutable atomic weights? Richards, one of the world's most accurate experimenters on atomic weights, had called them the "most significant set of constants in the universe." Scientists had believed all atoms of the same element, regardless of source or method of preparation, had fixed atomic weights. If the atomic weight of the element was not fixed, then the whole structure of chemical calculations was only a house of straw!

Was this all just a fabrication? Or was it a clue to the interpretation of the fractional atomic weights of chlorine, lead and neon? Perhaps chlorine, which chemists knew as a simple element, was in reality a mixture of isotopes, each of which possessed atomic weights of whole numbers. When mixed in less than identical amounts, these isotopes would yield a gas with an *average* atomic weight of 35.46. Was this the answer to the inconsistencies of Prout's Theory? Was the death knell of another dogma of chemistry to be heard?

The world again turned to the Cavendish Laboratory for the final answer. New methods of attack had to be devised. Here was the place for radical experiments. At about this time, J.J. Thomson and his "saints of Cambridge" were developing their "positive ray analysis." In this laboratory another of Thomson's brilliant students was at work on this perplexing problem. Francis William Aston came to Cambridge at about the same time that Moseley reached Rutherford in Manchester. The name of Aston had been heard at Cambridge long ago when Newton walked its halls. This Aston was descended from the distinguished family which held the Manor of Tixall in Staffordshire from 1500. Newton had written to a Francis Aston regarding the transmutation of lead into gold. This new Aston was immersed in a problem of modern alchemy—just as baffling as that of his ancestor. He was to solve the riddle of the isotopes.

"Positive rays" were first clearly described in 1886 by E. Goldstein. He obtained these rays by introducing a small quantity of gas in a Crookes tube containing a perforated cathode. Besides the usual cathode rays there formed, behind the perforated cathode, a stream of positively charged particles. Thomson realized that this stream was composed of nothing else but positively charged atoms of the gas, that is, of atoms which had lost electrons and had become ions.

The great English scientist saw in these positive rays a possible vindication of Soddy's isotopes with which he had just ruffled the chemical world. He argued that if these ions came from atoms of the same element having different atomic weights, then some means could be found to separate the element into its various isotopes. A powerful electromagnetic field could sort them out very neatly since the lighter ions would be deflected most.

Aston mastered this new approach to an extremely delicate analysis of the chemical elements and developed it with surprising accuracy. A narrow beam of positive rays was passed into an electromagnetic field which bent the stream of ions. This deflected beam of rays was then photographed on a sensitized plate. If the stream of ions was composed of atoms of equal mass only one band of light appeared on the plate. Positive rays consisting of atoms of different masses, however, were split into an electric spectrum, the number of bands depending upon the number of isotopes. Even the relative proportion of the isotopes could be determined from the size and darkness of the bands on Aston's "mass spectrograph."

Aston began the examination of those elements whose atomic weights were not integers. He worked first with neon. By 1919, definite proof of the physical separation of the two isotopes of the gas neon was established. He had found neon to be a mixture of 90% of neon with atomic weight of 20, and 10% neon, atomic weight 22—hence its accepted fractional weight of 20.2. Here was the first conclusive proof of the existence of isotopes, and the explanation of fractional atomic weights.

A few weeks later the occurrence of the six isotopes of mercury was similarly proved when W. D. Harkins and his students at the University of Chicago fractionally distilled mercury vapor and separated it into six isotopes. Here was another startling demonstration. In laboratories all over the world scientists followed the lead of Aston and his Master. Rapid progress was made the following year. By March the isotopes of argon, krypton and xenon were sorted out. By July came indisputable proof of the existence of isotopes of boron, silicon, bromine, sulfur, phosphorus and arsenic. Then in December of that year Arthur Jeffrey Dempster, a Canadian working at the University of Chicago, announced his separation of the three isotopes of magnesium.

The proof was overwhelming. Ingenious schemes were invented. Brönsted and Hevesy evaporated liquid chlorine and successfully isolated a minute volume of its isotopes. W. D. Harkins played with a series of porous clay pipes, ten feet long, through which, by a process of diffusion, he separated from a ton of HCl, lighter chlorine atoms from heavier ones, the lighter atoms passing through fastest. It was a long, slow, tedious process which took more than ten years. Chlorine was found to consist of two isotopes of atomic weights 35 and 37, so mixed as to give a mean value of 35.46. The evidence kept mounting. There was no question of the atomic weights of the elements being whole numbers. In 1922 Aston received the Nobel Prize for this epochal work. Soddy, speaking of the tremendous effort that had been put into the accurate determinations of atomic weights even to the fourth decimal place by pioneers such as Theodore Richards of Harvard University, declared that with the discovery of isotopes, "something surely akin to if not transcending tragedy overtook the life work of that distinguished galaxy of nineteenth century chemists.

The Unitary Theory of Prout began to be taken seriously. Scientists were arguing upon solid ground. The evidence was conclusive. Moseley had shown the way by determining the exact number of positively charged units in the nuclei of atoms. Rutherford later proved the existence of nothing but hydrogen and helium in these nuclei. And now Aston and his followers presented convincing evidence of the presence of isotopes, all of which had atomic weights of whole numbers. The overthrow of the old conception of the Daltonian atom was complete, and Aston declared, "Let us fix the word element precisely now and for the future, as meaning a substance with definite chemical and spectroscopic properties which may or may not be a mixture of isotopes." In other words, he associated it exclusively with the conception of atomic numbers rather than with the old idea of constant atomic weights.

Moseley had builded better than he knew. It is hard to say what this youthful genius might have accomplished had he lived the normal span of life. Had not that Turkish bullet cut him down in the fullness of his powers at Gallipoli, Moseley would undoubtedly have contributed to the great chemical harvest that was to come. It is safe, however, to say that he could never have outdone his greatest research—the discovery of the Law of Atomic Numbers which solved the riddle of the Periodic Table and the intimate relationship of all the elements.

> The beat of the harp is broken, the heart of the gleeman is fain
> To call him back from the grave and rebuild the shattered brain
> Of Moseley dead in the trenches, Harry Moseley dead by the sea,
> Balder slain by the blindman there in Gallipoli.
> Beyond the violet seek him, for there in the dark he dwells,
> Holding the crystal lattice to cast the shadow that tells
> How the heart of the atom thickens, ready to burst into flower,
> Loosing the bands of Orion with heavenly heat and power.
> He numbers the charge on the center for each of the elements
> That we named for gods and demons, colors and tastes and scents,
> And he hears the hum of the lead that burned through his brain like
>   fire
> Change to the hum of an engine, the song of the sun-grain choir.
>
> Now, if they slay the dreamers and the riches the dreamers gave,
> They shall get them back to the benches and be as the galley slaves.[15]

COMPLETE TABLE OF THE CHEMICAL ELEMENTS (1974)
(arranged according to atomic numbers)*†

At. No.	Element	Symbol	Atomic Weight	At. No.	Element	Symbol	Atomic Weight
1	Hydrogen	H	1.008	54	Xenon	Xe	131.30
2	Helium	He	4.003	55	Cesium	Cs	132.91
3	Lithium	Li	6.939	56	Barium	Ba	137.34
4	Beryllium	Be	9.012	57	Lanthanum	La	138.91
5	Boron	B	10.811	58	Cerium	Ce	140.12
6	Carbon	C	12.011	59	Praseodymium	Pr	140.907
7	Nitrogen	N	14.007	60	Neodymium	Nd	144.24
8	Oxygen	O	16.0000	61	Promethium	Pm	147.*
9	Fluorine	F	19.000	62	Samarium	Sm	150.35
10	Neon	Ne	20.183	63	Europium	Eu	151.96
11	Sodium	Na	22.989	64	Gadolinium	Gd	157.25
12	Magnesium	Mg	24.312	65	Terbium	Tb	158.924
13	Aluminum	Al	26.981	66	Dysprosium	Dy	162.50
14	Silicon	Si	28.086	67	Holmium	Ho	164.930
15	Phosphorus	P	30.973	68	Erbium	Er	167.26
16	Sulfur	S	32.064	69	Thulium	Tm	168.934
17	Chlorine	Cl	35.453	70	Ytterbium	Yb	173.04
18	Argon	A	39.948	71	Lutecium	Lu	174.97
19	Potassium	K	39.102	72	Hafnium	Hf	178.49
20	Calcium	Ca	40.08	73	Tantalum	Ta	180.948
21	Scandium	Sc	44.956	74	Tungsten	W	183.85
22	Titanium	Ti	47.90	75	Rhenium	Re	186.2
23	Vanadium	V	50.942	76	Osmium	Os	190.2
24	Chromium	Cr	51.996	77	Iridium	Ir	192.2
25	Manganese	Mn	54.938	78	Platinum	Pt	195.09
26	Iron	Fe	55.847	79	Gold	Au	196.967
27	Cobalt	Co	58.933	80	Mercury	Hg	200.59
28	Nickel	Ni	58.71	81	Thallium	Tl	204.37
29	Copper	Cu	63.54	82	Lead	Pb	207.19
30	Zinc	Zn	65.37	83	Bismuth	Bi	208.980
31	Gallium	Ga	69.72	84	Polonium	Po	210*
32	Germanium	Ge	72.59	85	Astatine	At	210.*
33	Arsenic	As	74.921	86	Radon	Rn	222.*
34	Selenium	Se	78.96	87	Francium	Fr	223.*
35	Bromine	Br	79.909	88	Radium	Ra	226*
36	Krypton	Kr	83.80	89	Actinium	Ac	227*
37	Rubidium	Rb	85.47	90	Thorium	Th	232.038
38	Strontium	Sr	87.62	91	Protactinium	Pa	231*
39	Yttrium	Y	88.905	92	Uranium	U	238.03
40	Zirconium	Zr	91.22	93	Neptunium	Np	237.*
41	Niobium	Nb	92.906	94	Plutonium	Pu	242*
42	Molybdenum	Mo	95.94	95	Americium	Am	243.*
43	Technetium	Tc	99.00*	96	Curium	Cm	247*
44	Ruthenium	Ru	101.07	97	Berkelium	Bk	247*
45	Rhodium	Rh	102.905	98	Californium	Cf	251*
46	Palladium	Pd	106.4	99	Einsteinium	Es	254*
47	Silver	Ag	107.870	100	Fermium	Fm	257*
48	Cadmium	Cd	112.40	101	Mendelevium	Md	258*
49	Indium	In	114.82	102	Nobelium	No	255*
50	Tin	Sn	118.69	103	Lawrencium	Lr	256*
51	Antimony	Sb	121.705	104	Rutherfordium	Rf	259*
52	Tellurium	Te	127.60	105	Hahnium	?	260?
53	Iodine	I	126.904				

*Most abundant or most stable isotope.
†Based on the relative atomic mass of ^{12}C = 12.000

XVI

BOHR AND LANGMUIR

PRESENTING NEW MODELS OF THE ATOM

MOSELEY had just been born when another lad six years old was growing up in Brooklyn, New York. Unlike the English boy, he could pride himself on no scientific forebears. His grandfather was a minister who had emigrated from Scotland to Canada and then brought his family to Connecticut. On his mother's side, too, there seems to have been no hereditary promise of scientific wizardry.

As a child, Langmuir asked many simple questions about the world in which he lived even as Clerk-Maxwell, at three, kept insisting: "Show me how it does." "Why does water turn to ice?" "Why does water boil in a kettle?" "Why does rain fall?" were some of the questions his parents kept answering. He also bombarded his brother Arthur, who was studying chemistry, with more queries, and when Irving was only nine they built a small workshop in the cellar of their home.

Arthur later graduated from Columbia College and was planning to continue his science studies at the University of Heidelberg. His parents decided to remain near Arthur while he was abroad, and, at the age of eleven, Irving was taken to Paris. And while Arthur was doing research in chemistry, Irving spent three years in a Paris boarding school under French tutors. He looked forward to the occasional visits of Arthur, and would listen breathlessly to his tales of research. Irving was only twelve, yet he wanted his own laboratory, and with his brother's aid he built a small one adjoining his room.

On one occasion Arthur took Irving mountain-climbing in Switzerland, and Irving became so enthusiastic over this sport that he wanted to climb everything in sight. Irving pleaded with his father and finally received permission to climb a mountain alone on condition that he would follow certain rules. He was to stay on a distinct trail, use the same trail going and returning, and make certain of returning at six o'clock by allowing as much time for descending as for ascending. Besides, he had to make sketches, maps and notes of the trails before starting. In this way the boy climbed several mountains about seven thousand feet high, often requiring several days of repeated effort before he could discover a route that led to the top. Such was the careful training his father gave him, which was to stand him in good stead years later both in his lifelong hobby of mountain-climbing and in his scientific adventures.

The Langmuirs spent three years in Europe. Arthur successfully completed his studies at Heidelberg, and in the fall of 1895 they sailed for the United States, but not before Irving had attended the public funeral of Pasteur in Paris—a scene he never forgot. At fourteen, he entered the Chestnut Hill Academy in Philadelphia under Dr. Reid. He knew all the chemistry they taught here and more. At this time he came across a book on calculus, became interested in the subject and at the end of but six weeks mastered it. "It was easy," he told Arthur.

The next year he was in Brooklyn again, attending Pratt Institute where his brother was now teaching chemistry. At eighteen he matriculated in the Columbia School of Mines, from which he received the degree of metallurgical engineer, and then left for Germany to do post-graduate work under Professor Nernst at Göttingen—the University made famous by Woehler.

Three years later, Langmuir returned a Doctor of Philosophy to teach chemistry at Stevens Institute in Hoboken. He remained here until 1909. That summer he made a visit to the Research Laboratory of the General Electric Company at Schenectady which had been established eight years before in a small shedlike structure. He planned to spend his ten weeks' vacation doing research work.

In charge of the Research Laboratory was Dr. Willis R. Whitney, former President of the American Chemical Society and a pioneer in the development of a new type of industrial research. He was an unusual leader. Instead of assigning him immediately to some definite piece of research, he suggested that the young teacher spend a few days watching the staff of researchers at work. Gulliver saw no stranger sights at the storied Academy of Lagado than Langmuir witnessed here.

One problem in particular attracted him. It was baffling a number of the research workers. They were trying to make a tungsten wire which would not break so easily in an electric light bulb. Hundreds of samples of this wire had already been prepared, but most of them were short-lived, once an electric current was passed through them.

He went to Whitney and asked to be assigned to this piece of research. He wanted to investigate the behavior of these imperfect wires when heated to incandescence in evacuated bulbs. Why did only three wires behave so perfectly? What was wrong with the rest of the samples? Langmuir saw the invisible trouble-makers before he began to investigate. He had an idea that the weakness lay in certain gases which they had absorbed.

Whitney agreed and placed the equipment of the General Electric Company and its entire staff of research workers at his disposal. Whitney dropped in occasionally to watch him. Years later he recalled those first few weeks.

> There is something in Langmuir's work that suggests by sharp contrast an oriental crystal-gazer seated idly before a transparent globe and trying to read the future. In my picture an equally transparent and more vacuous globe takes the place of the conventional crystal sphere. It is a lamp bulb, a real light source. Langmuir boldly takes it in his hand, not as some apathetic or ascetic Yogi, but more like a healthy boy analyzing a new toy. There might have been nothing in that vacuum, but he was driven by insatiable curiosity to investigate and learn for himself.

Langmuir expected to find a small volume of gas issuing from the heated wires in the glass bulbs. But what astonished him almost beyond belief was the tremendous quantities of gas given off by the hot tungsten wires—more than seven thousand times their own volume of gases.

His summer vacation had rushed by, and Langmuir must return to the comparative monotony of the classroom at Stevens. He had not discovered the cause of emission of the tremendous volumes of gases, but he suspected the reason. The glass bulb of the incandescent lamp, he surmised, gave off water vapor which, reacting with the glowing tungsten wire, produced immense volumes of hydrogen gas. This chemical action weakened the tungsten wire and shortened its life in the lamp.

Dr. Whitney had watched Langmuir at work. It would be a pity to lose this man who could "hold his theories with a light hand and keep a firm grip on his facts." Whitney made Langmuir a tempting offer to stay with him as a member of the research staff. His place in the classroom could be filled by some other less gifted instructor. Langmuir hesitated at first. Would it be fair, he asked Whitney, to spend the money of an industrial organization like the General Electric Company for purely scientific work which might never lead to any practical application? "It is not necessary for your work to lead anywhere," replied Whitney. Langmuir then and there made up his mind—he would remain in Schenectady.

It was a great day for Langmuir. He decided to continue his work on the incandescent lamp. Dr. Whitney believed with every other lamp engineer in the country that the solution of the lamp problem lay in obtaining a more perfect vacuum in the bulb. Langmuir would not admit this. On the contrary, he was going to fill the electric light bulb with different gases. By studying the bad effects of these known gases, he hoped to learn the causes of the early death of the incandescent lamp. This principle of research he found very useful. "When it is suspected," he declared, "that some useful result is to be obtained by avoiding certain undesired factors, but it is found that these factors are difficult to avoid, then it is a good plan to deliberately increase each of these factors in turn so as to exaggerate their bad effects, and thus become so familiar with them that one can determine whether it is really worth while avoiding them."

First Langmuir got rid of the immense volumes of gases which the tungsten wire had absorbed. Then, instead of working for a more perfect vacuum so that no oxygen would be present in the lamp to attack the wire, he filled the lamp with inactive gases. He chose nitrogen and argon, gases which would not attack the tungsten filament even at the temperature of incandescence. For years he worked persistently with his lamps. He was given the freedom of an academician, plenty of assistants, and tens of thousands of dollars to continue his work. Whitney was convinced that most of the practical applications of science had sprung from pure scientific curiosity. History had proved this over and over again. Clerk-Maxwell's work on light, for example, undertaken in the unalloyed spirit of philosophical inquiry, had ushered in the age of wireless.

In the meanwhile, the number of research problems had multiplied. Whitney needed more room for his enlarged staff of researchers. The General Electric Company met the situation by erecting a seven-story building into which the little laboratory soon moved. In each of its one hundred and thirty workrooms every facility of modern scientific research was available—water, illuminating gas, compressed air, vacuum pumps, high pressure hydrogen, oxygen, live steam, distilled water, standard equipment to supply amperage up to twenty thousand and voltage to two hundred thousand, and any temperature between two hundred degrees below zero and three thousand degrees above zero Centigrade. In the building was a library of more than three thousand volumes at the disposal of the staff. This entire plant was maintained by one of the largest annual subsidies devoted to the pursuit of pure and applied knowledge at that time.

Three summers passed without Langmuir's finding a single practical application to repay the huge sums of money he was spending. He continued to investigate the problem which had first attracted him, until finally the modern nitrogen and argon-filled tungsten lamps were developed. Langmuir, the theorist, saved America in those days a million dollars a night on its light bill of over a billion dollars a year. But that was not the purpose of his labors at Schenectady. "The invention of the gas-filled lamp," he assured the champions of applied science, "was nearly a direct result of experiments made for the purpose of studying atomic hydrogen (a purely theoretical problem). I had no other object in view when I first heated tungsten filaments in gases at atmospheric pressure."

His study of atomic hydrogen, carried through a period of fifteen years, led, in 1927, to his invention of the atomic hydrogen flame for welding metals which melt only at extremely high temperatures. A stream of hydrogen gas is sent through an electric arc. The molecules of the gas dissociate into hydrogen atoms which, on recombining, burn with a heat sufficient to melt metals which withstand the high temperature of even the oxyacetylene flame. These fifteen years of experimentation on purely theoretical problems brought a harvest of important applications in applied science.

When Langmuir first began the study of the tungsten filament heated to incandescence, he dreamed of the disembodied electricity of Thomson. He realized that the scientific progress of the future would be written in terms of electrons. He planned to set the reliable little electron to work. Langmuir was the first electron engineer. His bridges were the molecules, his locomotives were atoms, and his great power machines were the tiny law-abiding electrons rushing through the speedways of copper wires. He saw the heat and light of the electric bulb as the work of electrons passing through the filament.

To help him in his work on the tungsten gas-filled lamp, Langmuir invented the most perfect vacuum pump of his day—the high vacuum mercury pump. This invention was not a chance discovery, but the result of his remarkable powers to see the invisible. From his later researches on vacuum tubes for radio came the pliotron and kenotron used in radiotelegraphy; the magnetron, the thyraton, the dynatron, and the pliodynatron, in all of which he set the infinitesimal electron to work. During World War I he also found time to work for the government on submarine detection devices at the Naval Experiment Station at Nahant, Massachusetts.

From the beginning of his studies Langmuir was especially interested in the structure of the atom. He followed up every bit of research on this topic, and added mightily to the store of knowledge. Finally, he postulated a theory of the nature of chemical activity on the basis of the number and position of electrons in the atom. Langmuir brought together the loose threads of physical and chemical research and wove them into a beautiful fabric. The clear imagery of this man clarified much that was hazy in our conception of the outer structure of the atom. Out of the mass of theories, experiments and facts concerning the outer world of the nucleus of the atom, he constructed what appeared to be a sound and beautiful edifice.

The nature of the atom's structure was still very much in doubt. Many had crossed swords with nature to wrest this secret from her. Lord Kelvin had pictured the atom as consisting of mobile electrons embedded in a sphere of positive electrification. J. J. Thomson had developed this same idea but his model, too, had failed because it could not account for many contradictory phenomena. Rutherford's nuclear theory of the atom as a solar system was also regarded as incomplete. The greatest difficulty to the acceptance of these models was that they all lacked a consistent explanation of the peculiar spectra of gaseous elements when heated to incandescence.

Even before the discovery of the electron, Hendrik A. Lorentz of Amsterdam had come to the conclusion that these spectral lines were due to the motion of electrified particles revolving around the nucleus of the atom. He boldly predicted an effect which was later found by his countryman, Pieter Zeeman, in 1896. Zeeman showed that when incandescent gases were placed in powerful magnetic fields their spectral lines were split. Seventeen years later, a German physicist, Johannes Stark, exhibited a similar undeniable split when the incandescent gases were placed in strong electrical fields. Something electrical in the atom was the cause of the spectrum of elements.

Back in 1885, Johann J. Balmer, a Swiss schoolteacher and physicist, had made another interesting observation. The positions of the lines in the complicated spectrum of hydrogen were not chaotic. Their wavelengths could be represented by a simple mathematical formula. Nature again seemed to proclaim its mathematical proclivity. He noticed that the wavelengths of certain lines of the hydrogen spectrum could be accurately expressed by a number obtained by subtracting two figures. Subsequently other similar remarkable numerical relationships, such as the Rydberg constant, named after a Swedish spectroscopist, were discovered.

Still another puzzle which baffled the electronists was this. If electronic motion was the cause of spectral light then Rutherford's atom ought to radiate this light *continuously*. Stationary electrons were inconceivable. Such electrons would be attracted by and fall into the nucleus of the atom unless their stupendous speed around the center of the atom counteracted the powerful pull of the atom's kernel. What the spectroscope had revealed, scientists could not explain by all the known laws of classical electrodynamics. Here was a mighty impasse that challenged the world of science.

"There are times," said Professor W. F. G. Swann of Yale, "in the growth of human thought when nature, having led man to the hope that he may understand her glories, turns for a time capricious and mockingly challenges his powers to harmonize her mysteries by revealing new treasures." Among those who accepted the challenge was a young Danish scientist, Niels Henrik David Bohr. His father was a professor of physiology, his mother, Ellen Adler, a member of a prominent Danish Jewish family, and his brother a distinguished mathematician. Niels was to excel both father and brother in the field of science. He was eager to learn at first hand the latest developments in the structure of the atom.

Bohr first encountered Rutherford in the autumn of 1911 in J. J. Thomson's Cavendish Laboratory at Cambridge, where Bohr was studying the electron theory of metals. He was completing a year's study abroad after receiving a Ph.D. from the University of Copenhagen for an investigation of the passage of charged particles through matter. Rutherford had come to speak at the annual Cavendish dinner, and although there was no personal meeting at this time, Bohr was immensely impressed by his knowledge, vigor and charm. When a few weeks later he went up to Manchester to visit a friend of his deceased father who had also been a friend of Rutherford, he made sure to drop in at the Manchester laboratory. The warm reception he received captivated the young Dane and they talked at length about Max Planck and the German-born physicist Albert Einstein whom Rutherford had recently met in Brussels for the first time.

Bohr was eager to join Rutherford and his associates, and Rutherford, quite impressed by the young man, invited him to come at any time. Bohr appeared in the spring of 1912 and was at once placed in the hands of the team of Geiger, Makower, and Marsden to take that famous training course in radioactivity. Although essentially a theoretical physicist, Bohr was able to handle experimental equipment with competence. He soon got to know his young teachers, as well as Moseley and the other members of this bubbling fraternity.

The Dane was a youth of unusual imaginative power and intellectual courage. He bravely abandoned his classical mechanics and seized hold of a new key—Planck's conception of quantum of energy. Max Planck had enunciated a revolutionary theory. Energy, he said, is emitted not in a continuous way but only in tiny, finite bundles called quanta. Energy, he insisted, was atomic in structure. Bohr was not afraid to use this unorthodox idea.

In the summer of the following year Niels Bohr published in the *Philosophical Magazine* an article *On the Constitution of Atoms and Molecules* (*Part I*). For weeks at a time he had lived in his study scarcely ever emerging from it. Using Rutherford's conception of the atom as a miniature solar system, he boldly postulated a conception of the dynamic hydrogen atom, the simplest of all atoms, having but a single electron outside its nucleus.

He pictured the single electron of the hydrogen atom as revolving in an elliptical orbit around the nucleus unless disturbed by some outside force like cathode rays, X-rays or even heat. When thus disturbed, the electron would jump from one orbit to another orbit closer to the nucleus. When electrons leaped in this way light or some other form of radiation was produced. The transfer to each different orbit represented a distinct spectral line. "For each atom," he wrote, "there exist a number of definite states of motion called stationary states, in which the atom can exist without radiating energy. Only when the atom passes from one state to another can it radiate light."

Bohr believed that electrons may "jump" to a level of higher energy when the atom *absorbs* energy. When an electron drops to a level or state of *lower energy*, the atom *emits* energy. The electron cannot stop in between these levels—these are forbidden zones. Since these jumps can only occur between definite levels, definite amounts of energy are involved. In this way, atoms give definite line spectra of emitted energy. Every single line in a spectrum represents the transition from one energy state to another.

Professor E. E. Free at this time drew a beautiful analogy to explain Bohr's theory of radiation. He said:

> Imagine a series of race tracks one inside the other. Imagine these tracks are separated by high board fences. Put a race horse in the outermost track and instruct him to run around it until, when he happens to feel like it, he is to jump the inside fence into the next

track, run around it for a while, and then jump the next fence, and so on until he reaches the innermost track of all. If, then, you watch this procedure from the field outside the outermost fence, you will not see the horse at all as long as he is running in a single track. The fences hide him. But whenever he jumps from one track into the next, you will see him for an instant as he goes over.

Using this method of attack he cleverly explained Balmer's formula and even deduced this mysterious mathematical relationship by subtracting the calculated energies of electrons before and after jumping. He intimately associated the amount of energy required to move a single electron from one orbit to another with Planck's quantum of energy. He went further and explained that the spectrum of hydrogen was so complex because every sample of hydrogen gas used during any experiment consisted of a large number of atoms in different stages of equilibrium. Sommerfeld verified his conclusions and even succeeded in explaining the "fine" lines of the hydrogen spectrum (two or three lines very close together) by employing Einstein's mass-energy law which postulated electrons having varying masses as their speeds change.

Niels Bohr provisionally determined the position of the electron of hydrogen, its spectrum and the character of its orbit. The other more complex atoms still defied accurate analysis. He had made use of all the theories and discoveries known to science. By coordinating them he had finally postulated a fairly probable explanation of these phenomena. His thesis brought him undying fame and in 1922 the Nobel Prize. Einstein, his very close friend, had been awarded the Nobel Prize the previous year and Bohr had quickly dispatched his congratulations. Einstein responded with "Dear—or rather, beloved Bohr! Your heartfelt letter made me as happy as the Nobel Prize. I found especially charming your fear that you might have received the prize before me. That is genuinely Bohr-like (bohrisch)."

Bohr showed the microcosm of the atom to be a strange world. If we magnify the atom to the size of a football, the nucleus would be but a speck in its center and the electron, still invisible, would be revolving around its surface. Similarly, if we picture the atom as large as the New York City Trade Center tower, the electron, the size of a marble, would be spinning around the building seven million times every millionth of a second. There is relatively more empty space in the atom than between the planets in the solar system. Bertrand Russell, one of England's greatest mathematicians and philosophers, had expressed this idea rather fancifully. "Science compels us," he wrote, "to accept a quite different conception of what we are pleased to call 'solid matter'; it is in fact something much more like the Irishman's net, 'a number of holes tied together with pieces of string.' Only it would be necessary to imagine the strings cut away until only the knots were left." The only thing that gives porous matter the appearance of solidity is the rapid swarming of its electric particles.

Then came Gilbert Newton Lewis of the University of California who worked with mathematical formulas and complicated theories with uncanny precision. He had made outstanding contributions in the field of classical thermodynamics for which he was awarded the Davy Medal of the Royal Society of England. Lewis, born six years before Langmuir, was a lively character. One of his intimate friends had said that "with or without the stimulus of ethyl alcohol, in any company he is the focus of the liveliest discussion, and the center of the merriest group." Born in Weymouth, Massachusetts, ten years before Niels Bohr, he had studied at the Universities of Nebraska, Harvard, Leipzig, and Göttingen, where, like Langmuir, he received his doctor's degree. While lecturing at Harvard he was known to "pace back and forth across the platform covering miles at breakneck speed." His was a versatile personality who attracted many brilliant students.

Lewis, one of the first Americans to discover Einstein, was a theorist of the highest order. As early as 1902 he was pondering over the structure of matter and had already conceived the *cubical* atom. Lewis loved to speculate and play with fanciful theories and intricate formulas. Just before he left for France in 1916 as head of the Defense Division of the Chemical Warfare Service, he published a famous paper which laid the basis of the theory known as the static atom, as more fully developed a few years later by Langmuir. In every atom, said the California professor, is an essential nucleus which remains unaltered. Around this nucleus are cubical shells containing varying numbers of electrons which occupy fixed positions. Every atom tends to get one electron on each of the eight corners of its cube. Lewis based this *octet* theory upon the vast store of chemical and crystallographic properties of substances which had accumulated, and not, as Bohr had done, on the incomplete physical data then at his disposal.

In his 1916 paper "The Atom and the Molecule" he elucidated his ideas of nonionic compounds being the result of the sharing of electrons among atoms. In the formation of a molecular compound, there was a sharing of electrons of a pair of atoms by the two atoms. Lewis called such a sharing a *covalent bond*. This led to further theories of chemical *bonding*.

According to Lewis the following represented the atoms of

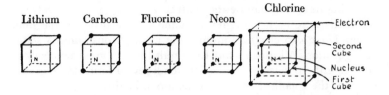

This was the state of our knowledge of the structure of the atom when Langmuir, the modern scientific conquistador, attempted to invade the tiny world of the atom. There was an unmistakable conflict between Bohr's theory of the hydrogen atom and the conception of Lewis. Chemists were unhappy with the Bohr atom. They wanted an atom which would explain chemical reactions. World War I over, Langmuir undertook to reconcile the two theories by publishing his own modified *concentric shell* theory of atomic structure.

About two hundred years ago Lavoisier tried to find the cause of the different behaviors of the elements. Why, for instance, was chlorine so violently active, while nitrogen and gold were almost completely inactive? Like thousands of other scientists, Lavoisier failed to explain this strange phenomenon. "The rigorous law from which I have never deviated," he wrote, "has prevented me from comprehending the branch of chemistry which treats of affinities or chemical unions. Many have collected a great number of particular facts upon this subject. But the principal data are still missing."

The great Berzelius half a century later was still puzzling over this question. "We ought," he wrote, "to endeavor to find the cause of the affinities of the atoms," and he suggested a possible method of attack. "Chemical affinity," he believed, "is due to the electrical polarity of the atoms." With the tremendous strides made in theoretical and applied chemistry, the solution of this important question still remained undiscovered.

Irving Langmuir, dreamer and practical engineer, saw in his conception of the tiny cosmos of the atom a probable explanation. Moseley's table of atomic numbers was his starting point. The inert gases of the atmosphere which had led him a merry chase in his researches on the gas-filled tungsten lamp were to furnish the clue to the cause of chemical activity.

The elements helium (atomic number 2) and neon (atomic number 10) were two stable elements. In these atoms the electrons outside their nuclei must therefore represent stable groups which rendered their atoms incapable of chemical activity. Langmuir pictured helium as containing a nucleus of fixed hydrogen ions (protons) and cementing electrons, and two additional electrons revolving in a shell outside the central core. The distances between the shells were made to agree with the various orbits of the Bohr atom. These two electrons around the nucleus constituted a stable configuration. All atoms, said Langmuir, have a great tendency to complete the outermost shell. This tendency to form stable groups explains the chemical activity of the atom. Hydrogen is very active because its shell, containing but one electron, is incomplete and needs another electron to form a stable group of two electrons, as in helium.

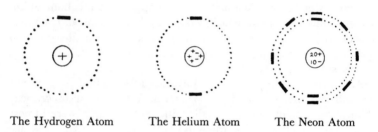

The Hydrogen Atom   The Helium Atom   The Neon Atom

Neon, with ten electrons outside its nucleus, represents another stable configuration having two electrons in its first shell and eight more in a second larger shell concentric with the first. (Walther Kossel, three years before in Germany, in attempting to construct a ring system which would correlate electronic shells with chemical valencies had also assumed a configuration of eight electrons as stable.) All the elements with atomic numbers between 2 and 10 are, therefore, active to an extent depending upon the completeness of their second shells. For example, lithium, atomic number 3, possesses only one single electron in its second shell, and hence in its eagerness to have its outside shell complete will readily give away this third electron to another element, and thus have left but two electrons in the first shell—a stable group. This tendency to lose electrons from the outermost incomplete shell makes lithium an extremely active substance. Fluorine, atomic number 9, shows two electrons in its first complete shell, and seven additional electrons in its second shell. It needs but a single electron to complete its second shell of eight electrons. Hence it, too, shows a violent tendency to capture an electron, thus manifesting extreme chemical activity.

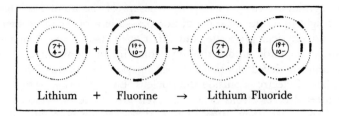

Lithium + Fluorine → Lithium Fluoride

Atoms, said Langmuir, differ from each other in chemical activity only because of their tendency to complete their outside shells and thus render the atom stable. Argon, the third inert gas in Moseley's Table of the Elements, has an atomic number of 18. Its first shell is complete with two electrons, its second shell is also complete with eight additional electrons, while its third shell likewise contains eight electrons, showing once more a stable configuration. Hence argon is inert. Chemical affinity is thus a condition dependent upon the nature of the outermost shell electrons. When the outside shell of an atom contains very few electrons, its tendency is to lose them. Such an atom is a metal. If, on the other hand, the outermost shell of an atom contains an almost complete ring, it will strive to borrow some electrons from other atoms which are anxious to lose them. Such an atom is a non-metal. Metals are lenders of electrons and non-metals are borrowers. Hence metals and non-metals will combine energetically with each other and both, by an exchange of electrons, assume the stable condition. Chemical affinity or union, therefore, depends upon this transfer of electrons. In a *polar union* said Langmuir, a positive atom loses its valence electrons to a negative atom and the two atoms are held by electrostatic attraction. In a *non-polar union*, on the other hand, electrons are not actually transferred—the two atoms approach each other so that one or more valence electrons of one atom occupy the vacant positions in the valence shell of the second atom. Octets are thus formed by a process of *sharing pairs of electrons*.

This modified concentric shell theory of Langmuir seemingly solved other riddles. It explained valence—the tendency of elements to combine with one or more atoms of hydrogen. Valence had baffled chemists ever since Frankland, an English chemist, had introduced the idea in 1852. Valence, according to Langmuir, is the number of electrons which the atom borrows or lends in its effort to complete its outside shell. Thus chlorine, which borrows but one electron, has a valence of one, which means that it combines with but one atom of hydrogen.

Langmuir's conception of the structure of the atom also threw some light upon the meaning of isotopes—atoms of the same chemical and physical properties but differing in mass. Since chemical affinity depends upon the electrons in the outermost shell, Langmuir believed chlorine isotopes, for example, to have the same number of electrons outside the nucleus. Each chlorine isotope has seventeen free electrons of which seven are in the outermost shell. Since, however, they differ in weight, Langmuir postulated quite rationally that the nuclei of isotopes differ by having different numbers of particles in their central cores. Isobars, *different* elements with *identical* atomic weights, were also explained. The last three even-atomic-numbered isotopes of selenium have the same atomic weights (78, 80, 82) as the first three even-atomic-numbered isotopes of krypton, yet the physical and chemical properties of these isobars are totally different from each other. This is because the number of electrons in their outermost shells are different.

What a simple, if very incomplete, conception of the atom Langmuir gave us. Chemical science, thought Langmuir, began to approach prophecy, for, according to him, knowing the atomic number of an element, we could with certainty, arrive at its properties without even seeing the element.

Langmuir gave a high place to intuition in scientific advance. Retiring as president of the American Association for the Advancement of Science in 1942 he stated,

> In almost every scientific problem which I have succeeded in solving, even those that have involved days or months of work, the final solution has come to mind in a fraction of a second by a process which is not consciously one of reasoning. Such intuitive ideas are sometimes wrong. The good must be weeded out from the bad, sometimes by common sense or judgment, other times by reasoning.

But Langmuir had by no means answered all the questions relating to the structure of the atom. There were many more knotty problems ahead. Both the brilliant formulations of the Copenhagen school led by Bohr and the Lewis-Langmuir picture of the atom developed in the United States contained dark spots that no amount of experimentation could illuminate. The experimental facts connected with the intensities of the spectrum lines of the elements, for example, proved an insurmountable objection to the acceptance of these theories.

From the field of light had already come rumblings of a violent upheaval in the definitions of science. Isaac Newton and his predecessors had believed light to consist of discrete particles too small and light to be measured. This corpuscular theory of light had been generally accepted until the opening of the nineteenth century when Thomas Young and later Auguste Fresnel, working independently, showed that under certain conditions rays of light could be made to interfere with each other to produce darkness. This phenomenon of *interference* could not be adequately explained by any corpuscular theory. Interference, however, made sense when light was considered as being of the nature of waves transmitted on a hypothetical medium, the ether. Just as the crest of one sea wave falling in the trough of a second fills it and makes the surface of the water level at that position, so, maintained Young, light waves meet and interfere with each other to produce darkness.

This undulatory theory of light, entrenched for a full century, in turn received a rude jolt. Hallwachs, in 1888, found that a beam of light falling on a negatively charged body produced a rapid loss of the charge, while a non-electrified body gradually became positively charged in the presence of light. No adequate explanation could be offered for this strange behavior until two years after the discovery of the electron. Both J. J. Thomson and Philipp Lenard showed that the emission of negatively charged particles or electrons from the surface of the metal was the cause of the phenomenon which Hallwachs was the first to observe. Three years later, Lenard proved that the speed of the emitted electrons depended only upon the color or wavelength of the light no matter how feeble. Further experimentation demonstrated that the *number* of electrons thrown out from the metal depended upon the *intensity* of the light. Here was another enigma for science to unravel. How could light, made up of waves in the ether, release electrons from the atoms of metals? The accepted concept of light was powerless to answer this baffling question.

In 1905, Albert Einstein, employed as examiner of patents in the patent office of Berne, made a bold attack upon this problem. In his now famous paper, this German-Swiss scientist of twenty-five theorized that the emission of electrons from metals by light could be explained by assuming that light consisted not of waves but of *photons* or concentrated bundles of energy traveling with the speed of light. Each photon or particle of radiation as it struck the metal surface gave up its energy to an electron and dislodged it from its atom.

In attempting to confirm Einstein's picture of this queer photo-electric effect, scientists delved once more into the possible structure of the atom and the nature of its electrons. If light is of a dual nature, exhibiting at one time the properties of a wave and at another the characteristics of a particle, then, perhaps, all the stuff of the chemist might show this same duality. Perhaps matter itself was of a wave nature, and the paradox of light would find its consort in the paradox of matter.

In a private laboratory in Paris worked Louis Victor, Prince de Broglie. One of his direct ancestors was Marshal of France under King Louis XV, while his great-grandfather, Victor Claude, Prince de Broglie, was a member of the Constituent Assembly and served as Lafayette's chief lieutenant during the American Revolution. This Jacobin, who admonished his young son to be faithful to the principles of the Revolution, however unjust it might be to him, was guillotined a month after the execution of Lavoisier.

Louis Victor, Prince de Broglie, at thirty-two had been admitted to the French Academy of Sciences. He had been trained under the excellent guidance of his older brother, Maurice, Duc de Broglie, who for more than two decades had made notable contributions in the fields of X-rays, electrons, and electricity. Louis de Broglie, in his study of black radiation and light quanta, was gradually led to what seemed to him an inevitable conclusion that light and electrons were similar. Seeing an imperative need for a new concept of matter which would be applicable to the structure of the atom as conceived by both chemists and physicists, and at the same time permit of an explanation of the paradox of light, he brilliantly enunciated a theory of the electron as revolutionary as the most sweeping hypothesis of the century. In a paper published in 1924 he suggested that the electron was not altogether a particle of electricity. The electron, he thought, was composed of, possessed, or perhaps was attended by, a group of waves which guided its path. He associated the motion of a particle with the propagation of a wave and related the energy and momentum of a particle with the frequency and velocity of the wave. By using the terminology of radiation for an atom of electricity and considering the electron as the wave itself, he calculated a definite wavelength for his electron.

If de Broglie's electron was of the nature of Young's rays of light then the electron ought to exhibit the peculiar characteristics of ether waves, namely interference and diffraction. For two years after de Broglie's hypothesis had been advanced no direct proof of its validity could be mustered. But in 1927 the experiment which men of science either scouted or awaited with bated breath came from America.

This most important discovery of the year was made in the research laboratories of a private industrial corporation, the Bell Telephone Company of New York. This was the probing ground of two men — Dr. Clinton J. Davisson and Lester H. Germer, a graduate student of Columbia University. Davisson had been graduated from the University of Chicago in 1908, and later, at Princeton University, under the direction of Professor Owen W. Richardson, had been drawn to thermionics. This branch of science which deals with the emission of ions and electrons from heated bodies was Richardson's own particular field, the laws of which he had discovered. Langmuir, too, had been attracted to this promising sphere of research and had made notable contributions to it.

Davisson had acquired a remarkable technique in his study of ions and electrons, and after de Broglie had astonished the world with his concept of electron waves he set to work to investigate it. For three years, with the aid of young Germer, Davisson experimented on the scattering of electrons from metal surfaces when bombarded both with ionized gas molecules and streams of electrons. Finally one day, from their graphs and figures, they discovered that a sharply defined stream of electrons, after striking a small plate cut from a single crystal of nickel, left the crystal in the direction of regular reflection. The angle of incidence of the stream of electrons was equal to its angle of reflection.

How could this action be accounted for? The surface of a crystal was made up of atoms very much larger than the tiny material electrons which struck the surface of the metal. How could these electrons rebound from such a surface with such regularity? "It is," said Germer, "like imagining a handful of birdshot being regularly reflected by a pile of large cannon balls. A surface made up of cannon balls is much too coarse to serve as a regular reflector of particles as small as birdshot." The reflection of the narrow beam of electrons from the nickel crystal indicated that the electrons were striking the crystal and were being reflected from the crystal in the same way as rays of light behave. Davisson and Germer drew their conclusion: "Our experiments establish the wave nature of moving electrons with the same certainity as the wave nature of X-rays had been established."

Here was the first fruit of de Broglie's daring assumption which had served as the guiding principle in this historic experiment. The scientific world carefully weighed the results of the Davisson-Germer discovery. Albert Einstein, the most profound scientific thinker since Newton, could not refrain from exclaiming: "We stand here before a new property of matter for which the strictly causal theories hitherto in vogue are unable to account." In 1929 Prince de Broglie was honored with the Nobel Prize for his work which culminated in the theory that the stuff of matter has wave properties like light. Physics and chemistry, turning an unexpected corner, had met and found themselves in need of new moorings.

Sir Joseph John Thomson had given the electron to the world and had spent a lifetime of research in its clarification. Then came his son, George Paget Thomson, trained by his father at Trinity College in the complicated field of electron mechanics. This younger Thomson, who was born in the same year as Louis de Broglie, had shown, even as a student, unmistakable promise in the field of science. He had won First Class honors in the Mathematical Tripos and shared honors in the Natural Science Tripos. Turning from aerodynamics, into which he had been led as an officer in the British Army during World War I, he tackled the problem with which de Broglie had wrestled. Perhaps, after all, his father's electron was not altogether what the world believed it to be.

While Professor of Natural Philosophy at the University of Aberdeen, he started to investigate the nature of the electron in order to strengthen the old concept or demolish it. If the electron consisted of waves it ought to exhibit the optical phenomenon of diffraction, that is, a change of direction as it passed from one medium like a solid metal to another like a vacuum or air. In the summer of 1927, G. P. Thomson sent rapidly moving electrons, the kind that Davisson was using, against extremely thin sheets of gold and other metals. He took hundreds of photographs of the area behind the metallic films and found unmistakable evidence of the diffraction phenomenon of waves.

A few months later the son of the master of the electron told the Royal Society of England something more startling perhaps than his father had told it thirty years back. The old picturesque conception of the electron as an atom of negative electricity was an outgrown image. There was no such thing as a material electron such as his father had described. "One may picture the free electron," declared the younger Thomson, "as something like a gossamer spider floating through the air at the center of a number of radiating filaments which control its flight as the air wafts them about, or as they are caught by solid objects." He claimed for this new conception of the electron as a wave motion "more prospect of accounting for the facts of chemistry than have the orbits" of Bohr and his school.

In 1925, Arthur H. Compton of the University of Chicago disclosed the fact that hard X-rays, which exhibited all the properties of light, showed a *lowering of frequency* after striking a metal plate. His photographs indicated that the X-rays bounced like balls rather than acted like waves. Two years later he received the Nobel Prize for this work and, soon after, from India came the announcement of the discovery of another somewhat similar effect. Chandrasekhara Venkata Raman, a Hindu scientist, detected that a beam of monochromatic light from a mercury vapor lamp was scattered by benzene, and new lines of *greater wavelength* than the incident beam were revealed by the spectroscope. Both the Compton and Raman effects pointed to the probability that both Einstein and de Broglie were right.

Recognizing the dual nature of light, X-rays and the electron, G. P. Thomson predicted that the proton also would eventually show this duality of wave and particle. Before long, many left less urgent researches to join in this gripping adventure. Among them was Arthur J. Dempster who had first shown positive proof of the existence of isotopes of magnesium. In the Ryerson Physical Laboratory of the University of Chicago, he plunged into the problem of discovering the nature of the proton. Hydrogen atoms were stripped of their electrons in a vacuum tube and then directed against a small calcite crystal. By their rebound he was convinced that the positively charged hydrogen atoms or protons were striking the crystal as waves rather than as tiny balls. This fulfillment of Thomson's prophecy made such a strong impression that Dempster was immediately awarded the annual thousand-dollar prize of the American Association for the Advancement of Science in 1930.

In the meantime, there had begun an orgy of speculation in the field of the physical sciences. Mathematical specialists entered the arena with a new attack. Physical and mathematical theories of the structure of the atom came "crowding on each other's heels with an increasing unmannerliness." In the forefront stood Erwin Schroedinger, an Austrian scientist. He introduced a new mathematical treatment known as *wave mechanics*, which looked upon the atom as a region permeated with waves. His technique confirmed the work of de Broglie and even predicted some facts connected with spectrum lines that were not foreseen by Bohr's ingenious approach.

At the same time an equally profound and suggestive idea was being evolved by a young theoretical physicist of Munich, Werner Heisenberg. This young man of twenty-three expressed the structure of the atom by means of mathematical formulas directly connected with the frequencies and intensities of spectrum lines—phenomena which could be observed and measured. His system, the *new quantum* or *matrix mechanics*, bothered even men accustomed to the twists and turns of serpentine theoretical physics. Wolfgang Pauli, a young Viennese who studied at Munich and Zurich, came to the conclusion that electrons have rugged individualities, and that there are never two or more equivalent electrons in the same atom. This idea, known as the Pauli exclusion or equivalence principle, became an important addition to atomic knowledge. Heisenberg, Pauli and Schroedinger all received Nobel Prizes for these contributions. In 1927 Heisenberg published in the *Zeitschrift fur physik* another paper in which he enunciated what is known as the *Principle of Indeterminacy*. He postulated that "a particle may have a definite position or it may have a definite velocity but it cannot in any sense have both."

These new principles were used with great success by many, including Linus C. Pauling, then head of the department of chemistry of the California Institute of Technology. Pauling, who was born in Portland, Oregon, in 1901, was one of the first chemists to interpret chemical behavior by the quantum theory. He cleared up many of the mysteries of the physical and chemical structure of organic compounds in several papers on the nature of the chemical bond and the electronic theory of valence. Experimentally, too, he developed with great brilliance a method of taking pictures of electron diffraction of gaseous molecules to elucidate the structure of organic molecules. His later contributions not only in the fields of chemistry and medicine, but also in the world peace movement, brought him two Nobel prizes—one in chemistry in 1954 and the other for peace in 1962, an extremely rare achievement.

In spite of all these new approaches which illuminated the outer regions of the atom, the center or nucleus of the atom continued to remain a bundle of uncertainties. Something of the composition of the nuclei of a few elements was already known. This information came from a study of the spontaneous disintegration of radium and other radioactive elements, such as thorium, polonium, uranium, and ionium. These elements break down of their own accord into simpler elements. Soon after the Curies' discovery of radium, Rutherford and Frederick Soddy, his student and collaborator, had found that the spontaneous breaking down of radium resulted in the emission of three types of rays and particles. Radium ejected alpha particles (ionized helium atoms), beta particles (electrons), and gamma rays (similar to X-rays). In radioactive elements, at least, it was believed that the nucleus contained electrons, protons, and electrified helium particles. (The gamma rays given off are energy rays rather than matter.) Was this true for other elements as well?

American science was ready to take another step forward in the realm of theory. In a room close to Millikan's laboratory at the

University of Chicago, William D. Harkins in the winter of 1914-15 attempted to find some characteristic of the atomic nucleus which would serve as a basis for a new classification. The characteristic chosen was the stability of the nucleus. He offered the theory that the nuclei of all elements were compounds of hydrogen and helium. Atoms having *even* atomic numbers were more stable than those possessing *odd* atomic numbers, and hence were more abundant in nature. He also assumed that the heavier elements were built from lighter elements by a step-by-step process in which hydrogen and helium groups were gradually added. So far as radioactive elements were concerned, Harkins' hydrogen-helium theory was acceptable, for they give off both electrons and helium. The disintegration of radium, for example, takes place in the following stages:

Radium	↦Radon	↦Radium A	↦Radium B	↦Polonium	↦Lead
at. wt. 226	at. wt. 222	at. wt. 218	at. wt. 214	at. wt. 210	at. wt. 206
loses one	loses one	loses one	loses	loses one	which is
charged	charged	charged	electrons	charged	the end
helium	helium	helium	and one	helium	product of
atom of	atom of	atom of	charged	atom of at.	radium
at. wt. 4	at. wt. 4	at. wt. 4	atom of	wt. 4 and	disinte-
and	and	and	helium and	*changes to* —	gration.
changes to —	*changes to* —	*changes to* —	*changes to* —		

But would Harkins' theory hold for other atoms? To be sure, Rutherford had already shown that nitrogen (atomic number 7), as well as all the odd atomic elements (5, 9, 11, 13, 15) which he had bombarded, liberated hydrogen but, so far as was known, no helium. On the other hand, not a single even-atomic-numbered element could be disrupted. Harkins decided to repeat Rutherford's classic experiment of 1919 which produced the first artificial transmutation in history by bombarding nitrogen with helium nuclei. He modified the Wilson cloud-chamber apparatus to suit his own procedure, and in 1921 took thousands of fog-track photographs of helium nuclei (obtained from thorium) shooting through nitrogen gas.

One of the pictures was a very strange one. It showed something new—a double deflection in a fog track with one line of the fork about ten times thinner than the other usual deflected line. To Harkins this solitary picture, one bull's-eye from among a hundred thousand shots (good marksmanship in the subatomic world), indicated that Rutherford's interpretation of the ejection of an electrified hydrogen particle from the nitrogen nucleus in his 1919 experiment was an incomplete story. In this destruction of an atomic world Harkins saw, also, the synthesis of a new one, for not only was hydrogen (thin line) ejected, but another element, oxygen (thick line), was in this case formed. He interpreted the picture as showing the union of the helium nucleus (alpha particle) with the nitrogen nucleus to form an atom of fluorine. This fluorine atom at once disintegrated to form an electrified hydrogen particle and an atom of oxygen. In other words, oxygen had been synthesized or built up from nitrogen and helium.

Harkins represented this result of seven years of intense work graphically and in equation form as follows:

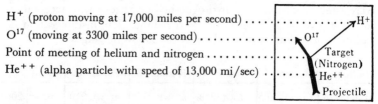

H^+ (proton moving at 17,000 miles per second) . . . .
O^{17} (moving at 3300 miles per second) . . . .
Point of meeting of helium and nitrogen . . . .
He^{++} (alpha particle with speed of 13,000 mi/sec) . .

H^+
O^{17}
Target (Nitrogen)
He^{++}
Projectile

helium	+	nitrogen	→	fluorine	→	hydrogen	+	oxygen
He^{++}	+	N	→	F	→	H^+	+	O
mass 4	+	mass 14	→	mass 18	→	mass 1	+	mass 17

The oxygen formed was an atom of atomic weight 17 instead of 16, the accepted atomic weight of the only kind of oxygen atom then known. In 1931, however, this new form or *isotope* of oxygen of atomic weight 17 was discovered, thus confirming what until then was only a possibility.

Harkins' picture of a nucleus containing nothing but helium and hydrogen nuclei and electrons, however, contained paradoxes. All of the protons of an atom are in the nucleus, but not all of its electrons are outside its nucleus. Some of its electrons must, therefore, be within its nucleus to help neutralize the positively charged protons, since normally elements are electrically neutral; they give no electrical shock when touched because they lack an excess of either positive or negative electricity. But how can negatively charged electrons and positively charged protons exist side by side in the nucleus? In other words, what prevented the negative electron and the positive proton from joining together since they were so closely situated in the tiny nucleus? Harkins saw the anomaly. Speculations were no longer unfashionable in twentieth-century science, and he was audacious enough to advance a seemingly preposterous theory of the existence of another entirely new unit in the nucleus. On April 12, 1920, he had written to the *Journal of the American Chemical Society* that in addition to the protons and alpha particles in the nuclei of atoms, there is also present "a second less abundant group with a zero net charge." He suggested the name *neutron* for this non-electrified particle of atomic number zero, made up of a single proton and a single electron very close together.

This was a bold prediction and an accurate one. Twelve years later, in the winter of 1932, the particle was actually discovered, not by Harkins but by an Englishman, James Chadwick, working in Rutherford's laboratory. Chadwick had shot helium bullets (from old radium tubes sent to him by the Kelly Hospital of Baltimore) against beryllium, a metal lighter than aluminum, and noticed that something of great penetrating power was knocked out of the target. To account for the high energy of this unknown something which was thrown out of the beryllium, and to save the law of the conservation of energy, Chadwick said that these new "rays" were really not rays at all. They must, he believed, be made of particles of the mass of protons, but unlike protons, they were not electrically charged. Since these neutrons were electrically dead, they could not be repelled by the impregnable electric walls of the atom, and hence they had a terrific penetrating power. Two and one-half inches of lead were capable of stopping only half of them. What had really happened could be expressed in the following equation:

Beryllium	+	Helium	→	Carbon	+	Neutron
at. weight 9	+	4	→	12	+	1

Where $\begin{smallmatrix} 4 + \\ 5n \end{smallmatrix}$ represents the nucleus of the beryllium atom containing 4 protons $(+)$ and 5 neutrons (n). Four electrons $(-)$ spin around this nucleus in two distinct orbits.

The announcement of this discovery was followed by feverish activity to learn more about this new member of the family of fundamental entities. John R. Dunning, Nebraska-born physicist, and George P. Pegram, at Columbia University, following the discovery of the Italian scientist, Enrico Fermi, in 1934, slowed down a stream of neutrons by passing them through paraffin. Then Isadore I. Rabi, their brilliant colleague, calculated from their experimental data that the diameter of the neutron was very much smaller than the diameter of the nucleus of the hydrogen atom. For its infinitesimal size, it was so tremendously dense and heavy that a lady's thimble tightly packed with them would weigh many tons. This compactness could explain the high densities of certain stars now known to consist of neutrons. Harkins was right, and once again the value of pure theory and creative imagination in science was demonstrated. This particle, the neutron, was a remarkable new key for science, one destined within a few years of its discovery to shake the whole world to its foundations.

What was the scientist's new picture of the structure of the atom back in 1932? The idea of planetary electrons outside the nucleus remained unchanged. But his conception of the nucleus was different. There were no longer any *free* electrons in the nucleus. Only protons and neutrons were found there. The atomic weight or *mass* of an element was equal to the total number of protons and neutrons in its atom. The *atomic number* of an element was defined either as the number of planetary electrons in its atom, or as the number of free protons in the nucleus of its atom. The two isotopes of chlorine whose atomic number is 17 were, therefore, represented as:

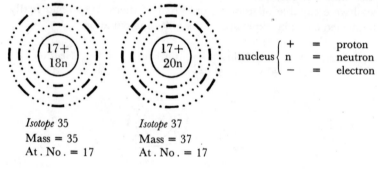

Isotope 35	Isotope 37
Mass = 35	Mass = 37
At . No . = 17	At . No . = 17

In the same year in which the neutron was discovered, the Swedish Academy of Science recognized the fundamental work of this man and awarded him the Nobel Prize for his researches in chemistry—the first American *industrial* chemist to be so honored. He went to Sweden to receive the award, did some skiing and mountain climbing in Switzerland, and visited Bohr in Denmark before he returned from Europe.

Langmuir, who in 1929 was president of the American Chemical Society, later turned to other fields of investigation, notably to the problems of surface phenomena. He pioneered in studying the question of how certain substances are adsorbed on the surfaces of other chemicals, and how molecules arrange themselves in thin layers on such surfaces. He prepared very thin layers of material, only one or two molecules thick in many cases, and investigated their unusual physical and chemical behavior. This field of two-dimensional chemistry was extremely important because the action of enzymes, toxins, antitoxins, vitamins and hormones, often distributed over membranes, walls and the like, is tied up so closely with human health. This approach opened up a new avenue of attack to some of the most baffling and stubborn problems concerning health and disease.

The study of films as thin as 0.000004 inch led Langmuir and his assistant, Dr Katherine B. Blodgett, to an immediate and practical application. They succeeded in rendering glass almost perfectly transparent by covering its surface with very thin films of fluorine compounds which eliminated reflections.

Later, Langmuir found himself immersed in the problems of weather control. A carefully planned series of experiments was undertaken to determine whether clouds could be made to change into rain or snow at the command of man rather than at the whim of nature. Such an achievement would be of tremendous practical importance. It could prevent huge losses to farmers when drought threatened to destroy their crops. It could "punch holes in the sky," and clear the atmosphere of fog and cloud quickly enough to permit pilots to take off and to land in perfect safety. It would encourage men to further investigations leading to more types of weather controls, and even to climatic changes.

At the General Electric Flight Test Center near Schenectady, New York, Langmuir and his assistants went up in planes in 1947 and seeded supercooled clouds from above with pellets of dry ice (solid carbon dioxide). Vincent J. Schaefer, one of his associates, had produced man-made rain and snow by this method for the first time in 1946. Langmuir's group succeeded in causing huge clouds to condense into rain by this method, and also by using silver iodine crystals instead of dry ice. On the basis of the unexpected behavior of certain clouds that had been attacked with dry ice, Langmuir predicted before the National Academy of Science that it would even be possible soon to get rain out of some clouds at will by the use of water itself dispersed at the right time. Late in 1947 he expressed even more optimistic ideas. He thought it would be possible, in the not too distant future, to change the general cloud formation over wide areas such as the northern part of the United States, and thus produce profound changes in the weather of thousands of square miles of territory.

More than a quarter of a century later his hopes were far from fulfilled. Numerous attempts have and are still being made to modify weather. For example, more than fourteen years of randomized experiments near Climax, Colorado, showed that seeding clouds in which the temperature ranged between $-11°$ and $-20°C$ at the cloud tops could increase the snowfall in this mountainous region by 10 to 30%. Australia, Israel and the U. S. S. R. have also experimented with some positive results.

Among the many problems still unsolved are the dangers of flood damage and the illegal use of this operation. During the Vietnam War rainmakers and weather modification were secretly used by the American Army for military purposes in North and South Vietnam, Loas and Cambodia. Its effectiveness for hampering enemy troop movements, commando operations and enemy missile fire was difficult to evaluate and has never been made public.

Our National Academy of Science has given this subject considerable attention, but after many years of study found some progress but even more problems. "Progress comes in the form of more statistically valid evidence that cloud seeding can increase precipitation, but only under some circumstances. Problems have to do with lack of research funds." The Academy proposed in 1973 three national goals with a target date of 1980. These include more experimentations, more attention to its hazards, and further study of the effects of man-made pollutants on global climate.

Irving Langmuir had great faith in the young research worker in science. He believed with that great protagonist of evolution, Thomas Huxley, who over half a century ago exclaimed: "I would make accessible (to the scientist) the highest and most complete training the country could afford. I weigh my words when I say that if the nation could purchase a potential Watt, a Davy, or a Faraday, at the cost of a hundred thousand pounds down, he would be dirt cheap at the money." Testifying in 1945 before a Senate subcommittee investigating the advisability of creating the present National Science Foundation, as suggested by Franklin D. Roosevelt and recommended by the Bush Committee, Langmuir called for the creation of a research foundation as an independent agency of the Federal Government with safeguards against political control. This National Science Foundation was finally created five years later by President Truman. Alan T. Waterman who had taught physics at Yale University was named its first director. It was authorized "to develop and encourage the pursuit of a national policy for the promotion of basic research and education in the sciences, to initiate and support basic research in the mathematical, physical, biological, engineering and other sciences by making contracts or other arrangements (including grants, loans and other forms of assistance) for the conducting of such scientific research." This would include, for example, money for building astronomical telescopes and particle accelerators. A phenomenal growth of expenditures for scientific research and development in this country resulted.

During World War II Langmuir again volunteered his services to the government. He was a consultant on the Manhattan Project which made the first atom bomb, and worked on other scientific problems connected with the war effort. He became more actively interested in social and political problems connected with the advance of science both here and abroad. As one of the American delegates he attended in 1945 the two-hundred-twentieth anniversary of the founding of the Russian Academy of Sciences in Moscow, and spent eighteen days in the Soviet Union. Among the new machines he was shown was a 40-ton cyclotron in the final stage of completion. He came back with great admiration for the strides which Soviet science had made and for the spirit of the Russian research workers he had met. "I found," he reported, "that the Russian scientists talked freely about their work and showed me through their laboratories. These men were working on problems that had been planned by scientists without undue political control." He had some good words to say about the Russian practices of wide differentials in pay, elimination of unemployment, "suppression" of strikes, and broad encouragement of science. He warned that Russian science was booming and "in ten or twenty years from now, the Russians may be far ahead of us." He called upon America to copy some of the ways of Russian scientists, and through science to make America an even more dynamic democracy and a more vital leader in international affairs. Late that year at a joint meeting of the American Philosophical Society and the National Academy of Science, he urged understanding with Russia to insure a durable peace. "It is highly desirable," he declared, "that there should be frank discussion between the Russian government and the American government regarding the troublesome effects caused by differences between our concepts of democracy and freedom of the press."

Langmuir believed that the future of mankind depended upon international peace guaranteed by a strong United Nations Organization, as well as upon the immediate application of the newer knowledge of physics and chemistry and the new mathematical systems and theories which had emerged within the past decades.

Langmuir retired in 1950 as associate director of the General Electric Laboratory and continued as a consultant to the laboratory until his death. This occurred in the summer of 1957 as the result of a heart attack which he suffered while visiting his nephew at Falmouth, Mass. He was seventy-six years of age.

Later that same year Niels Bohr was the first man to be honored as the first recipient of the Atoms for Peace Award. A check for $75,000 was handed to him by the President of the United States, Dwight D. Eisenhower, in Washington, D. C., in recognition of his great contributions to the peaceful uses of atomic energy. For many years he had worked very hard to turn the energy of the atom from war to peaceful uses. In 1952 at his suggestion representatives of 14 European nations met at Copenhagen to plan an international physics center, the Conseil Européenne Reserche Nucléaire, thereafter known as CERN. Seven years later the center was completed at a cost of $30,000,000 at Geneva, Switzerland.

In spite of his heavy involvement in his own scientific research, Bohr found time to do an enormous amount of work in many humanitarian and educational projects. He eagerly joined a committee to raise funds for the creation of a science institute honoring the chemist and statesman Chaim Weizmann whom he had known as a young man and who later became the first President of the State of Israel.

He travelled and lectured very widely. In the summer of 1962 he attended a conference in Germany where he suffered a mild cerebral hemorrhage. He recovered rapidly and by November 16 of that year was able to preside at a meeting of the Danish Academy of Science and Literature. Two days later he died quietly at home. Thus passed one of the two great giants in science of the century.

XVII

LAWRENCE

HIS NEW ARTILLERY LAYS SIEGE
TO THE ATOM'S NUCLEUS

WHILE Bohr and Langmuir had circumnavigated the atom and penetrated into its outer lines of electron defenses, the inner core or nucleus still remained very much a no man's land. What was now needed to enter the guarded citadel of the atom's nucleus were high-speed particles even more powerful than the alpha particles of radium disintegration. In 1932, within the space of less than a month, two new particles were revealed, two brand-new projectiles ready to be brought up with the artillery that was once more to lay seige to the subatomic world. The neutron was one of these bullets, deuterium or heavy hydrogen was the other.

Everywhere researchers close to the problem realized this need. A friendly yet spirited race had already begun in many laboratories of the world to build mighty armaments—new atomic siege guns which would hurl thunderbolts of staggering power to shatter the tiny nucleus into fragments that could be picked up and studied. High-potential drops were to send every kind of submicroscopic bullet available crashing into the subatomic defenses.

The greatest of these early ordnance builders was Ernest Orlando Lawrence. He was born, soon after the opening of the present century, in a little town in South Dakota. His paternal grandfather, Ole Lavrensen, was a schoolteacher in Norway who came in 1840 to Madison, Wisconsin, during the Norwegian migration, to teach school along the frontier of America. He had his name anglicized to Lawrence immediately upon arrival in the new land. Ernest Lawrence's maternal grandfather was Erik Jacobson, who came twenty years later to seek a homestead in South Dakota when that area was still a territory. Both his grandmothers were also natives of Norway. Ernest's father, Carl G. Lawrence, was graduated from the University of Wisconsin and became president of Northern State Teachers College at Aberdeen, South Dakota. His mother was born near Canton, where Ernest, too, was born on August 8, 1901, only twelve years after South Dakota became a state. As a young boy he was sent to the public schools of Canton and Pierre, South Dakota. Later he attended St. Olaf College and the University of South Dakota. He had become attracted to science through an experimental interest in wireless communication, but for a while it seemed that he might pursue his hankering for a medical career. Finally, however, he threw in his lot with the physics of the atom and radiation.

Lawrence did some graduate work at the University of Minnesota, where he came under the influence of Professor W.F.G. Swann. He followed Swann to Yale, did graduate work under him, took his doctorate there in 1925, and stayed on, first as National Research fellow, and later as assistant professor of physics. When Swann left to head the Bartol Research Foundation of Philadelphia, Lawrence answered a call from the University of California.

Late one evening in the sping of 1929, Lawrence quite accidentally came across an excerpt from the dissertation of Rolf Wideroe, an obscure Norwegian investigator working in Switzerland. Attracted to a diagram of a piece of apparatus used by this physicist, he never finished reading the paper. Wideroe had managed to impart to electrified potassium atoms in a vacuum tube energies equal to twice the energy of the initial voltage he used. A small voltage could thus give high velocities to projectiles if the voltage could be applied repeatedly to the bullets at just the right time. The idea was not an altogether new one. This scheme of multiple acceleration of atomic projectiles has been likened by Karl T. Compton, then president of the Massachusetts Institute of Technology, to "a child in a swing. By properly synchronizing the pushes, the child may be made to swing very high even though each individual push would lift him only a short distance."

Lawrence, who according to Swann "had always shown an unusual fertility of mind and had more than his share of ideas," picked up and nourished this fertile seed. He had been casting about for a means of side-stepping the difficulties of sustained or intermittent high voltage necessary for effective attacks on the kernel of the atom. He was searching for a technique which would require no high-tension currents of electricity and no elaborate vacuum-tube equipment, and yet enable him to get immense speed with his projectiles. Now he thought he had a valuable clue. Within a few minutes after he had seen that diagram he began sketching pieces of apparatus and writing down mathematical formulas. The essential features of his new machine came to him almost immediately. The next morning he told a friend that he had a new idea and was going to invent a new tool of science.

With this new instrument Lawrence planned to whirl an electrical bullet in a circle by bending it under the influence of a powerful electromagnet. As it passed around one half the circumference of a highly evacuated tank shaped like a covered frying pan, he was going to give the particle repeated electrical kicks which would send it racing in ever-widening circles at greater and still greater speeds, until it reached the edge of the evacuated tube, where it would emerge from a slit and be hurtled into a collecting chamber. Here he would harness it as a mighty projectile against the nucleus of an atom. He was going to adjust the magnetic field so that the particle would get back just as the initial alternating current changed direction and at the exact moment it was ready for another kick. The particle was to be speeded on its way by oscillating electricity of high frequency. He hoped in this way to get the same effect by applying a thousand volts one thousand times as he would by applying a million volts all at once.

"High frequency oscillations," said Lawrence, "applied to plate electrodes produce an oscillating electrical field. As a result, during a one-half cycle the electric field accelerates ions into the interior of one of the electrodes, where they are bent around on circular paths by the magnetic field and eventually emerge again into the region between the electrodes." It was a very bold scheme. Would it work? By January, 1930, Lawrence had built his first magnetic resonance accelerator, which later became commonly known as the *cyclotron.* Between the poles of an electromagnet was a vacuum chamber only four inches in diameter. In this were two D-shaped insulated electrodes connected to a high-frequency alternating current. Down the center ran a tungsten filament. The rest of the machine was constructed of glass and red sealing wax. With the help of N.E. Edlefsen, his first graduate-student assistant, he succeeded in getting actual resonance effects. The idea worked, and Lawrence made his first public announcement of the machine and method in September of that same year at a meeting of the National Academy of Sciences in Berkeley. Now at the age of twenty-nine he was made a full professor.

Lawrence had created a new tool for science and had added a new word to the dictionary. Said *The New York Times,*

The pioneers in experimental physics have always had to devise their own instruments of investigation. Men like Faraday, Hertz, and Helmholtz are not listed among the great inventors. For the servants

of science invent as a matter of course, rarely take out patents, and
concentrate on research . . . If Lawrence were what is called a *practical* inventor and his cyclotron were of any immediate commercial use,
he would take his place beside Watt, Arkwright, Bell, Edison and
Marconi, which would probably exasperate rather than flatter him.

Lawrence's cyclotron, at the beginning, was essentially a new tool
of theoretical research into the nature of the atom's structure. After
the first glass model of a cyclotron, Lawrence, with the help of M.
Stanley Livingston, another of his graduate students, made a metal
cyclotron of the same size. He was able with this new machine to
generate with a current of only 2000 volts a beam of hydrogen ions
with energies corresponding to those produced by 80,000 volts. By
February of 1932 Lawrence had built a model costing $1000. This
eleven-inch merry-go-round device was able to speed protons, obtained by ionizing hydrogen gas, with energies equivalent to 1,200,
000 volts. He was getting now into the big figures. With this
instrument he disintegrated the element lithium in that summer of
1932—the first artificial disintegration of matter carried out in the
Western Hemisphere, thirteen years after Rutherford had blazed
the trail in England.

There was great excitement in Lawrence's laboratory. Everybody
there was sure they could reach voltages only dreamed of before. A
huge magnet casting had been lying around idle in California since
World War I. It had been built by the Federal Telegraph Company
for use by the Chinese government in a radio-transmission installation, but had never been shipped across the Pacific. Lawrence spoke
to Professor L. F. Fuller, who was on the staff of the University of
California and also, at the time, vice-president of the Federal
Telegraph Company. Could he have this magnet? The immediate
answer was yes, and the seventy-five-ton monster was drafted for
research work. It was immediately set up and wired with eight tons
of copper in the newly established Radiation Laboratory of the
University, of which Lawrence was made director.

This $27\frac{1}{2}$-inch cyclotron was calculated to deliver several microamperes of 5,000,000 electron-volt deuterons and 10,000,000 electron-volt helium nuclei. The south pole of its huge magnet rose
flat-topped from the floor as high as a kitchen stove and had a
diameter of forty-five inches. The machine was operated from a
control board forty feet away, and the operator was further protected from the penetrating radiation of the machine by suitable
absorbing material, such as tanks of water and paraffin, placed
around the cyclotron. When all was ready, Lawrence and his

energetic group of very young assistants lost no time in trying this whirligig atom gun encased in the "frying pan" placed between the poles of the Gargantuan electromagnet. Every available projectile was hurled against every available target in the hope of breaking into and shattering the nuclei of every atom.

Protons and helium nuclei, as well as the nuclei of the newly discovered heavy hydrogen atom (deuterium), were hurled with crashing effects. Lawrence, at the suggestion of G.N. Lewis, called the heavy hydrogen bullets *deutons*, against the advice of Rutherford, who preferred *diplons* because, he thought, "deutons were sure to be confused with neutrons, especially if the speaker has a cold." Later by agreement of scientists here and in England, the nucleus of heavy hydrogen was named *deuteron*.

The discovery of deuterium had been predicted by Ernest Rutherford in England and by Gilbert N. Lewis and Raymond T. Birge of the University of California. Deuterium is a double-weight hydrogen atom; that is, an isotope of hydrogen of atomic weight one. Its structure may be represented as:

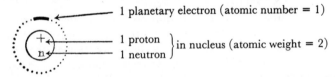

The prediction of deuterium's discovery came true in a chemical laboratory of Columbia University. Harold C. Urey, born in Indiana, had received his doctorate at the University of California and had studied under Bohr. Early in his career he had suspected the presence of *heavy hydrogen* as a result of his analysis of the spectrum of ordinary hydrogen. In the fall of 1931 F.G. Brickwedde of the United States Bureau of Standards evaporated a quantity of liquid hydrogen and sealed the last few remaining drops in a glass tube which he sent to Urey for examination. The Columbia scientist passed an electric discharge through the tube, scrutinized its spectrum lines, and announced the presence of the heavy isotope of hydrogen, which he named *deuterium* (D), from the Greek word meaning second. It occurs in ordinary hydrogen to the extent of about one part in four thousand.

This discovery was hailed as one of the most important of the century. It not only supplied atom smashers with a new type of bullet, but it also opened up a brand-new world for further research. When it is realized that the human body contains almost seventy per cent of water, the physiological importance of *heavy water* substituted in the body for ordinary water could hardly be exaggerated. With the three isotopes of hydrogen (*tritium*, atomic weight 3, is also found in hydrogen to a very minute extent), and the three isotopes of oxygen, as many as eighteen different kinds of water may be formed, each having different properties. Some scientists foresaw almost limitless possibilities of new compounds, since hydrogen occurs in more than a million organic compounds, alone. Most important of all, the new tool was put to use by researchers in physiology and medicine almost at once to tag atoms passing through the body. Several scientists drank water containing deuterium to determine its effects. The Hungarian Nobelist, Georg von Hevesy, recalled a conversation he had had with Ernest Rutherford over a cup of tea. The Englishman wondered how long it actually took that tea to pass through and out of his body. Hevesy thought we would never know since it was impossible at the time to follow the course of a fluid through the body. When, however, heavy water was discovered he at once saw the possibility of using it as a tracer, and tried to find the answer to Rutherford's question. He did: it was between 8 and 14 days.

Heavy hydrogen and, later, tritium were substituted for ordinary hydrogen in certain fats; then the course and changes which these "tagged" fat molecules underwent on their way through the animal body were studied. This new tool of research enabled scientists to attack many practical problems relating to human health. Here is another astonishing illustration of how theoretical problems in science may turn out to be of tremendous practical significance to mankind. In 1934 Urey earned the Nobel Prize for this work. Since then he has isolated other very important isotopes, such as those of carbon and nitrogen. These are proving to be additional weapons in man's fight against disease.

By 1935 Lawrence had shot deuterons against the element lithium and obtained helium, and had effected many other similar transmutations. The way was now clear for the transmutation of every element in the table of atomic numbers—including even the transformation of baser metals into the gold of the alchemists' dreams.

Lawrence's fame spread rapidly. At the age of thirty-two he was elected to membership in the National Academy of Sciences. By this time the center of gravity of scientific talent in the United States had definitely moved westward. From Lawrence's laboratory streamed an army of young researchers who were put in charge of construction or maintenance of new cyclotrons built by other universities and several industrial laboratories. By 1940, thirty-five cyclotrons were in operation in as many laboratories both here and abroad. One of these, a 200-ton job, was shipped to Dr. Yoshio Nishina's laboratory at the Tokyo Institute of Physical and Chemical Research. At least another twenty more were then under construction. Lawrence was constantly called upon to offer expert advice on these new installations. Much of his time was also taken up by supplying many centers of research both in this and foreign countries with new radioactive products of his huge cyclotron.

Honors, too, came flowing his way. These were finally capped by the award of the Nobel Prize in 1940 for the invention of the cyclotron and especially for the results attained by means of this device in the production of artificially radioactive elements. Hitler, in the meantime, had overrun much of Europe, making it quite impossible for Lawrence to go to Stockholm to receive the award personally from Sweden's king. Instead, the presentation was made at Berkeley, with the Consul General of Sweden present to represent his government. The prizewinner's colleague, Raymond T. Birge, made the presentation address, and reminded his audience of the splendid example of co-operative effort represented by Lawrence's Radiation Laboratory. Lawrence's first remark on hearing of the award was, "It goes without saying that it is the laboratory that is honored, and I share the honor with my co-workers past and present." When in 1942 the Academy of Sciences of the Soviet Union elected its first foreign members, three Americans were included — G.N. Lewis, Walter B. Cannon, and Ernest O. Lawrence.

Other generals were operating in the atomic field. In the laboratory of the Department of Terrestrial Magnetism of the Carnegie Institution in Washington, D.C., Merle A. Tuve, another scientist of Norwegian ancestry and a boyhood playmate of Ernest Lawrence, worked out a different method for obtaining high voltages by means of a modified Tesla coil and huge glass plate condensers. Tuve, guarded by thick plates of lead and aided by the biological research of his wife, who studied the effects of these penetrating radiations on rats, drove projectiles by intermittent excitation to their speed limit and produced momentary voltages as high as five million.

Robert van de Graaff, an Alabaman, while a Rhodes scholar at Oxford, got the notion that perhaps a return to simple principles might help solve the perplexing problem of high voltages. He gave up the idea of using transformers to increase voltage and devised a machine which built up a high potential by gathering large quantities of static electricity from a friction machine, such as was first produced by Otto von Guericke in 1671. Karl T. Compton, then head of the department of physics of Princeton University, where van de Graaff served as a National Reasearch fellow, helped the Alabaman construct the first working model, which supplied energies as high as 1,500,000 volts. In 1933 van de Graaff was able to build a Big Bertha consisting of two units, each weighing sixteen tons. A highly polished aluminum shell fifteen feet in diameter and one-quarter of an inch thick was mounted on the top of a twenty-five-by-six-foot insulating cylinder. Inside this hollow cylinder a rapidly moving silk belt sprayed static electricity onto the surface of the sphere. The machine was capable of supplying 7,000,000-volt sparks. Later the Westinghouse Electric and Manufacturing Company constructed a huge pear-shaped atom-smasher built on the same principle.

W.H. Keesom at Leyden, Holland, tried for years to realize the dream of wrenching atoms apart with a fourteen-ton electromagnet immersed in pure liquid helium at a temperature of 272.29° below zero Centigrade (just a few tenths of a degree above the lowest theoretically possible cold). Peter Kapitza, brilliant Soviet physicist, employed a different strategy. He tried to rip the atom apart by subjecting it to tremendously powerful momentary currents strong enough, he hoped, to overcome the terrific magnetic forces (estimated at 7,000,000 gauss) which hold it intact. He went to Ernest Rutherford with his idea, and was given laboratory facilties to test his theories. Later he was made director of the Mond Laboratory of the University of Cambridge. This laboratory, an adjunct of the Cavendish Laboratory, was built for Kapitza with the aid of the Rockefellers. Here he struggled with huge electromagnets, an electric alternator giving 20,000-ampere current, ingenious and elaborate systems of switches, and especially designed cables that carried great surges of magnetic pulls through coils which would melt if the surges were of longer duration than a mere hundredth of a second. The shock produced when the circuit was closed for a fraction of a second resembled a minor earthquake, but the jar reached the other end of the laboratory, eighty feet away, where the delicate measurements were being taken, only after the experiment was over. Kapitza pitted here all the knowledge of science and all the skill of man against the atom, whose symbol, a dragon, was carved over the entrance to his laboratory. In 1935 he returned to Russia, where the Soviet government appointed him director of the Institute of Physical Research of the Leningrad Academy of Sciences. Here a new laboratory was equipped with a forty-ton magnet with which Kapitza resumed his experiments begun in England. In

1946 Kapitza was elected to membership in our own National Academy of Sciences.

The atom was also attacked in the Kellogg Radiation Laboratory of the California Institute of Technology. Here C. C. Lauritsen, a Dane, who gave up a promising career as a sculptor to design electrical equipment and join the men arrayed against the atom, sat in the center of a great concrete block while he controlled a million-volt X-ray tube. Deuterons, protons, neutrons, and electrons were speeded up into powerful bullets for atom smashing. Here, too, a young man who had wandered into physics and taken his doctorate under Millikan photographed an unexpected curved fog track which turned out to be the face of another newcomer from the atom's nucleus—a strange wanderer that stirred the scientific world. The young man was Carl D. Anderson, who was born in 1905 in New York City of Swedish parents. He was graduated from the Los Angeles Polytechnic High School and then started to study electrical engineering at the California Institute of Technology. In his sophomore year he suddenly switched to physics.

In the spring of 1930 Millikan was searching for a way to determine the energy of cosmic rays, a highly penetrating form of radiation which he believed was intimately connected with the building up of the atoms of matter. He set Anderson at work on a machine which might succeed in bending these rays by means of strong magnetic forces. Three years before, Skobelzyn, a Soviet scientist, had for the first time obtained photographs of tracks of cosmic-ray particles. But Anderson had a tougher job on his hands.

Between the poles of a powerful magnet capable of maintaining a magnetic field of 24,000-gauss strength was placed a vertical Wilson cloud chamber 15 cm. in diameter and 2 cm. deep (the first of its kind ever built). Photographs were taken through a hole in the pole piece of the magnet along the lines of force, thus making possible the revealing of a particle that might be reflected by the magnetic field as an arc of a circle. On moving-picture film thousands of photographs were taken of the effect of cosmic rays, some of 3,000,000 electron-volts energies, striking atoms of gas in the cloud chamber.

On the afternoon of August 2, 1932, one film, exposed and developed by Anderson, showed the image of a blasted atom with a graceful track which had never before been observed. "He at once realized its importance," wrote Millikan, "and spent the whole night trying to see if there were not some way of looking at it from what was already known about atomic nuclei." At first Anderson thought an electron had suffered a reversal of direction due to a sudden scattering, or that possibly it was a proton, for the direction of the curve was opposite to that formed by a negative electron. This indicated that it possessed a positive charge. Its ability to pass through a lead plate 6 mm. thick indicated a tremendous power of penetration possessed by no electron known. The length of the path of its fog track was ten times greater than the path of a proton of this curvature, proving that it could not be the positively charged proton. It seemed, in fact, to belong to a particle of positive charge, yet of mass equal only to that of a negative electron. It seemed to belong to a new particle—a positive electron.

Anderson repeated his experiments, obtaining a multitude of new photographs and confirming the result, which he published in September 1932. Like the neutron, this particle, too, had been predicted. The great English theoretical physicist, P.A.M. Dirac, later a Nobel laureate, had theorized its existence. Anderson's discovery, however, was not guided by this theory. The new arrival from the subatomic world was christened by its discoverer the *positron*. The name *oreston* had been suggested because Orestes, brother of Electra, came to an early and tragic end even as the positron is quickly annihilated when it encounters an electron (or negatron) and the two are changed into gamma rays having about half a million electron-volts energy.

In the month of Anderson's discovery of the positron, the same particle was obtained by Patrick M. S. Blackett at Rutherford's Cavendish Laboratory, by Skobelzyn at Leningrad, and by a young French couple in the Radium Institute of the University of Paris. Irene Curie, a tall, shy, serious-looking woman who had inherited the Slavic features of her mother, was walking in the footsteps of the immortal Marie Curie, discoverer of radium. Irene had married Jean Frédéric Joliot, whom she had met in the laboratory even as Marie had met Pierre Curie, and like her parents they were working side by side on the problem of atomic structure. The Joliot-Curies obtained their positrons by using gamma rays from radioactive elements instead of from cosmic rays.

After further study, Anderson declared, "It looks as though it is a general property of electromagnetic radiation to give rise to positrons when the radiation penetrates matter." However, positrons are not released from atoms if the radiation possesses energy of the order of less than one and one-half million electron volts. Positrons have been found, when alone, to have an extremely short life. For the discovery of the positron, the first *anti-matter* reported, Anderson went to Stockholm in the winter of 1936 to receive the Nobel Prize.

The year 1932 was a real *annus mirabilis* in the history of science. Four great discoveries and inventions were made in that single year. The neutron, heavy hydrogen, and the positron were all brought to light for the first time, while the first practical cyclotron was added to the key instruments of physical science. It is interesting to note in this connection that three of these four milestones were first reached by young American scientists, while the fourth, predicted by an American, was discovered by an Englishman. The tempo of first-rate American contributions in science had definitely reached a new high.

The positron soon appeared in another way. When radium, recklessly throwing away electrons and alpha particles until it changes by easy stages to lead, was first described as an element that spontaneously disintegrated, forming different elements, the first authentic case of transmutation—the change of one element into another—was actually recorded. Over this sort of transmutation, man had, however, no control. He was like an astronomer witnessing the galaxies in the heavens spinning their mysterious tales. But this discovery gave man hope that perhaps some day he would find the mechanism of transmutation and bring it within his own sphere of influence.

More than two decades were to pass after the discovery of radium before Rutherford, chief of the early atom-smashers, actually succeeded in 1919, with the help of ionized helium bullets, in breaking nitrogen into hydrogen. A great ferment arose and physicists everywhere strove to develop the method. Another decade passed, and Harkins submitted photographic proof not only of the breaking down of a heavier element into a lighter one (nitrogen into hydrogen) but of an even more unexpected feat, the building up of oxygen from the lighter element nitrogen. Still the skeptics sneered. In every reported case of a transmutation, the means used, the alpha particle, had been itself obtained by the breaking down of a naturally radioactive element.

Scientists still hoped for a transmutation which could be effected by means of a simple tool not fashioned from spontaneously degenerating elements. The day of triumph came on April 28, 1932, a few weeks after the discovery of the neutron, to two of Rutherford's young lieutenants, J. D. Cockcroft and E.T.S. Walton, working in his laboratory. Instead of alpha particles they used protons stepped up by means of a transformer-rectifier to a velocity of about 6000 miles per second—equivalent to a voltage of 600,000. They aimed these projectiles at lithium fluoride crystals and obtained visual flashes of helium atoms striking a special zinc screen placed in a position to receive them. For the first time in history a method of transmuting elements by means other than radioactive products had been accomplished. It was a startling announcement.

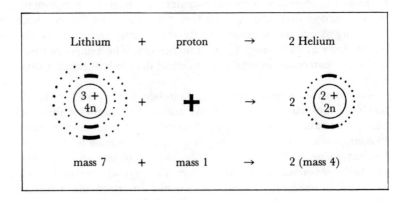

Not only had this new kind of transmutation been achieved but another phenomenon had been observed—*artificial radioactivity,* or radioactivity in nonradioactive elements—a new and major contribution. The announcement of this discovery was made on January 15, 1934. Fate had waited for the Joliots to open up another room in that mansion of radioactivity first unlocked to the world by the Curies. The Joliots accomplished this discovery by bombarding boron with alpha particles from polonium (discovered by Marie Curie), producing a neutron and a form of *radioactive nitrogen* which continued to disintegrate after the bombardment was stopped. Half of this radioactive nitrogen changed within fifteen minutes into an inactive form of nitrogen and a positron. Half of the remainder disintegrated within the next fifteen minutes, and so on progressively. (We say that the *half-life* of radioactive nitrogen is, therefore, fifteen minutes.) The Joliots explained the new phenomenon as a capture of the electrified helium particle, forming a radioactive nitrogen nucleus and a neutron. The radioactive nitrogen nucleus then disintegrated, forming a stable nucleus of carbon, and emitting a positron.

FIRST STEP	Boron + Helium ion	→	radioactive Nitrogen	+ neutron
	mass 10 + mass 4	→	mass 13	+ mass 1
SECOND STEP	radioactive Nitrogen	→	non - radioactive Carbon	+ positron
	mass 13	→	mass 13	

The experiment was repeated by the Joliots with other elements, such as magnesium and aluminum, and similar results were obtained. Other laboratories took up the work. Within eight years science had added 360 artificially produced radioactive substances, of which 223 were discovered by means of the cyclotron, 120 of them in Lawrence's own laboratory. One of the most interesting of these products produced by Lawrence was *radioactive sodium*, which has a half-life of about fifteen hours. This was obtained by bombarding sodium with deuterons shot with energies of 1,750,000 electron volts. The explanation of this change is based on the fact that a deuteron is composed of a neutron and a proton. As the deuteron approaches the nucleus of the atom at which it has been hurled, it is broken up. The strong electric field around the atom's nucleus stops the electrified proton, but the electrically neutral neutron goes right on and enters the nucleus, producing a radioactive atom, thus:

$$Na^{23} \quad + \quad H^2 \quad \rightarrow \quad Na^{24} \quad + \quad H^+$$

non-radioactive + deuteron → radioactive + proton
sodium   sodium

This radioactive sodium produces gamma rays with energies of 5,500,000 electron volts, almost three times as penetrating as the gamma rays formed during the natural disintegration of radium. The use of radioactive sodium for producing this new type of gamma rays was tried out in cancer therapy as a substitute for both X-rays and radium. Ernest's brother, Dr. John H. Lawrence of the Department of Internal Medicine of Yale, used both radioactive sodium and radioactive phosphorus in the study of chronic leukemia in mice and men. The half lifetime of radium is 1700 years, that of radioactive sodium is only about fifteen hours. This is an advantage in cancer therapy, since it reduces the danger of burns. In addition, radioactive sodium gives only gamma rays, whereas radium produces other products which must be filtered out. The amount of radioactive sodium obtained by Lawrence after his giant slingshot machine has been in continuous operation for twenty-four hours bombarding with 5,000,000 electron-volt deuterons (twenty microamperes) was only about one-fifth of a gram, that is, an amount of radioactive sodium having a gamma-ray activity equivalent to that of one-fifth of pure radium. This, at the time, was regarded as a considerable amount.

The radioactive elements supplied by Lawrence were also used as *tracer* elements or chemical sleuths in physiological and medical research. For example, radioactive sodium present in common table salt (sodium chloride) is taken into the body in foods. In about twenty-four hours it has completely changed to a new element and has ejected a high-speed particle. This ejection can be recorded by means of a Geiger counter placed next to various parts of the body. In this way the itinerary of a tracer atom can be followed to find the answer to some health problem. Or fluids may be removed from certain parts of the human body and then tested for their radio-activity without disturbing the metabolic process.

Hevesy was the first to use this technique. It was back in 1923 when he was working on plant metabolism. He used lead and bismuth, which are slightly radioactive in their natural state. He used it once, he told a friend, at a boarding house where he suspected the quality of the food that was served. One day he brought to the table a millionth of a millionth of a gram of a radioactive compound and dropped it on a small scrap of meat which he left in his plate. The next day he appeared at his usual place in the dining room armed with a Geiger counter. As the meat dish—hash—was placed before him the Geiger counter clicked the warning. It was the same meat that had been left on his plate the day before. That settled it. Hevesy changed his boarding house.

Still another technique, called radioautography, was used in this connection. When it was found necessary to determine the route and rapidity of concentration of phosphorus in the bones of a rat, for example, the animal was fed with a radioactive phosphorus compound. At intervals sensitive films were placed against its bones, and the radioactive phosphorus which had collected made its own image on the film. By this same technique Herbert M. Evans in 1940 tagged radioactive iodine in the bodies of sixty-three of his experimental tadpoles. He was attempting to determine the way in which iodine is extracted from the blood and stored in the thyroid gland. Sections of the tadpole's thyroid tissue were placed in contact with X-ray film, and the radioactive iodine, which has a half-life of about eight days, registered its presence on the film. At the Westing-house Electric and Manufacturing Company this technique was also used to show the distribution of phosphorus, carbon, silicon, and other elements in steel. A. V. Hill, the eminent English physiol-ogist and Nobel prize winner, told Lawrence that it was his belief that "the use of such tracer elements will some day be recorded in history as a technique of equal importance with the use of the microscope."

New and more sensational developments were being reported so rapidly that research workers were left almost breathless in the effort to study in detail the ground already won. On June 4, 1934, the world of science was again shaken. At a meeting of the Academy dei Lincei in Rome, experimental science stirred once more in the direction heralded years before.

The announcement from Italy told of the synthesis of a new element, No. 93, by Enrico Fermi, a young academician of the Institute of Experimental Physics of the University of Rome. The new element was similar to manganese, fitted into its proper place in Moseley's table of atomic numbers, and was radioactive, half of it disintegrating spontaneously within thirteen minutes, thus accounting for its absence on earth. Fermi was among the scientists who had been completely absorbed in the new developments of atomic research. While bombarding uranium (element No. 92) with neutrons, he found that the neutron insinuated itself in the nucleus of uranium and built up the heavier, transuranium element, No. 93. Two scientists, Grosse and Angruss, thought that Fermi had only obtained a new isotope of protactinium. But Fermi made a chemical test to convince himself that his interpretation was plausible. Later investigations showed that he had made uranium 239 which changed into element 93. Here was another proof of the building-up theory of Harkins. Before the outbreak of World War II Fermi left Fascist Italy to breathe the freer air of American laboratories, and became professor of physics at Columbia University.

American genius in science was to be heard from again. In 1936 another major discovery was made by the same Carl D. Anderson who had given the positron to the world. This discovery, too, came in connection with cosmic-ray research. Robert A. Millikan and Arthur H. Compton of the University of Chicago had been carrying on a friendly rivalry which sent each of them to many parts of the earth and high into the sky to collect data in the hope that he might be the first to find out definitely the manner of creation of cosmic rays and the nature of their effects.

For six weeks during the summer of 1935 Anderson worked on top of Pikes Peak, taking 10,000 photographs of possible collisions of *cosmic-ray* particles with atoms in the upper atmosphere. Associated with him was Seth H. Neddermeyer, who was taking his doctorate under him in this field. Upon their return to California they made a careful study of their photographs. They measured the range, curvature, and penetrating power of the fog tracks. In August, 1936, they reported, among other findings, that, "About one percent of the exposures reveal the presence of strongly ionized particles which in most cases seem to be neither electrons nor protons. They arise from a type of nuclear disintegration not heretofore observed. Their source is in the cosmic rays." The following year, after further experiments, in which they used a 1-cm. platinum plate across the center of the cloud chamber, they obtained another 6000 photographs. After an analysis of these pictures, they announced "the existence of particles of a new type which cannot readily be explained except *in terms of a particle of mass greater than that of an ordinary electron!*" This new type was named by its discoverers the *mesotron*, meaning intermediate particle, since its weight is intermediate between that of an electron and a proton. The discovery of the mesotron had been predicted by a Japanese scientist, Hideki Yukawa, and for a while was humorously known as the *Yukon*, at J. Robert Oppenheimer's suggestion. The mesotron was later renamed the *meson*.

Just before this important news was printed in the May, 1937, issue of the *Physical Review*, J. C. Street and E. C. Stevenson reported the same phenomenon to the American Physical Society. Soon after, Blackett and others also confirmed the discovery. The mass of Anderson's new negative particle had been estimated to be 200 times that of an electron. Anderson also reported that it travelled almost as fast as light and that, like positrons, mesotrons may be created in pairs by protons. Five years later, Bruno Rossi, then at Cornell University, estimated that the life of this new entity was of about the same order as that of the positron. He found it to be about one-half millionth of a second. The mesotron was regarded from the very beginning of its discovery as an extremely important particle. Lawrence believed that a thorough understanding of mesotron forces might help uncover some new secrets of atomic energy.

In 1939 Lawrence had jumped from his 37-inch cyclotron to a 60-inch cyclotron, which was built for medical research and daily cancer treatment. The funds for this machine were supplied by Francis P. Garvan of the Chemical Foundation, the Rockefeller Foundation, and the National Cancer Advisory Council. William H. Crocker gave the money for the construction of the building, named the William H. Crocker Radiation Laboratory of the University of California, situated on the campus in Berkeley. The 60-inch cyclotron, weighing 220 tons, gave 100 microamperes of 16,000,000-volt deuterons and one microampere of 32,000,000-volt helium ions. The heavy hydrogen "death-ray" beam, several inches in diameter, shot 600,000,000 individual atoms per second a distance of about five feet out of the window of the machine at a speed of 25,000 miles per second. This is equivalent to the effects of disintegration of thirty tons of pure radium worth, in 1940, about $32,000,000 per pound!

After this cyclotron had proved itself very efficient, Lawrence thought of an even more powerful machine. When in 1932 he had jumped from his tiny cyclotron to a machine containing a magnet with a pole diameter of $27\frac{1}{2}$-inches, he showed that he was more than an academic physicist. He had invaded the field of engineering. He could handle big projects. A 5000-ton cyclotron did not frighten him. He was ready to make one great leap from the 60-inch medical cyclotron to a 184-inch machine, even as George Ellery Hale had jumped from his 100-inch telescope on Mount Wilson to the newer 200-inch giant eye on Mount Palomar. Lawrence drew the designs for a new cyclotron which would be capable of furnishing deuterons flying at a speed of more than 60,000 miles per second, and other atomic bullets of energies well above 200 M.E.V. (200,000,000 volts). This machine, he hoped, would become a national institution and the mecca of hundreds of research men.

In reply to the presentation to him of the Nobel award on February 29, 1940, Lawrence humorously remarked, "Perhaps the difficulties in the way of crossing the next frontier in the atom are no longer in our laboratory; we have handed the problem over to the president of the university!" The president lost no time. Money was found. The International Education Board appropriated $1,150,000 for the project. A site on Charter Hill near the Berkeley campus was chosen, and steel and concrete began to roll in for the new atom-smasher, largest in the world. The mammoth core was buried in the hill. For remote-control operation, an underground power cable for carrying current from Gilman Hall over "Tightwad Hill" to the cyclotron building was completed, and in July, 1941, work was started on the twenty-four sided building, 90 feet high and 160 feet in diameter, which was to enclose the cyclotron.

The bombing of Pearl Harbor on December 7 of that year did not put a halt to the project, which was 70 per cent complete. The machine was finished in May, 1942. It was agreed that research with this new tool should be continued for the sake of the war effort. Its 3700-ton magnet was temporarily removed for some special war work, and in addition to the many immediate practical problems there still remained that old, tantalizing question of whether it would ever be possible to pick the lock that holds the enormous reservoir of energy inside the nucleus of the atom. Every Nobel Prize winner in the physical sciences believed that goal possible of attainment. William Bragg, of crystal structure fame, had stated his belief: "A thousand years may pass before we can harness the atom, or tomorrow might see it with the reins in our hand."

Almost half a century had passed by this time since the bubble of the invisible Daltonian atom was first burst. Some of the fragments of that explosion—electrons, protons, alpha particles, neutrons, deuterons, positrons, mesons—lay about for examination. Other fragments were destined to be picked up among the debris of atom-smashing in the years to come.

The fact that some of several predicted particles had not been found had not prevented theoretical physicists from employing them in attempting to explain not only the nucleus but also the universe as a whole. In fact, Einstein in his last attempt to unify all the phenomena of both stars and atoms into one all-embracing system, made use in his mathematical equations of the then still undiscovered neutrino and of still another, very queer, undiscovered particle—an electrical atom of zero mass.

Until all the parts of the extremely intricate machinery of the atom's nucleus could be located and the forces that held them together understood, the release of the stupendous energy tied up in the atom would remain, it was believed, in the future. Physical science was still groping its way through the atom's nucleus. Lawrence to the end remained vigorous, optimistic, enigmatical. Referring to his newest cyclotron at that time, he remarked: "There lies ahead for exploration a territory with treasures transcending anything thus far unearthed. It may be the instrumentality for finding the key to the almost limitless reservoir of energy in the heart of the atom." There was a great deal more that Lawrence knew at this moment. But it was top secret.

For the next sixteen years Lawrence was a key figure in the most unique and daring project ever undertaken by science. It was destined to change the course of world history. In 1957 he received the second Enrico Fermi Award of $50,000 presented by the United States Atomic Energy Commission. This event took place hardly a year before he succumbed, while undergoing surgery to correct an ulcerated colitis condition, at Palo Alto, California. He was fifty-seven years old.

XVIII

THE MEN WHO HARNESSED
NUCLEAR ENERGY

LATE in 1938, in Berlin-Dahlem, an experiment in nuclear chemistry touched off a wave of excitement throughout the world which even reached the front pages of the most conservative newspapers. At the Kaiser Wilhelm Institute for Chemistry, only a few miles from Hitler's Chancellery, three researchers had proceeded to repeat some experiments first performed by Enrico Fermi in Rome in 1934. The Italian scientist, in an attempt to produce the Curies' artificial radioactivity in the very heavy elements by bombarding them with neutrons, believed he had created an element (No. 93) even heavier than uranium.

Two of these scientists in Berlin-Dahlem, Otto Hahn and Lise Meitner, had already confirmed Fermi's results, and by 1936 it was proposed to name this new element *Ansonio* after an ancient name of Italy. Later, Fritz Strassmann joined the team and together they continued with these experiments. On January 6, 1939, they observed a strange result which they published two months later in *Die Naturwissenschaften*. According to Hahn and Strassmann, the bombardment of uranium with neutrons had split the uranium atom almost in half! The smash-up had produced what they had reason to believe were two different and lighter elements, isotopes of barium and krypton ($U^{92} \rightarrow Ba^{56} + Kr^{36}$). Hitherto only bits of the heavier atoms had been chipped away. No new theory or explanation was offered for this unexpected effect.

What was even more startling than this transmutation was the announcement of the three scientists that during this spectacular change their oscilloscope recorded a release of energy equivalent to 200,000,000 electron volts. The Germans were completely at a loss for a logical explanation of this phenomenon which they had witnessed. Hahn was overcome by his interest in the chemical change, and the problem of the energy change escaped him. Dr.

Lise Meitner, a mathematical physicist, knew, however, that something new and tremendously important had happened in the subatomic world of the nucleus of uranium. She pondered deeply over the probable mechanism of this change. In the meantime, however, the purge of non-Aryans and other intellectuals from German universities under Hitler caught up with the sixty-year-old woman scientist.

Born in Vienna, the daughter of a lawyer, Lise chose scientific research as a career and came to Berlin before World War I to assist both Max Planck and Fritz Haber. By 1917 she was made head of the physics department of the Kaiser Wilhelm Institute at the time that Otto Hahn was head of its chemistry department. Working with him in that year she had discovered and isolated a new element No. 91—protactinium.

In spite of a lifetime of distinguished scientific work in Europe, during an intensified period of purges and atrocities Lise Meitner was finally marked by the Nazis as a Jew for arrest and a concentration camp. Early in 1939, therefore, she "decided that it was high time to get out with my secrets. I took a train for Holland on the pretext that I wanted to spend a week's vacation. At the Dutch border I got by with my Austrian passport, and in Holland I obtained my Swedish visa." Meitner, of course wanted to escape the concentration camp but, even more important, she desperately needed to get out of Germany because she felt that she had an interpretation of the Hahn-Strassmann experiment—an explanation whose implications might change the course of history. On the basis of mathematical analysis Meitner saw in the Berlin experiment a splitting or *fission* of the nucleus of uranium into two almost equal parts. This atomic fission was accompanied by the release of stupendous nuclear energy resulting from the actual conversion of some of the mass of the uranium atom into energy in accordance with Einstein's mass-energy law. This was her greatest contribution.

Back in 1905, Albert Einstein, in developing his theory of relativity, announced that there was no essential difference between mass and energy. According to his revolutionary thinking, energy actually possessed mass and mass really represented energy, since a body in motion actually possessed more mass than the same body at rest. Instead of two laws—the law of the conservation of mass, and the law of the conservation of energy—there was only one law, the law of the equivalence and conservation of mass-energy. Einstein advanced the idea that ordinary energy had been regarded as weightless through the centuries because the mass it represented was so infinitesimally small as to have been missed and ignored. For example, the amount of heat energy required to change 300,000 tons of water into steam is equivalent to less than $\frac{1}{30}$ ounce of matter.

Einstein published a mathematical equation to express the equivalency of mass and energy. The equation is

$$E = MC^2$$

where E represents energy in ergs, M is mass in grams, and C is the velocity of light in cm/sec. This last unit is equal to 186,000 miles per second. When this number is multiplied by itself as indicated in the formula, we get a tremendously large number; hence, E becomes an astronomically huge equivalent. For example, one pound of matter (one pound of coal or uranium) is equivalent to about 11 billion kilowatt hours, if *completely* changed into energy. This was roughly equivalent to the amount of electric energy produced by the entire utility industry of the United States in a period of one month at that time. Compare this figure with the *burning* (chemical change rather than nuclear change) of the same pound of coal, which produces only about 8 kilowatt hours of energy. In terms of energy produced, therefore, a chemical change such as burning is an extremely inefficient energy-producing process in comparison to the release of the energy locked up in the nucleus of an atom (nuclear reaction). In other words, the available *nuclear* energy of coal is about 2 billion times greater than the available *chemical* energy of an equal mass of coal.

These ideas of Einstein were pure theory at the time. He had no experimental data to confirm the truth of his equation, but suggested that research in radioactivity and atom-bombardment might furnish the proof. If the tremendously great electrical forces, the binding energy, that held the different particles inside the nucleus of the atom of radium or other elements could be suddenly released, Einstein's ideas might be shown to be true. The first bit of confirmation came in 1932. That was the year in which the neutron was discovered. But the neutron was not the bullet used in the bombardment experiments. High-speed hydrogen ions or protons were the battering rams employed. J. D. Cockcroft and E. T. S. Walton, working in Rutherford's laboratory, ionized hydrogen gas by removing their electrons and then accelerated the resulting protons in a transformer-rectifier high-voltage apparatus to an energy of 700,000 volts. The very swift protons then were made to strike a target of lithium metal. The lithium atom was changed into helium ions with energies many times greater than those of the proton bullets employed. This additional energy apparently came as a result of the partial conversion of some of the mass of lithium into helium in accordance with the following nuclear reaction:

LITHIUM	+	HYDROGEN	→	HELIUM	+	energy
$_3Li^7$	+	$_1H^1$	→	$_2_2He^4$	+	17 M.E.V.
Mass 7.0180	+	Mass 1.0076	→	2 (Mass 4.0029)		
		Mass 8.0256	→	Mass 8.0058		

This equation (8.0256 → 8.0058) seems to show a condition of imbalance, for the whole is less than the sum of its parts. There is an approximate loss of Mass 0.02—a fatal decimal that was to shake the world. This loss of mass is accounted for by its conversion into the extra energy of the swiftly moving helium nuclei produced. This energy turns out to be the exact mass equivalent as determined by Einstein's energy-mass equation mentioned above. However, the method used by these experimenters was extremely inefficient; only one out of several billion atoms actually underwent the change. There was, therefore, no great excitement over this bit of scientific news.

But the publication of the energy release in the Hahn-Strassmann experiment not only revived the old interest, but raised it to a fever heat. Before Meitner had reached Stockholm in her flight from the Nazis the Joliot-Curies had obtained the same effect independently of the German investigators. In Stockholm, Lise Meitner communicated her thoughts regarding uranium fission to Dr. Robert O. Frisch, another German refugee who was then working in the laboratory of the world-famous atom-scientist, Niels Bohr, in Copenhagen. On January 15, 1939, Bohr's laboratory confirmed the Hahn experiment. Frisch was terribly excited. He sent the news immediately to Bohr, who had just reached the United States for a stay of several months to discuss various scientific matters with Einstein at the Institute for Advanced Study at Princeton, New Jersey.

Bohr, too, became excited at the news and communicated it to other scientists. Within a few days three American research groups confirmed the experiment. On January 25, 1939, Fermi, John R. Dunning, and associates repeated the Hahn experiment with the help of the Columbia University 75-ton cyclotron located in the rock-hewn vault underneath the Physics building. They obtained the violent splitting of uranium, and their photographs showed high peaks of discharge of 200,000,000 electron volts. M. A. Tuve, L. R. Hafstad and R. B. Roberts, working in the Department of

Terrestrial Magnetism of the Carnegie Institution of Washington, repeated the historic reaction on the 28th, and on that same Saturday morning, workers at the Johns Hopkins Laboratory obtained the same results.

On January 26th of that year Bohr attended a Conference on Theoretical Physics at George Washington University in Washington, D.C. Atomic fission had electrified the many scientists gathered there. There was much discussion and speculation over this new phenomenon. Among the top-flight atomic artillerymen present was Enrico Fermi. He was now professor of physics at Columbia University. He had just then arrived from fascist Italy, with his scientist wife, Laura, who had worked with him in some of his early experiments. When Mussolini embraced racism, Fermi, an anti-fascist, thought the time had finally arrived when he must leave his native country and try the free air of America. During his talk with Bohr, Fermi mentioned the possibility that nuclear fission might be the key to the release of colossal energy by the mechanism of a chain reaction. He speculated that the fission of the uranium atom might liberate additional neutrons which might be made to fission other atoms of uranium. In this way, there might be started a self-propagating reaction, each neutron released in turn disrupting another uranium atom just as one firecracker on a string sets off another firecracker until the whole string seems to go up like a torpedoed munition ship in one mighty explosion. Subatomic energy could thus be released and harnessed, producing from a single pound of uranium energy equivalent to that produced by 40,000,000 pounds of TNT.

That possibility was really something to think about. But nothing of this sort had, as yet, been seen in either Berlin or New York. Atomic energy of this order still remained fettered. Fermi at that moment may not have realized that there was the genesis of a profoundly disturbing drama.

The possibility of a chain reaction still obsessed nuclear physicists. Why had not the chain reaction of uranium fission actually occurred? Niels Bohr and a former student, John A. Wheeler of Princeton University, puzzled over this question. At a meeting of the American Physical Society at Columbia University on February 17, 1939, they advanced a theory of uranium fission which postulated that not all the uranium employed as target actually fissioned. They believed that less than one percent of their uranium target disintegrated because only one of the three isotopes of uranium was actually capable of fission. This fissionable isotope has an atomic weight of 235 instead of 238 which is the atomic weight of 99.3% of the uranium mixture found in nature. U-238 is extremely stable; its half-life has been estimated to be one hundred million years. It behaves like a wet blanket over U-235. (There is another isotope which has the atomic weight of 234. This is found in uranium only to the extent of a negligible 0.006%.) Isotopes, as you know, are atoms of the same element having the same atomic number and similar chemical properties, but differing in their atomic masses.

Bohr and Wheeler reasoned that a chain reaction could be obtained only from pure U-235. They also proposed that the chain reaction could be initiated by bombardment with the *slow* neutrons that Fermi had first used in 1934, and Fermi suggested that graphite could be used as the slowing-down agent or moderator. Neutrons normally emitted are very fast (10,000 miles per second). Such fast neutrons are easily captured by U-238, but no fission occurs. When forced to hurdle some retarding agent such as graphite or heavy water, fast neutrons collide with it and lose some of their energy, which may slow down their speed to a pace no greater than 1 mile per second. The slow neutron may bounce around from one U-238 nucleus to another until it strikes the nucleus of a U-235 atom and splits it. The effectiveness of the slow or *thermal* neutron has been compared to the slow golf ball which rolls along slowly and drops gently into the cup on the green while the fast moving golf ball simply hops past the cup.

The first researcher to separate a minute quantity of U-235 from the isotopic mixture of natural uranium was Alfred O. Nier of the University of Minnesota. He sent this microscopic quantity of U-235 (about 0.02 micrograms) to Fermi and others at Columbia University. This bit of U-235 and another speck from the General Electric laboratory were bombarded with slow neutrons in the Columbia cyclotron and the prediction of Bohr and Wheeler was confirmed in March, 1940.

Nier had worked hard and long to separate the tiny bit of U-235 by means of the mass spectrometer, but the process was extremely slow. At the rate at which he was separating the U-235 from the rest of the mixture it would have taken 75,000 years to manufacture just one single pound of this key isotope. Thus the possibility of releasing huge quantities of atomic energy still remained a dream. Fantastic stories went the rounds to the effect that Hitler had ordered his scientists to redouble their efforts to supply him with several pounds of the powerful element whose terrific destructive powers would bring world domination for Nazi arms. But, for the moment, it continued to remain as devastating a secret weapon as the rest of his threats.

Nevertheless, a great deal of research in this field continued. In 1939, more than one hundred papers on nuclear experiments were published. But the consensus of expressed opinion was against any early solution of the problem. "I feel sure, nearly sure," Einstein had observed, "that it will not be possible to convert matter into energy for practical purposes for a long time." On February 2, 1939, Fermi delivered himself of the opinion that "whether the knowledge acquired of the possibility of a chain reaction will have a practical outcome, or whether it will remain limited to the field of pure science cannot at present be foretold." On his return from the Washington meeting of the American Physical Society, Fermi was asked on the radio how soon the world would blow up. He remained silent on this question. Eighteen months later the journal *Electronics*, summing up all the work published to that time, declared, "The matter stands at present waiting a conclusive demonstration that the chain reaction of U-235 is indeed a reality . . . . In the meantime U-235 is an isotope to watch. *It may be going places.*"

But there was nothing for the general public to watch. In March, 1939, Bohr returned to Denmark. A year later he and the American nuclear scientists voluntarily agreed not to publish any more of their findings in this explosive field. In June, 1943, Byron Price, U. S. Director of Censorship, sent a confidential note to 20,000 news outlets asking them "not to publish or broadcast any information whatever regarding war experiments involving production or utilization of atom smashing, atomic energy, atomic fission, atom splitting, or any of their equivalents, the use for military purposes of radium or radioactive materials, heavy water, high voltage discharges, equipment, cyclotrons, and the following elements or any of their compounds, namely, polonium, ytterbium, hafnium, protactinium, radium, rhenium, thorium, and deuterium."

Thus the security blackout of all U-235 news left the world speculating as to whether atomic energy could actually ever be harnessed for practical use. When the news of triumph finally came on August 6, 1945, it surprised even the most optimistic scientists. The great marvel, said President Truman, "is not the size of the enterprise, its secrecy or cost, but the achievement of scientific brains in putting together complex pieces of knowledge held by many men in different fields of science into a workable plan." The controlled release of atomic energy was not only the most spectacular but also the most revolutionary achievement in the whole history of science. Within the short span of five years a handful of scientists, standing on the shoulders of thousands of others who had been probing the heart of the atom for fifty years, uncorked a torrent of concentrated energy that can improve the world immeasurably or blot it out completely.

The manufacture of the atom bomb is another example of the oneness of pure and applied science. Out of the purely theoretical investigations relating to the composition and heat of the sun, the nature of radiation, and the structure of the atom came remarkable inventions such as the photoelectric cell or magic eye, television, and the electron microscope. Men who never dreamed of having a hand in the construction of a practical gadget were supplying concepts and mathematical equations which eventually made possible the most devilish war weapon ever dreamed up by man.

The thousands of scientists of every race, nationality, religion, and motivation had, except for the last chosen few, no idea of the monster they were fashioning. They knew only that they were adding just another bit to human knowledge. Science is an international activity. The widespread dissemination of the findings of researchers in hundreds of laboratories throughout the world makes possible the cooperation of all peoples in the hunt for new principles and new machines. Men and women from almost every corner of the earth played their parts in the drama of atomic energy. Only a very few of these actors were aware that, near the close of the drama, there would emerge an atomic bomb. William Roentgen, the German who discovered X-rays in 1895, could not have dreamed of it. The Frenchman, Henri Becquerel, who noticed the effect of the uranium ore, pitchblende, on a photographic plate in a darkroom, could not have guessed it. The Polish-born scientist, Marie Curie, caught a glimpse inside the spontaneously disintegrating world of the radium atom, but could not foresee the harnessing of subatomic energy. J. J. Thomson of England and Ernest Rutherford of New Zealand, who gave us the electron and the proton, considered controlled atomic energy both too expensive and too far distant.

Scientists working in the field of nuclear physics included Niels Bohr, a Dane, Enrico Fermi, an Italian, Wolfgang Pauli, an Austrian, Georg von Hevesy, a Hungarian, Peter Kapitza and D. Skobelzyn of the Soviet Union, Chadrasekhara Raman of India, and H. Yukawa, a Japanese who as early as 1934 foreshadowed the presence of a new nuclear unit, the mesotron, which was later discovered by California Tech's Carl D. Anderson, son of a Swedish immigrant.

When the curtain that hid the work on atomic fission during the war was partially lifted after Hiroshima, a thrilling story was revealed. After the reality of atomic fission had been demonstrated early in 1939 and the possibility of a chain reaction had been partially proved in the spring of that year by Frédéric Joliot and his collaborators in France, the whole world knew these results. Among those interested in this new milestone were, of course, several German nuclear physicists who saw the possibility of manufacturing a super-high explosive on the basis of the concentrated energy locked up in the heart of the atom's nucleus. There were two laboratories in Germany in 1939 capable of nuclear research. Neither of them had a cyclotron, but in the autumn of 1940 the first "pile" was set up at Berlin-Dahlen consisting of layers of uranium oxide and paraffin. It failed. Hitler kept rattling his sword more and more menacingly over Europe, as his scientists continued their investigations.

It was very different in the United States. "American-born nuclear physicists," wrote Dr. Henry D. Smyth in the official Army Report on the Atomic Bomb released six days after Hiroshima, "were so unaccustomed to the idea of using their science for military purposes that they hardly realized what needed to be done. Consequently, the early efforts both restricting publication and getting government support were stimulated largely by a small group of foreign-born scientists in this country." The first of these scientists, who included Edward Teller and Victor F. Weisskopf was Enrico Fermi, who was put in touch with Navy officials by George B. Pegram, Dean of the Graduate Faculty of Columbia University, as early as March, 1940. The European-born scientist suggested to the Navy the possibility of producing terrific explosions with the aid of uranium and neutrons. The Navy showed interest.

Another refugee scientist working on atomic energy at this time was forty-year-old Leo Szilard, visiting experimental physicist at Columbia University. This Hungarian had served his native country in World War I, had continued his studies in Berlin, and then moved to England. On March 3, 1939, he and Walter Zinn, a Canadian attached to the College of the City of New York, were working on the seventh floor of the Pupin Building of Columbia University. They were attempting to confirm the reality of atomic fission. They set up the necessary apparatus, turned a switch, watched the screen of a television tube for the telltale sign. "That night," wrote Szilard, "I knew that the world was headed for sorrow." After Hahn's epochal experiment, Szilard, a very active anti-Nazi, had come to the United States with some apparatus he had constructed in England to continue his experiments at Columbia. Szilard was seriously frightened by the implications of atomic fission. In September, 1939, the German Heereswaffenamt had organized a research group to examine its possibilities. It consisted of W. Bothe, Hans Geiger, Otto Hahn, P. Harteck, and C. F. von Weiszaecker, son of the German Undersecretary of State. Suppose Hitler's scientists went to work in dead earnest to construct a bomb on the atomic energy principle? They might succeed and enslave the whole world! He rushed to Princeton to talk the matter over with his friend Eugene P. Wigner, who had come here from Hungary in 1930, and possibly with Einstein. (In 1937 Wigner had become an American citizen, and Szilard followed him in 1943.)

Another American of foreign birth who was scared almost to distraction was Alexander Sachs. He had come here from Russia as a boy, had been educated at Harvard and Cambridge, had become informal economic adviser and industrial consultant to Franklin D. Roosevelt, who in 1933 had appointed him first chief economist and organizer of the NRA (National Recovery Administration). Sachs agreed with Szilard that something had to be done quickly. He knew the tremendous prestige of Einstein. When the latter suggested that Sachs go to see Franklin D. Roosevelt, he lost no time. On October 11, 1939, he delivered a letter from Einstein to the President at the White House. Hitler's armies were already on the march. Poland had been crushed.

The letter read in part as follows:

> In the course of the last four months it has been made probable through the work of Joliot in France as well as Fermi and Szilard in America—that it may become possible to set up a nuclear chain reaction in a large mass of uranium, by which vast amounts of power and large quantities of new radium-like elements would be generated. Now it appears this could be achieved in the immediate future.
>
> This new phenomenon would also lead to the construction of bombs, and it is conceivable—though much less certain—that extremely powerful bombs of a new type may thus be constructed. A single bomb of this type, carried by boat and exploded in a port, might very well destroy the whole port, together with some of the surrounding territory. However, such bombs might very well prove to be too heavy for transportation by air.

Sachs reminded the President that Fermi, Szilard, and our American scientists were probably only one step ahead of the Nazi scientists. Germany had overridden Czechoslovakia, its precious uranium ores were in its hands; the most important source of uranium was the Belgian Congo; Belgium would undoubtedly be invaded by the German hordes, and this huge source of uranium would then be lost to the United States.

Roosevelt saw the danger at once. He had the vision and the courage to act promptly. He brushed aside those of his advisers who were hesitant and less alarmed. Hardly five weeks after World War II had broken out, President Roosevelt appointed an "Advisory Committee on Uranium." This committee consisted of Alexander Sachs, E. P. Wigner, Edward Teller of George Washington University, Enrico Fermi, Leo Szilard, and several Army and Navy men. They met on October 21, 1939. The military men on the committee felt that the federal government should not engage in atomic energy experiments but should leave this work to the universities. In November of the same year the committee recommended getting four tons of graphite to be used as moderator and 50 tons of uranium oxide. The first appropriation, a pitifully tiny sum of $6000, was made for this purpose on February 20, 1940.

In the meantime, Ernest Orlando Lawrence had been awarded the Nobel Prize on December 10, 1939, for his invention in 1932 of the cyclotron, the superb atom-smashing machine which he had, in the succeeding years, developed into a mighty tool for atomic fission work. This event served as a sharp reminder that others, too, had cyclotrons with which to carry out atomic fission experiments on a large scale. Early in 1940 Sachs and Einstein were dissatisfied with the snail-like progress of the uranium research. In April of that year Sachs was again at the White House pleading with Roosevelt for more haste and more money.

At this very moment England was getting nervous about the possibility of a German atom bomb. British scientists had learned that a large section of the Kaiser Wilhelm Institute had been set aside for nuclear research. They were now doing something to meet the crisis. A committee under the leadership of Nobel Prize winner Sir George P. Thomson, son of the discoverer of the electron, was appointed in April, 1940. It was first under the Air Ministry and later under the Ministry of Aircraft Production. The work was started by O. R. Frisch and J. Rotblat at Liverpool, and later was extended to the famous Cavendish Laboratory of Experimental Physics under N. Feather and E. Bretscher. French scientists, too, were aware of the danger of an atom bomb in the hands of the Nazis. When in June, 1940, France fell to Hitler, Frédéric Joliot-Curie, codiscoverer of artificial radioactivity and a leader in the field of atom chain reaction, sent his collaborators, H. H. Halban and L. Kowarski, to Cambridge to aid in the work of the British atom scientists. Kowarski took with him the 165 quarts of heavy water which the French government had brought from Norway before its invasion. The French scientists, pioneers in the field of slow neutron bombardment, thought of using this heavy water as a moderator in the slowing down of neutrons. Joliot remained in France to become an active worker in the resistance movement in his country and to organize the manufacture of munitions for the underground partriots in Paris.

At the same time that French scientists were leaving for England, President Roosevelt in June, 1940, set up the National Defense Research Committee (NDRC), and the original Committee on Uranium became a subcommittee of this new body. Before the end of that year Columbia University received $40,000 for study of a chain reaction. In the summer of 1941 Vannevar Bush, director of the NDRC, visited Roosevelt on the President's return from the Atlantic Charter meeting with Churchill. Bush gave Roosevelt a brief account of the work on atomic energy already under way and told him also of the reports which K. T. Bainbridge and C. C. Lauritsen, who had attended meetings of Thomson's committee in England, had brought back. The British scientists by the summer of 1941 had definitely come to the conclusion that an atom bomb was feasible. Roosevelt then suggested to Clement Attlee, then a member of the Churchill cabinet, that the British scientists working on atomic energy pool their knowledge and efforts with those of our own scientists working on nuclear fission. This proposal was eagerly accepted by the Churchill government.

Harold C. Urey, discoverer of heavy hydrogen, and George B. Pegram, physicist of Columbia University, were sent abroad in November, 1941, to confer with the British scientists. Just two months before, Churchill had asked Sir John Anderson to supervise the work of an atom bomb project, and the latter had brought together England's ace atom scientists, including Sir Charles Darwin, member of the fourth generation of Darwin's family; J. D. Cockcroft, M. L. Oliphant, N. Feather of Cambridge, first to split the oxygen atom, and two refugee scientists, Rudolph Peierls, Professor of Applied Mathematics at the University of Birmingham since 1937, and Franz E. Simon, reader in thermodynamics at Oxford since 1933. Sir James Chadwick, discoverer of the neutron and also a member of this original group, was terribly worried. During the summer of that year the Germans were experimenting in Leipzig with a small uranium pile using heavy water as a moderator. Their results were cloaked in secrecy. Werner Heisenberg, top German nuclear physicist and one of the world's greatest authorities in this field, was believed to be directing this work.

Urey returned to the United States in the week that terminated in Pearl Harbor Sunday. He shared the anxiety of Chadwick and brought home a sense of the utmost urgency. Heisenberg, on Feb. 26, 1942, had told Rust, German Minister of Education that "an atomic bomb could be produced in the pile on theoretical grounds." At this moment American and German scientists had arrived at similar results if we exclude our great success in isotope separation. What if the German scientists got the jump on us and the German High Command had agreed on an all-out effort to make atom bombs? This was altogether possible, for on November 6, only a month before, the NDRC had reported that "a fission bomb of superlatively destructive power results from bringing quickly together a sufficient mass of U-235 . . . . . If all possible effort is spent on the program, we might expect fission bombs to be available in significant quantities *within three or four years*."

Things now began to move much faster. Just a day before the attack on Pearl Harbor an "all-out effort" to manufacture an atom bomb was finally decided upon. Eleven days later the NDRC was reorganized under the Office of Scientific Research and Development (OSRD) headed by Vannevar Bush. A delegation of British scientists, including Peierls, Simon, and Halban, came to the United States to help in the coordination of the work. Finally, on August 14, 1942, the Manhattan Engineer District project was started by order of Secretary of War Henry L. Stimson. Major-General Leslie R. Groves, forty-two-year-old army construction engineer, a West Pointer, was made director of all army activities relating to the project. Richard C. Tolman, dean of the graduate school of the California Institute of Technology, was appointed his scientific adviser.

Long before this last step had been taken, various research groups had already been assigned to several crucial problems. One of these was the production of a controlled and self-maintaining nuclear chain reaction. As early as January, 1942, it was decided to concentrate this project at the University of Chicago where Arthur H. Compton was working with neutrons. Fermi's group working at Columbia, a number of scientists working at Princeton University, and several other researchers came to Chicago to team up in what became known as the Metallurgical Laboratory.

One of the essentials of this project was a good supply of pure uranium of which only a few pounds were actually available in 1941. There was plenty of the impure uranium ore obtainable from the Belgian Congo, and from the Eldorado pitchblende mines of the Canadian arctic wilds which had been taken over by Canada during the war. Getting the pure uranium metal from these ores was no simple matter. The first supply was delivered to the Chicago pile operators early in 1942. It came from a long wooden shed of the Iowa State College at Ames, where formerly coeds practiced archery. Here F. H. Spedding of the chemistry department supervised the purification of the uranium which was to be used for nuclear fission. At this time Arthur Compton also called upon Harvey C. Rentschler to make three tons of the metal for the Metallurgical Laboratory. Rentschler had worked with this metal at the Westinghouse Research laboratories in connection with electric lamp filaments, and had been supplying college laboratories with small amounts of this element for research. Rentschler got to work at once and before long had stepped up his yield from half a pound to 500 pounds daily. And instead of uranium costing $1000 per pound he pushed its production cost down to only $22 per pound.

A large doorknob-shaped structure called a *pile* was set up by Fermi on the floor of the squash-rackets court underneath the west stands of Stagg Field of the University of Chicago. The pile contained 12,400 pounds of specially purified graphite bricks with holes at calculated distances in which were embedded lumps of uranium oxide and pure uranium sealed in aluminum cans to protect the uranium from corrosion by the cooling water pumped through the pile. The bricks were arrayed in the form of a cubic lattice as suggested by Fermi and Szilard. The lattice structure was found to be the most effective arrangement of material for the slowing down of neutrons. The graphite bricks act as a moderator, to change fast neutrons into slow or thermal neutrons. The thermal neutrons produced then cause fission in U-235, producing a new generation of fast neutrons similar to the previous generation. Thus neutron absorption in U-235 maintains the chain reaction as a further source of neutrons.

A chain reaction will not maintain itself if more neutrons are lost than are produced. Just as coal will not continue to burn and the fire will be extinguished when the heat it generates is lost faster than it generates new heat, so U-235 will not fission so long as it loses neutrons faster than it generates them by fission. Neutrons may be lost by being absorbed either by U-238 or by impurities present in the uranium or in the graphite. U-238 absorbs neutrons but does not fission, hence these neutrons may be considered as lost. By careful purification of uranium and graphite, proper spacing of target and moderator to cut down further the losses of neutrons, and by accurate determination of the size of the pile, the chain reaction should, theoretically at least, be kept under control. That was the job assigned to Fermi and his assistants.

There was a great deal of theorizing, calculating, discussing, and changing of plans. There was a great deal, too, of piling and repiling of graphite bricks, hence the name *pile* for the uranium reactor. On the final day of trial Fermi, Compton, W. H. Zinn, and Herbert L. Anderson stood in front of the control panel located on a balcony ten feet above the floor of the court. Here stood George L. Weil, who was to handle the final control rod which held the reaction in check until it was withdrawn the proper distance. Another safety rod, automatically controlled, was placed in the center of the pile and operated by two electric motors which responded to an ionizing chamber. When a dangerously high number of neutrons were escaping, the gas in the ionizing chamber would become highly electrified. This would automatically set the motor operating to shoot a neutron-absorbing, cadmium-plated steel rod into the pile. As an added precaution an emergency safety rod called *Zip* was withdrawn from the pile and tied by a rope to the balcony. Norman Hilberry stood ready to cut this rope if the automatic rods failed for any reason. Finally, a liquid control squad stood on a platform above the pile trained and ready to flood the whole pile with water containing a cadmium salt in solution.

Fermi started the test at 9:54 A.M. by ordering the control rods withdrawn. Six minutes later Zinn withdrew *Zip* by hand and tied it to the rail of the balcony. At 10:37 Fermi, still tensely watching the control board, ordered Weil to pull out the vernier control rod thirteen feet. Half an hour passed and the automatic safety rod was withdrawn and set. The clicking in the Geiger counters grew faster and the air more tense. "I'm hungry. Let's go to lunch," said Fermi, and his staff eased off to return to the pile at 2 o'clock in the afternoon. More adjustments, more orders, and at 3:21 Fermi computed the rate of rise of neutron count. Then suddenly, quietly, and visibly pleased, Fermi remarked, "The reaction is self-sustaining. The curve is exponential." Then for 28 more minutes the pile was allowed to operate. At 3:53 P.M. Fermi called "OK" to Zinn, and the rod was pushed into the pile. The counters slowed down. It was over. The job that came close to being a miracle was completed. December 2, 1942, marked the first time in history that men had initiated a successful, self-sustaining nuclear chain reaction. Only a handful of men surrounding Enrico Fermi knew that on this wintry Wednesday afternoon mankind had turned another crucial corner.

Arthur H. Compton had witnessed this successful achievement and put through a long distance telephone call to James B. Conant, who was in charge of the project for OSRD. "The Italian navigator has landed in the New World. It is a smaller world than he believed it was," said Compton to Conant. To Conant the "smaller world," though no code had been prearranged, meant that the atomic pile was smaller and its fires were not as violent as had been expected. Conant then asked, "Are the natives friendly?" Compton took this to mean whether he was ready to go ahead full blast. The answer to that query was "Yes, very friendly," and Fermi and his Chicago group lost no time in following through.

Working with this uranium pile partly shrouded in balloon cloth to keep out neutron-absorbing air was a most dangerous business. Death and destruction threatened at almost every move. A chain reaction might get out of control and produce a super-explosive that could blast the researchers into kingdom come. And there were thousands of other people who stood in imminent danger of being crippled by atomic fission. People living near the university might have been blown to vapor one fine morning had not the men at the pile taken every conceivable precaution—and been lucky, too. It had been anticipated that the nuclear reaction would start from spontaneous fission caused by a stray neutron or other source (such as bombardment from a wandering cosmic ray) just as soon as the pile reached a certain size, known as the *critical size*. This condition is reached when the number of free fission neutrons just equals the loss of neutrons due to non-fission capture and escape from the surface. Since the rapid loss of neutrons into the surrounding space is a surface phenomenon, and nuclear fission of U-235 is a volume effect, it would be disastrous to build a pile so large that the number of neutrons produced by fission would be greater than the number of neutrons lost by non-fission capture and by escape from the surface of the pile. Such a pile would produce an uncontrolled fission chain reaction explosion. The critical size of the pile under construction had been calculated from all the available data, and it turned out that an error had actually been made, for the approach to critical size was later found to occur at an *earlier stage of assembly than had been anticipated*. Of course, Fermi and his men, working in an unfamiliar field, had taken every conceivable precaution. "This was fortunate," wrote Dr. Smyth, and these three words must remain one of the classic understatements in the long history of the hazards of scientific discovery.

If the men at the pile escaped sudden death, they might have still succumbed to a slow, painful destruction caused by the penetrating rays and poisonous radioactive particles emitted during nuclear fission. As a safeguard against the perils of such penetrating radiation and poisons, the pile was shielded very carefully by 5-foot-thick walls of absorbent material. No one dared come within reach of the pile, and manipulations had to be performed by ingenious devices permitting remote control.

The original purpose of the Metallurgical Laboratory project was the creation of a large and easily controlled chain reaction. This objective was achieved. In addition, however, the pile turned out to be a plant which efficiently manufactured a new element in large quantities. This element is plutonium. It is a brand new man-made chemical element which fissions just as easily as U-235. The story of the birth of this synthetic element goes back to a day in May, 1940, when two men using Lawrence's cyclotron at Berkeley, California, bombarded uranium with neutron bullets. The two men were Edwin M. McMillan and Philip H. Abelson, who was later decorated by our government for this achievement. After the bombardment of U-238 they detected traces of a new element, heavier than uranium. This new element, No. 93, was named *neptunium* by McMillan. It was a very difficult element to study, for its life span was very short. It threw out neutrons almost immediately and in a split second's time was no longer neptunium.

It was exciting enough to have made a new element, but what was even more thrilling was the discovery, before the end of that same year, of still another element which turned out to be even more interesting than neptunium. McMillan, Glenn Seaborg, A. C. Wahl, and J. W. Kennedy learned late in 1940 that neptunium actually changed into another element heavier than itself. This fairly stable element, No. 94, was sensitive to neutron bombardment and fissioned in a similar manner to U-235, emitting other neutrons capable of producing a chain reaction. This was a tremendously important fact, for here science had a substance which could be used instead of U-235 in the projected atom bomb. Furthermore, this new element, *plutonium*, could be separated from natural uranium much more easily than could U-235. This was true because it is an entirely different element and could be separated by chemical means rather than by the very difficult physical means necessary for separating the isotopes of uranium.

The nuclear reactions involved in the discovery of neptunium and plutonium, and in the fission of the latter element, may be represented by the four steps indicated on page 301.

If this newly discovered plutonium was to be manufactured on a scale large enough to meet the needs of the Manhattan District project, more of its chemical properties would have to be known. At the beginning the men at the Metallurgical Laboratory had only a trace of the element to work with and made some preliminary

investigations of its properties by a tracer technique. Because of the
radioactive nature of the element it was possible to gather some
information of its behavior by mixing it with other elements and
compounds.

A larger sample of the element, an amount that could be seen,
was needed if further information vital to the whole project was to
be obtained. The big cyclotrons of Lawrence's laboratory at
Berkeley, California, and of Washington University in St. Louis
went to work bombarding uranium with neutrons. For weeks the
big machines were kept operating until as much as one-thousandth
of a single gram of the element had been collected, enough for
direct observation. Years before, chemists had developed a branch
of analysis called microchemistry which could handle tiny amounts
of chemicals weighing as little as 0.001 gram. But not even such a
tiny bit of plutonium was available. So the chemists at the University of Chicago under Glenn Seaborg began in April, 1942, to develop a new method which could handle chemicals which weighed
no more than 500 micrograms (1 microgram equals one-millionth of
a gram) or about 1/5000 the weight of a single dime. (A human
breath weighs 750,000 micrograms.) This method is known as
ultra-microchemistry. Midget test tubes called microcones were
used. Ingenious devices were invented to handle these minute
quantities of chemicals, and very clever methods were introduced to
safeguard the health of the men handling these radioactive substances. Finally 2.77 micrograms of "the first pure chemical compound of plutonium was prepared by B. B. Cunningham and L. B.
Werner at the Metallurgical Laboratory at the University of
Chicago on August 18, 1942. This memorable day will go down in
scientific history," wrote Seaborg, codiscoverer of plutonium, "to
mark the first sight of a synthetic element, and, in fact, the first
isolation of a weighable amount of an artificially produced isotope
of any element."

The first self-sustaining chain reaction had been achieved in
December, 1942, only three months after the first weighable amount
of plutonium compound had been prepared and studied. A third
major problem in the preliminary work of Manhattan District was
also well on its way to solution. This was the preparation of pure
U-235, the first fissionable substance then known. Nier had first
separated this isotope from natural uranium, but his method was so
slow and laborious that it would have taken scientists several
thousand years to separate a single gram of pure U-235. Better and
faster methods of separating the precious U-235 had to be devised.
Several methods had been suggested and tried as far back as 1939.

(1) U-238 + neutron → U-239 (*no fission . . . nucleus contented*)
$\binom{92+}{146n}$ $\qquad\qquad$ $\binom{92+}{147n}$

(2) U-239 $\xrightarrow[\text{half-life}]{\text{23 min.}}$ Np-239 (*radioactive*) + electron
$\binom{92+}{147n}$ $\qquad\qquad$ $\binom{93+}{146n}$

(This change occurs by the breaking down of 1 neutron in the nucleus of U-239 into 1 proton and 1 electron which escapes. The neutron is here considered as a particle composed of 1 proton and 1 electron very tightly packed together.)

(3) Np-239 $\xrightarrow[\text{half-life}]{\text{2.3 days}}$ Pu-239 + electron
$\binom{93+}{146n}$ $\qquad\qquad$ $\binom{94+}{145n}$

(This change occurs by the breaking down of 1 neutron in the nucleus of Np-239 into 1 proton and 1 electron which escapes.)

(4) Pu - 239 $\xrightarrow[\text{\textit{slow} neutrons}]{\text{fissionable with}}$ U-235 + Helium^{++} (mass 4)
$\left(\begin{smallmatrix}\text{half-life}=\\ \text{24,000 years}\end{smallmatrix}\right)$ $\qquad\qquad$ (alpha particle)

The electromagnetic method, first employed by Francis W. Aston in 1919, was already being used by Lawrence at the University of California. This consisted of shooting ionized gaseous particles of a uranium compound through an electric field which accelerated them to a speed of several thousand miles a second. They then entered a strong field between the poles of a powerful electromagnet which curved them into a circular path. Molecules of the lighter isotope (U-235), being bent more than those of the heavier isotope (U-238), could thus be separated, and the pure U-235 trapped. Lawrence used the huge electromagnet of his dismantled 37-inch cyclotron for this job. The huge electromagnet machine employed by him in preparing pure U-235 was named the *calutron* from the two words California and cyclotron.

A second method, known as the gaseous diffusion method, was used by Harold C. Urey and Dunning at Columbia University. Urey had had considerable experience in separating the isotopes of hydrogen by this method and developed it further for this most pressing need. The diffusion method consists of passing a gaseous uranium compound (fiercely corroding uranium hexafluoride, UF_6) through barriers of very fine filters. The lighter vapor passed through filters faster than the heavier vapor, and thus by a continuous process a complete separation of U-235 from the other isotopes of natural uranium was effected. Both of these methods, as well as two other—thermal diffusion and centrifugal—were tried. The thermal diffusion method depends upon the fact that if there is a temperature difference in a vessel containing a mixed gas, one gas will concentrate in the cold region and the other in the hot region. All of these methods were slow, laborious, at times disappointing and discouraging. But new tricks were devised, new improvements were introduced, and new information was gathered during this long period of preliminary investigation.

Much of the preliminary work had now been completed by hundreds of scientists in dozens of laboratories around the country. More than 30 volumes of reports had been written. It was decided to go into large-scale production of both U-235 and plutonium for the making of atomic bombs. A bold step had to be taken. Instead of setting up a small pilot plant to test final manufacturing procedures, it was necessary to jump at once into large-scale production. "In peacetime," wrote Smyth, "no engineer or scientist in his right mind would consider making such a magnification in a single step." But there was no alternative. The Nazis were thought to be working on an atomic bomb. At the Norsk Hydro Company plant at Vermork near Rjukan, Norway, the Germans were already producing several hundred liters of "heavy water" to be used in slowing down neutrons. This was considered such a threat to us that the Allies with the help of Norwegian patriots decided to put the plant out of commission. In the fall of 1942 five Norwegians were parachuted from a British plane near the plant and awaited four others who came later. Under the leadership of the daring 26-year-old Sverre Haugen, the patriots succeeded in blowing up the heavy water apparatus. Two more attacks were made on it by commandos during the winter of 1942 and the spring of 1943, but the plant was repaired. Finally, on November 16, 1943, Flying Fortresses and Liberators of the U. S. Bomber Command attacked, and the Germans decided to dismantle the plant and send the heavy water and machinery to Germany. Norwegian patriots blew up the ferryboat carrying the heavy water in February of the following year.

The Germans were actually spending very little money on atomic-bomb work. Heisenberg revealed this in an article in *Nature* in 1947. In the early part of 1942 he had tried to get some of the top flight Nazis to a luncheon for the purpose of getting more help. Fortunately for us, he failed. The Germans expected a quick victory and thought a successful bomb project would take at least twenty years. In addition, the Germans were blinded by belief in their own superiority; many of their scientists, selected for political reasons rather than for their scientific abilities, had a contempt for so-called non-Aryan science. Nevertheless, they tried to get information about our own atom-bomb program. They knew that Niels Bohr would be valuable to the Allies. Hitler issued an order for the capture of the Danish scientist, who was working in Denmark at that time. The Danish underground learned about this, and smuggled Bohr and his family and Georg von Hevesy out of the country in a fishing boat which landed them in Sweden. Bohr was still not safe, and 19 days later a British Mosquito bomber picked him up for a flight to England. Because of fear that the Germans might intercept the plane, the pilot was instructed to drop Bohr by parachute through the bomb bay should his plane be attacked. Bohr almost died during the flight. As the plane rose into high altitudes the pilot instructed him to put on his oxygen mask. But Bohr did not get the warning, for he was not wearing the earphones which were too small for his rather large head. He became unconscious but fortunately recovered when the plane came down to lower altitudes.

The Nazis also sent spies to the United States to get information about our atom bomb plans. At least five of these German spies reached the country but they were all intercepted, and American alertness prevented any of our secrets being revealed or any sabotage being committed in our uranium plants.

In December, 1942, it was decided to proceed at once to the manufacture of sufficient quantities of both U-235 and plutonium to be used in atomic bombs. This meant increasing the available amounts of these elements many millions of times. It meant amplifying the ultra microchemistry of the Metallurgical Laboratory workers a billion times or more. It meant operating several huge uranium piles in plants designed from data collected during experiments with almost gossamer bits of plutonium. It meant constructing entire cities of many massive concrete structures to perform the colossal tasks ahead. Two sites were chosen and three plants were to be constructed.

One of these sites was Oak Ridge, Tennessee, near Knoxville. Here, early in 1943, the DuPont Company built a plant at the Clinton Engineer Works site, including an air-cooled uranium-graphite pile which was to manufacture plutonium. This pile was started in November of that year and within two months the first batch of plutonium, less than 1/5 gram, was delivered. It had come at the end of a long process of separation from the uranium from which it had been formed during the chain reaction initiated in the pile. Uranium slugs, after exposure in the pile, were transferred under water to hermetically sealed rooms surrounded by thick walls of concrete, steel and other absorbent material. Several of these cells, two-thirds buried in the ground, formed a continuous fortress-like, almost windowless structure, 100 feet long, called a canyon. The plutonium-rich uranium rods were passed along from one cell to another, chemically treated until all of the plutonium was separated from the uranium. All of these operations were performed by remote control because of the hazards to health and even life caused by the radioactive products of nuclear fission that included not only neutrons, penetrating radiation similar to X-rays, but also dozens of radioactive isotopes such as iodine and xenon.

A second plant was erected at Oak Ridge where pure U-235 was manufactured by separation from natural uranium by the electro-magnetic method developed by Lawrence. Mammoth electromag-nets, 250 feet long, containing thousands of tons of special steel were employed. The pull of these magnets on the nails in the shoes of workers in this plant made walking difficult. Another plant using the gaseous diffusion method developed by Urey was built by the Chrysler Corporation. It consisted of sixty-three buildings. A thermal diffusion plant was also built here.

The other site selected was on the Columbia River near the Grand Coulee Dam. It was known as the Hanford Engineer Works and located in an isolated region near Pasco, Washington. Construction started in 1943, and the job was completed in record time. Its first large water-cooled pile for the manufacture of plutonium was built by DuPont and went into operation in September, 1944. By the summer of the next year, just before the war ended, the entire plant was humming. All of its three piles, located several miles apart for safety, were heating the waters of the Columbia River. During the operation of a pile a large amount of heat is produced by atomic fission. The pile is cooled by water, which at the Hanford plant came from the Columbia River in volumes estimated to be big enough to supply a large city. The cold water of the Columbia River was first filtered, then treated before it was circulated through the pile.

Sixty thousand workers and their families, sworn to the strictest secrecy, poured into Richfield and the Hanford area. Out of a barren wasteland, a city which became the fourth largest in the state sprang up almost overnight—a city whose name had never even appeared on a map. The story of Oak Ridge was equally amazing. Oak Ridge did not exist in 1941. Within a short time farms and forests became a city of 75,000 people, the fifth largest city in Tennessee. At the peak of the Manhattan project 125,000 people were engaged, including 12,000 college graduates.

The radioactivity of a single pile was estimated to be equal to that of a million pounds of radium. Hence the health of the scientists, engineers and other workers had to be constantly and carefully watched. The white-blood-corpuscle count was used as the main criterion as to whether a person was suffering from overexposure to radiation. Men carried small, fountain-pen-shaped electroscopes in their pockets to indicate the extent of exposure to dangerous fission products. Later a small piece of film was placed in the back of every worker's identification badge. The film was periodically developed for signs of blackening, indicating exposure to radiation. Contamination of laboratory furniture and equipment by alpha rays was charted by another specially constructed instrument, euphemistically called Pluto, for the god of the underworld. Geiger counters were used to check contamination of almost everything, including mops and laboratory coats, before and after laundering. No person whose clothes, hair, or skin was contaminated by even traces of radioactive material could get by the exit gates of certain laboratories without being detected, for concealed in these gates were instruments which sounded an alarm. The water which left the factories was constantly analyzed for radioactive material, and the dust of the air was checked by another instrument, called "Sneezy." Factory stacks were built high enough to insure more adequate dispersion of gases, dust, and vapors.

Even before the mammoth plants at Hanford and Clinton had been completely designed, an atomic bomb laboratory was erected on an isolated mesa 7000 feet above sea level near Los Alamos, New Mexico, twenty miles from Santa Fe. At the foot of this mesa were the ruined cliff dwellings and mud huts of the Pueblo Indians. A winding mountain road led to the top of the mesa. Here, in what was once a boys' ranch school, almost completely hidden from the rest of the world, for safety and security, the first atom bomb was to be constructed. To this spot, which was soon to become the best equipped physics laboratory in the world, a young man in his late thirties was called from the University of California. This scientist was the gifted, versatile, and brilliant theoretical physicist J. Robert Oppenheimer, a New Yorker and graduate of Harvard University. "Oppy," as he was known to his many friends, was placed in charge of this laboratory, where he arrived in March, 1943. With tremendous energy, superb organizing ability, an uncanny insight into the multitude of problems confronting the project, and great personal charm Oppenheimer soon turned Los Alamos into a marvelous workshop staffed by the best brains in atomic physics.

S.K. Allison of the University of Chicago was his right-hand man.

Machines were soon installed. A cyclotron from Harvard, two Van de Graaff electrostatic generators from Wisconsin, a Cockcroft-Walton high voltage device from the University of Illinois, and a thousand and one other contraptions were made ready for a crucial experiment. Men were assigned to their tasks and frequent consultations were called to decide upon ticklish problems as they arose from time to time. The men who worked at Los Alamos or were consulted included the American Nobel Prize winners in physics and chemistry—Carl D. Anderson, Arthur H. Compton, Clinton J. Davisson, Albert Einstein, Irving Langmuir, Ernest O. Lawrence, Robert A. Millikan, Isidor I. Rabi, Otto Stern, and Harold C. Urey. Here, too, Niels Bohr (alias Mr. Nicholas Baker), who had eluded the Gestapo, was flown from England to help in the work. James Chadwick, discoverer of the neutron, H. A. Bethe of Cornell University, originator of the hydrogen-helium nuclear theory to account for the sun's heat, Robert F. Bacher of Cornell, Russian-born George B. Kistiakowski of Harvard, Enrico Fermi, and several other refugee scientists pooled their knowledge in the final problem of constructing an atomic bomb.

The first experiments were begun at Los Alamos in July, 1943. Arthur H. Compton at Chicago had bet James B. Conant that he would deliver the first major batch of pure plutonium to Oppenheimer in New Mexico on a specific date. The stake was a champagne supper for the members of the old executive committee of the uranium project. The plutonium for the atom bomb was actually delivered a month later and the bet was paid off.

More and more plutonium and pure U-235 arrived at Los Alamos. It was known by this time that a chain reaction would not complete itself unless the fissionable material reached a certain minimum size as the critical size. To prevent premature detonation of the bomb, therefore, several pieces instead of a single lump of U-235 and plutonium were used. The pieces could be brought together suddenly to form a large mass of critical size by shooting one fragment from a gun as a projectile against another piece as target. When the pieces met to form one single piece of critical size, the whole mass would explode in a split second. To reduce the critical size of the bomb, a covering (*tamper*) which reflects neutrons back was used.

That the scientists in the mesa laboratory did their job well was known to a handful of men who midst rain and lightning witnessed the first explosion of an atom bomb. This test explosion took place at 5:30 A.M., July 16, 1945, at the Alamogordo Air Base in a desert of New Mexico, 120 miles southeast of Albuquerque. The bomb, containing (according to the Smyth Report) between 4 and 22 pounds of plutonium and U-235 fissioned with a blinding glare, vaporized the steel tower on which it had been exploded, and left a crater half a mile in diameter covered with sand sintered to a green, glassy consistency. It was estimated that a temperature as high as 100,000,000°F. was reached in the center of the explosion. The age of nucleonics was here.

Finally, the whole world learned about the new terror fashioned by science when on the early morning of Monday, August 6, 1945, a single atom bomb dropped through the open bomb bay of the *Enola Gay* wiped out the city of Hiroshima, killing 50,000 people. The bomb had been carried in a B-29 piloted by Colonel Paul W. Tibbets, Jr. To avoid the possibility of premature detonation, an accident which would have wiped out all life on Tinian Island from which the plane took off, the bomb was not assembled until the plane was in the air and at a safe distance from the island.

The objective of Manhattan District project had been reached. Never before in the history of man had such a colossal task been completed in so short a time. A heritage of scientific brains unsurpassed in the annals of theoretical science, a reservoir of brilliant engineering and industrial talent, a life-and-death situation that compelled planned, coordinated, and accelerated action, and finally an expenditure of two billion dollars made this epochal achievement possible. The long search for the key to some of the energy locked up in the heart of the atom was ended. The goal of the ancient alchemists had not only been reached but had been left far behind. And in this triumph, modern man had created a new problem for himself, a problem as challenging as the bomb itself—the problem of survival in a world which might be foolish enough to remain divided into hostile camps. This is the life-and-death problem raised by science which must be solved by statesmen, scientists and other men of power, good sense and good will.

XIX

NUCLEAR ENERGY TODAY
AND TOMORROW

THE challenge was immediate. The facts were simple. Hiroshima was destroyed by a single atomic bomb. Others could make the bomb. Forty million Americans could be killed in a single night. The Hiroshima bomb was puny in comparison with the new ones already on hand and yet to be developed. A thousand of these new missiles could destroy civilization. The Smyth report made this abundantly clear. Oppenheimer, later director of the Institute for Advanced Study, said so. Einstein believed it. Other men closely involved in the manufacture of the first atom bomb made no bones about it.

There is no weapon in the hands of man to counter the atomic bomb. There would probably never be any physical device that could neutralize its devastating effects. Hans Bethe said so. Fermi could not deny this. Other scientists intimately connected with the application of atomic fission for war purposes were all agreed. "Unless we act now, we shall eat fear, sleep fear, live in fear and die in fear," warned Urey.

Men could not turn back the pages of history. The bomb was here to stay. Was mankind here to stay? That was the burning question raised by physical science and thrown into the laps of statesmen and social scientists. The destructive power of this new discovery was so overwhelming that the atom bomb became the all-absorbing problem of millions of people. Never before had such a frightening event occurred in a world where the dissemination of news was so rapid and so widespread.

There had been other revolutionary milestones in the progress of science, but none shook the world more promptly and more profoundly. For example, when Copernicus in 1543 destroyed the geocentric theory of the structure of the universe by establishing the heliocentric conception of the earth and the other planets revolving around the sun at the hub, the foundation of one of the most cherished human beliefs was shattered. It was a terrible blow to man's pride when it was shown that *his* was only a minor planet, and not the principal focus of all creation. It was a turning point in the history of civilization, but there was plenty of time for the world to adjust itself to this intellectual revolution.

Less than half a century later Galileo performed a simple experiment that chalked up another basic advance in human thought. Aristotle, centuries before him, had taught that objects fall at different speeds. Galileo refused to accept any such intuitive conclusion. He dropped objects of different weights from a high tower in Pisa. Despite the fact that all these objects reached the ground at the same time, "The Aristotelians," wrote Galileo, "ascribed this effect to some unknown cause, and preferred the decision of their master to that of Nature herself." But Galileo had won his fight—he had established the essential need for man to think for himself. This was another stirring revolution born of science. Again, however, its effects spread slowly; there was plenty of time for mental adjustment.

Other basic discoveries brought tremendous changes in the social, political, and economic life of large sections of the world. In 1769 the patenting of the first successful steam engine by Watt helped to usher in the factory system and the Industrial Revolution. In 1828 the first synthesis of an organic compound, urea, marked the birth of synthetic chemistry. It pointed the way to the chemical revolution which gave men more potent drugs, more nourishing food, more beautiful dyes, more serviceable rubber, plastics, and fuel than they had ever before possessed. In 1831 the simultaneous discovery of electromagnetic induction by Faraday in England and Henry in the United States made available a new torrent of power and led the world into the electrical age.

But none of these advances was as global, rapid, and revolutionary in its effects as the controlled splitting of the atom. The achievement of a controlled chain reaction in the first successful uranium pile at the University of Chicago on December 2, 1942, was tremendously new. It was vastly different from any and all of the previous triumphs of the laboratory. The new age of nucleonics had a deeper and more immediate impact on the daily lives of every member of society. No other collection of scientific discoveries of the past brought with it the terrifying possibility that mankind could, by the pressing of a button, commit mass suicide.

Nuclear fission differs from all previous achievements in science in still another way. For the first time in history the basic power of the universe is at our disposal. Science and technology have suddenly jumped at least a full century, and mankind has at its fingertips as much concentrated power as it needs to raise the well-being and living standards of every inhabitant of the earth. It was estimated that less than 500 tons of fissionable material were needed to replace the more than one billion tons of coal burned annually throughout the world at that time. Only ten freight cars would be needed to carry this fuel, for "It is quite feasible that a city the size of Seattle, Washington, should be completely heated from an atomic source in ten years." This was the overly optimistic opinion of Oppenheimer in 1946.

Even before the wounds of Hiroshima had been dressed, bills were proposed and resolutions were introduced in Congress to meet the challenge. Fear inspired the first moves. On September 5, 1945, hardly a month after the first atom bomb has been dropped on Japan, Representative Bender of Ohio introduced a bill in the House making it a capital offense to disclose information or impart knowledge with respect to the atom bomb. On the same day, Representative Ludlow of Indiana submitted a resolution urging action by the United Nations to ban the atomic bomb as an instrument of war. On the following day Senator McMahon of Connecticut brought forward a bill designed to conserve and restrict the use of atomic energy for the national defense, to prohibit its private exploitation, and to preserve the secret and confidential character of information concerning the use and application of atomic energy. On the eleventh of the same month Representative Harris of Arkansas proposed a resolution creating a joint committee of the House and Senate to study and investigate the control of the atomic bomb. Other bills followed. The most important of these was the May–Johnson Bill introduced by Representative May of Kentucky, and Senator Johnson of Colorado.

Scientists, too, were deeply concerned. They had unleashed a force that directly threatened the lives of hundreds of millions of people. What were they to do about it? What was their responsibility? This was really not a new question. The place of the scientist in society has been debated for many years. Nearly every scientist and engineer in the United States had for decades past interested himself mainly in the search for scientific truth, the creation of new inventions, the construction of better mechanical machines and engineering works. He had been satisfied to leave to big business, politicians, economists, and sociologists the application of the fruits of his scientific innovations in a free enterprise system of economy. He had been prone to shy away from serious consideration of such matters as health insurance, flood, drought, and erosion control, the stifling of new inventions, and even the muzzling of science in Fascist, Nazi, and other authoritarian countries.

Justification for this attitude at this time had been expressed in several ways. "We need specialists," said one scientist, "with no outside disturbance. Economists and sociologists working in the field are better qualified to handle such problems. The scientist would tend to oversimplify a social problem."

The detached view, however, was not shared by all of our men of science even before World War II. Many maintained that it was the duty of scientists to help in nationwide planning in an effort to raise the general level of living standards through the widespread application of scientific knowledge. They pointed to the fact that scientists are peculiarly fitted for such a responsibility, that to make the benefits of their advances in science available to the great masses of the people would transform the life of millions for the better. They believed that science and scientific thinking must fit into the whole pattern of our activities.

This broader outlook was gaining adherents even before 1939. At the close of 1937, for example, the Council of the American Association for the Advancement of Science resolved that it

> makes as one of its objectives an examination of the profound effects of science upon society; and that the Association extends to all other scientific organizations with similar aims throughout the world an invitation to cooperate, not only in advancing the interests of science, but also in promoting peace among nations and in intellectual freedom in order that science may continue to advance and spread more abundantly its benefits to all mankind.

This historic declaration came on the heels of a similar announcement by the British Association for the Advancement of Science at the 1936 meeting held at Blackpool, England. The British scientists declared that the social consequences of the advances of science are part of the business of scientists. The dread of a growing Fascism and the threat of an impending world war accelerated the emergence of this new point of view.

These statements of principles were shortly followed by the organization of the British Association of Scientific Workers. The new society included such eminent scientists as Julian Huxley, H. G. Wells, John D. Bernal, Frederick G. Donnal, and the Nobel Prize winners in science, Sir Charles S. Sherrington and Sir William Bragg. The president of this organization, Sir F. Gowland Hopkins, Nobel prize winner and president of the British Association for the

Advancement of Science, declared at that time, "In these days when science plays so great a part in every field of modern life it is essential for scientific workers to work for the better organization and application of science for the benefit of the community."

A similar society was then founded in the United States. The American Association of Scientific Workers was organized in 1938. Among its early members and sponsors were Franz Boas, Anton J. Carlson, Karl T. Compton, Watson Davis, Kirtley F. Mather, Harlow Shapley, and two American Nobel prize winners in science, Harold C. Urey and Arthur H. Compton. This society was affiliated with the American Association for the Advancement of Science. The purposes of the American Association of Scientific Workers were "to promote and extend the applications of science and the scientific method to the problems of human welfare, to promote better understanding of pure and applied science by the general public, to secure more effective organization and adequate financing of scientific work, and to safeguard the intellectual freedom and professional interests of science."

The unexpected release of nuclear energy from the atom of uranium accelerated this slow trend toward social consciousness among scientists to a dramatic spurt. Scientists were suddenly catapulted into world politics. There began a considerable descent of American scientists from their ivory towers. Our chemists and physicists saw at once the meaning of this revolutionary event. They realized the urgency of safeguarding mankind against the catastrophic power of the atom. They felt it their duty to assure its use for peaceful ends.

Even before Alamogordo and Hiroshima, small groups of atomic scientists had met secretly to discuss the social implications of controlled atomic fission. One such group at Chicago's Metallurgical Laboratory drew up a memorandum which was sent to Secretary of War Stimson as early as June 11, 1945. It opposed the use of a surprise atom-bomb attack on a Japanese city, and recommended instead a demonstration of its annihilating effects in some uninhabited region.

By the close of 1945 sixteen independent local groups of scientists had organized to study the question of what to do with atomic energy. The following month delegates from thirteen groups met to consider joint action. The result was a merger of these thirteen associations. The new body, called the Federation of American Scientists, had an original membership of more than two thousand scientists and an advisory panel of such top-ranking figures as Oppenheimer, Urey, Shapley, Smyth, Szilard, and Edward U.

Condon, who later became Director of the U. S. Bureau of Standards. W. A. Higinbotham was made chairman of the Federation's Administrative Committee.

The Federation was determined *"to meet the increasingly apparent responsibility of scientists in promoting the welfare of mankind and the achievement of a stable world."* Its objectives included:

> placing science in the national life that it may make its maximum contribution to the welfare of the people, urging that the United States help initiate and perpetuate an effective system of world control of atomic energy based on full cooperation among all nations, countering misinformation with scientific fact, and *especially disseminating those facts necessary for intelligent conclusions concerning the social implications of new knowledge in science.*

The Federation set itself a double task. Its first job was to work for the passage of domestic legislation to insure civilian rather than military control of atomic energy. A vigorous attack was made on the May–Johnson bill which had been drawn up by Army men even before Hiroshima and which would have put this new power under the control of the military. Urey, Morrison, Szilard, Smyth, and other atomic scientists sacrificed valuable research time to go to Washington to lobby against this measure. They talked to Congressmen and Senators, testified at public hearings as citizen-scientists, and joined with church, educational, farm, labor, professional, and youth groups in opposition. They backed the McMahon bill to place atomic energy in the hands of a five-man civilian commission with only a single military representative as the director of the division of military application.

After a hard battle "the House, which a week before did all it could to continue military control, changed its mind, accepted the McMahon bill, and did not even bother to take a roll-call vote on the final passage. The Senate shouted it through as though it were a pet post-office project in a colleague's home town." The McMahon bill became law on August 1. It was known as The Atomic Energy Act of 1946.

The battle was not quite won, however. President Truman named David E. Lilienthal head of the new commission, and Congress debated the advisability of confirming this choice. Accusations against Lilienthal ranged all the way from alleged injudicious appointments he had made as head of the TVA to sinister radical leanings. The Federation of American Scientists fought for Lilienthal's confirmation, and the good sense of the American people prevailed. The Senate early in 1947 confirmed the appointment. The other members of the Commission were: Robert F. Bacher, who was in charge of the final assembly of the core of the atom bomb; Sumner Pike, a veteran of the government service and business; Lewis L. Strauss, a former partner of Kuhn, Loeb and Company; and William Haymack, of the *Des Moines Register*.

The fight on the international front turned out to be even more difficult. It was apparent to everybody that the peacetime uses of atomic energy could not be separated from its threat as a military weapon. It was also immediately recognized that the control and development of atomic energy could not safely be left to any private industry or to any single nation. Some form of international control was absolutely essential if the peace of the world was to be preserved for all time and if the peacetime uses of this new force were to be rapidly expanded.

To most people the United Nations organization seemed the logical body for the solution of this vital problem. Some scientists, notably Einstein, Urey, and Oppenheimer, while supporting the United Nations, urged that countries surrender part of their sovereignty at once to a World Government which would be entrusted with the control and development of nuclear fission.

As sole owner of atomic bombs and, so far as was known, sole manufacturer of fissionable material in quantities large enough to be used in bombs, it was natural that the United States should suggest a plan of control to the United Nations. On June 13, 1946, at the first meeting of the United Nations Commission on Atomic Energy, the American delegate, Bernard Baruch, proposed a plan patterned after the so-called Acheson-Lilienthal Report written by a seven-man Board of Consultants appointed by the Secretary of State on January 7 of that year. Six days after Baruch's plea, the United Nations Atomic Energy Commission met for a second time and listened to the Russian proposal presented by Andrei Gromyko. The chief points of difference between the two plans were:

(1)   The United States plan provided for the establishment of an autonomous Atomic Development Authority which would exercise rigid control over all mining, processing, research, and manufacturing of fissionable material by a system of inspections, and certify cases of violations to the Security Council. After the establishment of this Authority we would yield our atomic bomb secrets step by step. The Russian plan called for an international pact forbidding the use of atomic weapons, and for all atom-bomb stockpiles to be destroyed within three months after the ratification of this international pact.

(2)   The United States plan demanded that the Security Council

give up its veto power in cases of violations certified by the Atomic Development Authority. The Russians opposed any effort to undermine the principle of unanimity of the permanent members of the Security Council. They opposed any relinquishment of the veto.

(3) The United States plan would punish individuals and nations for violations by a system of sanctions imposed by the United Nations. The Russian plan would let each government pass legislation to insure compliance with international agreements within its own borders. Neither plan was adopted. A stalemate resulted.

Since ultimately the people of the United States would have to decide on an American policy regarding international control of atomic energy, it was essential that they understand the meaning of the controlled release of nuclear energy. From the very beginning, the Federation of American Scientists was aware of this necessity and made a nationwide public education program its second major project. An independent publication under the editorship of Eugene Rabinowitch was founded in 1945. It was named the *Bulletin of the Atomic Scientists* and continues today under the name of *Science and Public Affairs* as a powerful educational tool.

On the first anniversary of the horrors of Hiroshima, the United Nations was still deadlocked over atomic energy, and time was rapidly running out. More effective action had to be taken. On August 12, 1946, an Emergency Committee of Atomic Scientists was incorporated at Princeton, New Jersey. It included officials of the Federation of Atomic Scientists as well as Hans Bethe, Linus Pauling, Selig Hecht, and Philip M. Morse. Its purpose was to conduct an educational program on the meaning of atomic energy, to make the people realize that they must live together peacefully or die. Einstein, chairman of the committee, declared, "We must carry the facts of atomic energy to the village square." The public had to be warned that atomic bombs can be made plentifully and cheaply, that they will become more destructive, that there can be no permanent monopoly on knowledge of how to make them, that preparedness against them is futile, that their general use in war would destroy civilization, and that there is no other solution than international control of atomic energy, and, ultimately, the elimination of war.

On September 22, 1947, this advice was repeated by Lilienthal. He made an impassioned plea for a "specific program whereby in Crawfordsville, Indiana [where he was speaking], and in all other American communities the people on their own can set about to learn the essential facts about atomic energy." He asked all Americans "to look upon this task as an obligation." He characterized as "dangerous nonsense, dangerous to genuine national security, the

growing tendency in some quarters to act as if atomic energy were
none of the American public's business." Finally, he warned the
American public: "You need to watch your public servants, to keep
an eye on us, whether in the executive branch, the military, or in
Congress. And to do so effectively your views and judgments must
be based upon some knowledge of the background of facts." Decades have passed but the warning needs to be repeated.

During the years of waiting for a settlement of the military
atomic energy problems, scientists in and out of the Manhattan
District Project continued their researches. Some restrictions, imposed for security reasons, impeded the progress of the work. Some
scientists, irked by measures and actions which interfered with their
freedoms, found employment in other research fields. Nevertheless,
even during this early period of far-from-ideal working conditions
further facts were unfolded, new instruments and techniques were
developed, and additional uses were made of the new atomic force.
The identification and properties of four previously but erroneously
announced naturally occurring chemical elements were finally and
definitely established, and a substantial number of brand-new elements were created, the so-called *transuranium* elements.

The first of these newly created elements, Nos. 93 and 94, were
born in the Berkeley cyclotron, as mentioned, in 1940. Nos. 95 and
96, were identified in 1944 and reported in the fall of 1945 by
Glenn Seaborg and his young co-workers, Ralph A. James, Leon O.
Morgan, and Albert Ghiorso. The public announcement of these
discoveries was one of the most curious in the history of science.
They were to be announced in the usual way before a meeting of
the American Chemical Society. But on the preceding Sunday,
Seaborg, a guest on a popular radio program called the Quiz Kids,
was asked by one of the youngsters if any new elements had been
discovered. Before he realized it, Seaborg told five million people
listening in about elements 95 and 96. These elements were first
obtained in ultramicroscopic amounts, with the aid of the University of California's 60-inch cyclotron at Berkeley, by bombarding
U-238 and Pu-239, respectively, with 40-million-electron-volt
helium ions. The radioactive properties of the new elements made it
possible to study these elements through the tracer technique. In
April, 1946, Seaborg named the newcomers *americium* (Am) after the
Americas, and *curium* (Cm) after the first pioneers of radioactivity.
Curium was finally isolated in September, 1947, by Isadore Perl-
man and L. B. Warner in quantities large enough to be seen with
the naked eye. It turned out to be the most violently radioactive
element thus far prepared, with a half-life of five months.

Another five years passed before two new births were announced by the same superb nuclear scientist. In 1949, element No. 97 was born of americium-241 and christened by Seaborg *berkelium* (Bk) after the city of Berkeley, home of the cyclotron that Lawrence had given to science. Then the state in which these deliveries were made was honored with *californium* (Cf), element No. 98, which was prepared from curium-242, again with the aid of alpha particles.

Another five crowded years went by before the announcements of still two more transuranium elements were made. Element No. 99 (atomic weight 247) was prepared by bombarding uranium-238 with nitrogen-14 projectiles. The great Albert Einstein, father of the nuclear age, had just died, and his illustrious name was given to this new element, *einsteinium* (Es).

The hundredth element of the expanding periodic table, together with the ninety-ninth, was first unexpectedly detected in the fallout of a hydrogen bomb explosion at the Eniwetok atoll in the Pacific. Samples of the dust were collected on large filter papers carried by drone (unmanned) airplanes flying through the radioactive clouds. The discovery of these new elements, kept a top secret for more than a year, was achieved by three teams of scientists stationed at Berkeley, Chicago, and Los Alamos, New Mexico. Element No. 100 also turned out to be radioactive, with a half-life of sixteen hours. Seaborg had, in the meantime, been selected to receive the Nobel Prize for his contributions to creative nuclear chemistry, and some whispered that it might be a good idea to name element No. 100 *seaborgium*. Seaborg protested. Then late in November, 1954, Enrico Fermi, one of the greatest of the pioneers of the nuclear age, died of cancer at the age of fifty-three. There was no doubt now in the minds of all the men who had taken part in the adventure of tracking down the new element. It was by common consent named *fermium* (Fm).

One hundred might have been a nice round number of chemical elements to satisfy any chemist, but in 1955 yet another element appeared in the sixty-inch cyclotron at Berkeley, again uncovered by Seaborg. Its parent was element No. 99, and this new element was an extremely active one. In less than an hour it had changed into a lighter element—that is, only half of it had changed. If it were really present on that cosmic occasion when the earth with all its chemical elements is believed by some to have been created in one big bang, this particular element must have broken down into one of the more stable and familiar elements soon afterward. Like most of its transuranium kin, it was too nervous and erratic to retain its individuality for any appreciable length of time. Its existence was incontestably established for the first time on the basis of the detection of only a single atom of this element.

When the moment came to give this newcomer a name, Seaborg and his colleagues thought it was high time, despite cold-war feelings, to dedicate it to the memory of the man who had given science the first practical catalogue of the chemical elements almost a century before. *Mendelevium* (Md) became the name of element No. 101.

Then came *nobelium*, element No. 102, first created by an international team of American, British, and Swedish scientists working at the Nobel Institute in Stockholm in 1957. It was slowly and painfully detected, one single atom at a time, after the bombardment of curium-244 with carbon-13 ions. Seaborg, who later became the chairman of the U.S. Atomic Energy Commission, did not believe this would be the last. He predicted that within the next few years elements Nos. 103 through 108 would come to light. This prediction was based partly on the fact that a powerful new particle accelerator had been built at the University of California. HILAC (Heavy Ion Linear Accelerator) would probably father these new elements.

In 1961 part of this prediction came true. A few atoms of element No. 103 were produced early that year with the help of HILAC by the bombardment of californium with nuclei of boron-10 having energies of 70 million electron volts. It was found to have a half-life of about eight seconds and was named *lawrencium* after the inventor of the cyclotron. This element rounded out the *actinide* series, a second inner group of transition elements starting with the element thorium and ending with lawrencium.

Then, in 1964, a group of scientists at Dubna, U.S.S.R., under the direction of G. N. Flerov, announced that they had made the isotope 260 of the element No. 104 by bombarding plutonium with neon-22. It had a half-life of only 0.3 seconds and resembled hafnium in chemical properties. They tentatively named it *kurchatovium* in honor of the Russian nuclear physicist Igor Kurchatov. Albert Ghiorso and his team at the Lawrence Radiation Laboratory failed to find the new element by Flerov's method, but in 1969 Ghiorso announced the discovery of element No. 104 (at. wt. 257) by the bombardment of californium-249 with carbon-12 nuclei in HILAC. It had a half-life of about four seconds, decaying into nobelium-253 by emission of an alpha particle. Despite what appears to be conclusive proof of Ghiorso's discovery no name for this element has yet been accepted by the International Union of Pure and Applied Chemistry, although *rutherfordium* has been suggested by the Berkeley team.

Finally Ghiorso and a team of four scientists announced in 1970 the creation of another new transuranium element, No. 105. In the same machine that gave birth to element No. 104 Ghiorso bombarded californium-249 with a beam of nitrogen-15 ions. Four neutrons were emitted as the nucleus of the target atom became an isotope of element No. 105 with an atomic weight of 260 and a half-life of 1.6 seconds. The team suggested that the new element be named *hahnium* in honor of the German scientist Otto Hahn, who won the Nobel Prize in chemistry in 1944 for his discovery of nuclear fission.

With this announcement, more than half a century had passed since Moseley gave science his table of atomic numbers. Perhaps some day, as would be both fitting and proper, the name of Moseley will be lifted to immortality by naming the still undiscovered element No. 106 after him, even as element No. 101 was named mendelevium in honor of the architect of the periodic table of the elements. Or, perhaps Moseley might be remembered by naming the atomic number the *Moseley number*, to be universally used, even as Amedeo Avogadro, the Italian scientist, was honored long after his death with the *Avogadro number*.

The conquest of atomic fission also brought rapid and significant activity in the field of the radioisotopes. The first artificially radioactive element was created as mentioned before on January 15, 1934, when Irene and Frédéric Joliot produced a radioactive isotope of nitrogen by bombarding boron with helium ions. Radioactive nitrogen with an atomic weight of 13 has a half-life of only about 15 minutes. That same year, the Joliots transmuted ordinary aluminum into radioactive phosphorus, and within five years only seven additional radioactive isotopes had been reported. By 1942, however, science had added 360 artificially produced isotopes of which 223 were discovered by means of the cyclotron, 120 of them in Lawrence's own laboratory. By 1948 more than 800 of these atomic nuclei were known, about half of which are stable. Many more were later created.

The cyclotron, however, was a very slow and inefficient manufacturer of radioisotopes. Uranium fission produced radioisotopes infinitely faster. For example, it would take Lawrence's 60-inch cyclotron *five years* of continuous operation to produce one millicurie (one unit of radioactivity equivalent to that produced by one mg. radium) of radioactive carbon of atomic weight 14 (C-14). In the Oak Ridge pile only *two and a half days* were needed to get the same yield. In other cases the pile out-produced the cyclotron one billion times! The cost of radioisotopes fell accordingly at a phenomenal rate. Made in the cyclotron, C-14 cost about $1,000,000, while the same amount of it was being turned out by the pile in 1947, at a cost of only $50. Today its production cost is even less.

Radioisotopes can be made in reactor piles by inserting chemicals and subjecting them to neutron bombardment, producing a transmutation. For example, to make C-14, a compound containing nitrogen such as NH_4NO_3 (ammonium nitrate) is placed in the pile. The nitrogen picks up a neutron emitted by uranium fission during the "cooking" process and discharges a proton from its nucleus. This changes a stable nitrogen atom into a radioactive carbon-14 atom. Some of these transmutations are very slow and it frequently takes days to produce appreciable quantities.

The radioactive isotope was a new and revolutionary tool. It opened a new avenue of approach. Laboratories all over the world were eager to use it in tracer-technique researches. It was a very delicate tool. Only tiny amounts were necessary. For example, carbon-14 can be diluted to one part in ten million, and radioactive sulfur to a ratio of only one part in one trillion and their presence can still be detected by the Geiger counter. The tagged atom can be detected among millions of ordinary atoms "just as a man with a ringing clock in his pocket can be picked out of a silent crowd of 100,000 closely packed people."

Up to August 2, 1946, these radioisotopes were either altogether unobtainable or prohibitive in cost. On that day, almost thirteen years after the Joliots' first creation of radioactive nitrogen, an important event took place in the pile building of the Clinton Laboratory at Oak Ridge. Eugene P. Wigner, director of research, turned over a pea-sized amount of C^{14} (in the form of $BaCO_3$) manufactured in the pile to a representative of the Barnard Free Skin and Cancer Hospital of St. Louis for cancer research. Four other bits of this precious substance were released to the University of Pennsylvania's Medical School for research in diabetes, to the University of Minnesota for studies on teeth, to the Medical School of the University of California for investigations of the liver, muscle tissue, and blood. The last piece went to James Franck of the University of Chicago where this Nobel Prize winner was conducting research in the mysteries of the photosynthetic process. Since the half-life of this isotope is 5,100 years it can be kept for a long time. The following year both deuterium and deuterium oxide were also made available to American researchers. Even the super-heavy water, tritium oxide, which is radioactive (half-life is 31 years), can now easily be obtained. The reactor pile that had produced the atom bomb was now beginning to work for peace and humanity.

Any isotope can now be shipped to any part of the world within one day. They are packed in lead containers, the walls varying in thickness depending upon the penetrating power of the radiation emitted. Thus, radioactive iodine, which emits gamma rays, requires a shield of about one hundred pounds, while cobalt must be shielded in a sixteen-hundred-pound container.

Radioactive cobalt-60, introduced in 1948, took the place of radium and X-rays in many cases of cancer therapy. Its half-life is 5.3 years and it is more than 300 times as powerful as radium. Phosphorus-32 has already proved of great value in agricultural research. It has given us clues as to the best way to add phosphate fertilizers to the soil for larger and healthier crops. Carbon-14, in Nobelist Melvin Calvin's hands, has unravelled many secrets connected with photosynthesis—the most important chemical reaction in the vegetable world.

Scientific discoveries often follow the most curious turns. Take the case of carbon-14, which has a half-life of about 5400 years. In 1946 Willard F. Libby, who later became a Nobel Prize winner, saw in this radioisotope a new method of determining the age of things that date back as far as 30,000 B.C. Wooden coffins and the clothing of ancient mummies, charred timber of ancient dwellings, organic relics, artifacts, bones and bits of charcoal found in prehistoric caves —and the Dead Sea scrolls, perhaps 1900 years old—have all been accurately dated by the number of electrons emitted by the carbon-14 they contain. C^{14} is formed by the action of cosmic rays on nitrogen atoms in the atmosphere. The C^{14} finds its way into plants as CO_2 and eventually is incorporated into animals which eat the vegetation. The amount of C^{14} in plants and animals ceases when the plant or animal dies. A carefully worked out formula relates the emission of electrons from the object under investigation with the time that its C^{14} ceased to increase. The method is uncanny in its accuracy. Archaeologists and other students of prehistory regard it as an exciting and reliable new tool of research for accurate dating back to 70,000 years.

Radioisotopes are also aiding industry. Radioactive iron and sulfur are being used to study the exchange of sulfur between slag and iron in the blast furnace. Metallurgists are using several radioisotopes to improve the quality of steel alloys, and to study the effects of friction and heat on many metals used in heavy machinery. During the manufacture of steel alloys the percentage and distribution of carbon and alloying metals such as titanium, molybdenum, and vanadium, can be quickly determined by using only a tiny amount of the radioactive forms of these substances.

Radioisotopes such as H^3 and C^{14} are also being employed in the petroleum industry to study the nature and action of catalysts, polymerization of hydrocarbons, the flow of underground water, oil and gas, and to map subterranean pools of crude oil.

The future uses of these chemical detectives and healers in the fields of human health, control of insect pests, plant and insect physiology, the nature and mechanism of chromosome changes, food sterilization, and the manifold areas of industrial research are practically limitless.

With the first announcement of the use of an atomic bomb came a widespread interest in *Atomics*. Fantastic stories were circulated about the new and revolutionary source of power which we were told would completely eclipse overnight our basic fuels such as coal, oil, and running water. Visions of the immediate bankruptcy of coal and oil companies in the midst of a far-reaching economic upheaval were conjured up by some. More than a whole generation has now passed, and yet none of these dire or optimistic predictions has materialized. Just what is the true picture with regard to nuclear fission as a rival of our present sources of energy?

Richard C. Tolman predicted in the Smyth Report that "a great industry would arise comparable with electronics. Nuclear power for special purposes could be developed within 10 years." C. A. Thomas of the Monsanto Chemical Company, which operated the Oak Ridge installation, also thought it should take ten years. On the other hand, Major-General Groves said "decades" will have to pass. J. R. Oppenheimer said, "Thirty years from now nuclear power will be common." Lillienthal in 1947 declared that it would be six to ten years before atomic plants generating electricity will be operating in the United States. Fermi, constructor of the first pile in history, was of the opinion that a large central installation changing atomic power into electricity successfully for local use would be a reality within the next twenty to thirty years. In 1948, Lyle B. Borst, chairman of the Nuclear Reactor project of the Brookhaven National Laboratory, declared it would be 10 to 20 years before atomic energy can compete favorably with coal as a source of industrial power.

Predictions that, within a few years, we would be able to heat a small, radioactive, lamp-lit home all year round with a single tiny uranium pill were nonsense. In 1946 Fermi declared, "It does not appear possible to design an atomic power unit light enough to be used in a car or plane of ordinary size." But there are always charlatans to seize on such fantastic schemes, and victims to fall prey to them. In the summer of 1946 a Mr. John Wilson attempted to float a company in London to manufacture an atom-powered car driven by a "uranium" engine. He picked up a considerable amount of money from innocents before the law caught up with him and gave him a 21-month jail sentence.

Research was soon under way on the problem of the propulsion of a large plane by nuclear energy. In 1946 the Fairchild Engine and Airplane Corporation of New York was awarded a contract to undertake such a project. This project, called NEPA (Nuclear Energy for Propulsion of Aircraft), was established at Oak Ridge as a joint effort of the Atomic Energy Commission, the National Advisory Committee on Aeronautics, and the Army Air Forces. The project, however, proved a failure and was dropped soon afterwards.

Submarines equipped with an atomic-energy installation independent of an air supply, since no oxygen is needed during the "burning" of atomic fuel, have, on the other hand, already been constructed and are roaming the oceans. Other naval vessels and commercial ships are also in operation. Nuclear explosions have been used to try to free a few huge reservoirs of gas and oil trapped deep in the ground (AEC's Project Ploughshare), and it has also been suggested that atomic bombs might be used to change weather and even climate. Atomic explosions might, it is argued, modify the normal movements of air masses so as to change weather conditions over fairly large land areas. Some, including Eddie Rickenbacker, the famous World War I air pilot, had even suggested that atom bombs might be used to blast away the polar ice caps. Such a venture, if successful, would loosen a flood of water which would inundate large land masses, change the direction of winds and ocean currents, and produce profound world-wide climatic changes. Even if such a highly questionable undertaking were considered worthwhile it would entail the use of perhaps several thousand atomic bombs. Such drastic landscaping of the earth is, however, still extremely questionable.

The predictions of the nuclear experts turned out to be completely overoptimistic. Electrical power from the energy of the atom's nucleus is still a minor factor in the world's energy consumption. For example, in 1975 some forty nuclear power plants were in operation in the United States. Yet only about 5 per cent of the nation's electricity was supplied from all of these plants when *in full operation*. It was hoped that by the 1980's one hundred and fifty nuclear plants would satisfy 20 per cent of our national electrical power needs. (In 1974, 30 nuclear plants supplied 10 per cent of the electricity of England and Wales, and in that year the U.S. Atomic Energy Commission projected about one thousand reactors at the turn of the century.)

The first nuclear reactor was built by the United States Government in 1943 at Oak Ridge, Tennessee. It resembled the Chicago atomic pile constructed by Fermi the previous year. Several nuclear reactions took place in this device as shown on page 301. The method for separating the Pu-239 from the U-238 in this pile had been first worked out by Seaborg, Segre and two associates. Because this procedure had preceded their employment by the United States Government on the bomb project, the Patent Compensation Board of the Atomic Energy Commission in 1955 awarded them $400,000 for their rights to this process. Many reactors followed both here and around the world.

Every nuclear pile is a potential electric power station. During its operation uranium is fissioning and large quantities of heat are being liberated. This heat changes water to steam, which operates a conventional turbine. Electricity is generated and distributed from the nuclear power plant to wherever it is needed. The three essential parts of a nuclear reactor are the fuel, the moderator, and the protective shielding. The main fuel is uranium-235. When this is bombarded with neutrons it fissions and produces heat. The moderator, which is usually either graphite or heavy water, slows down the neutrons liberated and makes them more effective for fissioning. The shielding of lead and concrete prevents the very dangerous fission products from leaving the reactor, thus safeguarding the health and lives of its operators.

Of the many reactor types tried or considered since 1943 in this country only about five have survived. The dominant reactor type uses enriched uranium oxide fuel and is moderated and cooled by water. Another type called the *fast breeder reactor* actually produces more nuclear fuel than it uses. Breeders use U-235 (which is in short supply) or plutonium-239 as fuel. Liquid sodium metal is the coolant. Neutrons produced by the fission of the U-235 or Pu-239 react in a "blanket" of nonfissionable U-238 (abundant but in dilute form in granite rocks) which surrounds the reactor core and create more Pu-239 than they consume. This extra Pu-239 is then periodically removed and made into fuel for other breeders or water reactors that can also run on Pu-239. A nuclear reactor operating for one year can produce as much as 700 pounds of *weapon-grade* Pu, enough to make several dozen A-bombs. The U.S.S.R., Britain, West Germany, and France have one or more of these *experimental* commercial breeders in operation. In this country active but rather slow development of this type was proceeding in 1975.

In 1973 a report by a panel headed by Hans Bethe recommended the liquid sodium-cooled fast breeder reactor as the most logical first choice for future reactor construction. Late that year a severe energy crisis confronted the world partly due to an Arab oil squeeze, just as a fast flux test facility of this type was rising in the middle of the Hanford reservation on the Columbia River in the State of Washington to test materials and fuel. It is called the Fast Flux Test Facility (first proposed six years before). A projected commercial breeder power plant was started along the Clinch River in Tennessee. This is a multi-billion dollar project intended to be ready by the late 1980's.

In 1975, however, it was still not certain whether this was the answer to the energy problem. The nuclear industry was still plagued with shortages, bottlenecks, late deliveries, sharp cost overruns and technological malfunctionings. Few reactors operate at intended capacity. Years are required to license, build and complete a nuclear power plant. There are other problems—thermal pollution of streams resulting in the death of fish, and the real dangers connected with the disposal and transportation of nuclear material (often in commercial planes, trains, and trucks) and nuclear wastes (the ashes of nuclear plants). These are potential hazards for hundreds and even thousands of years. Despite all the latest sophisticated equipment installed to insure security against theft, terrorism, and leakage and minimize other hazards, many people feel apprehensive and even worse about peril to their health and even lives. The Union of Concerned Scientists, the Committee for Nuclear Responsibility, and the Scientists Institute for Public Information among others have questioned the wisdom of the entire project.

A few can even remember an incident connected with the fast reactor at Los Alamos on May 21, 1946. A piece of equipment accidentally slipped and its fissionable material was ready to explode and send a stream of gamma rays, neutrons and other products through the laboratory. Louis Sloton, who had had a colorful career as a Canadian scientist, literally separated the dangerous material with his bare hands, prevented the explosion, and saved the lives of several other men working near him. He died nine days later, at the age of thirty-five, of radiation poisoning.

Alvin W. Weinberg, while director of the Oak Ridge National Laboratory at Oak Ridge, Tennessee, put the matter this way, almost thirty years later: "We nuclear people have made a Faustian bargain with society. On the one hand we offer—in the breeder reactor—an inexhaustible source of energy. But the price we demand of society is both a vigilance and a longevity of our social institutions that we are quite unaccustomed to."

The real terrors of the A-bomb and the nightmare of other and accidental nuclear explosions did not exhaust the horrors of the atom's nucleus. Another was not long in coming.

More than thirty years had passed since the tragedy of Hiroshima. Only a little progress has been achieved in reducing the dangers of atomic fallout and nuclear warfare. In 1963 a partial nuclear test ban was enacted by the treaty of Moscow which banned test explosions of atomic bombs in the atmosphere, on the surface of the ground, and under water, *but not under ground*. This treaty was initialled by the representatives of the United States, the Soviet Union, and Great Britain. However, France, the People's Republic of China, and India, which joined the exclusive nuclear power club in 1960, 1964, and 1974 respectively, continued to set off nuclear blasts in the atmosphere. In 1969 the United States ratified a nuclear non-proliferation treaty, this time prohibiting installation of nuclear weapons on the seabed beyond any nation's 12-mile coastal zone. It was signed by 63 nations. Again Peking, Paris, and New Delhi still insisted on their "legitimate" right of self-defense to continue nuclear testing and installing nuclear weapons beyond the limits proposed.

However, the nuclear dangers continued to increase. Five countries that are all permanent members of the Security Council of the United Nations now have frightening arsenals of nuclear weapons, enough in some instances to destroy not only each other but civilization. At least another five countries already have or will soon possess nuclear capability since nuclear reactors built for peaceful purposes can also create weapon-grade plutonium used in weaponry. These include West Germany, Canada, and Sweden. Nuclear engines continue to proliferate unrestrained in many parts of the world. The dangers continue to mount daily.

In addition to the perils of the A-bomb based on nuclear *fission*, a new terror has been introduced—the *hydrogen* or *fusion* bomb. Such a fusion or *thermonuclear* bomb is a million times more destructive than an A-bomb. Soon after the A-bomb, loaded with U-235 and plutonium, had been exploded in 1945, Enrico Fermi, together with Edward Teller, a refugee physicist from Hungary, and British-born James Tuck considered the possibility of achieving a bomb explosion by fusion rather than by fission. Edward Teller with fanatical zeal was all for manufacturing it as quickly as possible. Ernest Lawrence and Luis Alvarez, another Nobel laureate, agreed. But the consciences of some other scientists bothered them. "The physicists have known sin," declared J. Robert Oppenheimer; Fermi and Rabi were also reluctant to join in this new venture of creating an H-bomb. This attitude was partly responsible for the revocation of Oppenheimer's security clearance in 1954 even though he had been declared loyal to his country.

The work on the H-bomb went on. The principle of this weapon is somewhat different from that of the A-bomb. The destructive force of the H-bomb comes from the *fusion* of lighter atoms into a heavier one rather than from the *fission* of a heavy element into lighter ones. Two isotopes of hydrogen take part in the fusion process. Heavy hydrogen or deuterium has a mass of two, double that of ordinary hydrogen. Tritium or radiohydrogen, the heaviest form, has a mass of three. Heavy hydrogen is found in all water, including that of the oceans, to the extent of about one part in 6000. Tritium with a half-life of twelve years is seldom met in nature but can be manufactured in a nuclear reactor by bombardment of the isotope of the element lithium of atomic weight 6 with neutrons.

The nuclei of deuterium (D) and tritium (T) are made to merge or fuse. During this fusion, the hydrogen is transmuted into helium whose mass is four (He^4). One neutron is liberated and nuclear energy is produced in tremendous quantities, because in fusion, too, there is a loss of matter. This *thermonuclear* reaction may be expressed as follows:

$$D^2 + T^3 \rightarrow He^4 + n + energy$$

The energy liberated is equivalent to about 176,000,000 kilowatts per pound of fuel. (One pound of coal when *burned* produces only 8 kilowatt hours of energy.)

For such a nuclear reaction to take place, however, an enormously high temperature—at least 100,000,000° Centigrade—is necessary. Such a temperature is found only in the sun and other stars where it is generated by just such a fusion reaction, according to the originator of this theory, Hans A. Bethe, who later became a Nobel laureate. Edward Teller had been interested in this thermonuclear reaction as far back as 1935 when he came to George Washington University as professor of physics. There he had worked with George Gamow on nuclear reactions in the stars.

In the sun, according to Bethe, temperatures of 100,000,000 degrees Centigrade and pressures that stagger the imagination are ripping protons and neutrons out of atoms, and building heavier elements such as helium out of the lighter element hydrogen. It is generally agreed that the stupendous and age-old heat of the sun is generated by the release of nuclear energy during this building of helium from hydrogen. The transmutation is consummated after several intermediate changes known as the *carbon cycle* as follows:

$$C^{12} + H^+ \rightarrow N^{13} + energy$$

$$N^{13} \rightarrow C^{13} + positron$$

$$C^{13} + H^+ \rightarrow N^{14} + energy$$

$$N^{14} + H^+ \rightarrow O^{15} + energy$$

$$O^{15} \rightarrow N^{15} + positron$$

$$N^{15} + H^+ \rightarrow He^4 + C^{12} + energy \quad (\text{The } C^{12} \text{ nucleus is thus regenerated.})$$

Upon adding the above 6 equations, it is seen that the over-all reaction is $4H^+ = He^4$. This cycle in the sun is then repeated.

Can man imitate this cycle of nature's celestial furnaces? We have achieved this awesome heat in the explosion of the A-bomb. This terrifically high temperature is needed for only about one-millionth of a second. With the creation of the A-bomb such a temperature became available here on earth, for it is reached during an A-bomb explosion. The detonation of an A-bomb can thus be made to act as a trigger for the explosion of an H-bomb. Such a double bomb explosive can be constructed to provide almost unlimited destructive power. Ordinary A-bombs are in the kiloton or *thousand*-tons-of-TNT class; H-bombs are in the megaton class; that is, they can produce energy equivalent to as much as sixteen *million* tons of TNT.

The first hydrogen bomb was made and exploded by the United States in November, 1952, at the Atomic Energy Commission's Eniwetok proving grounds in the Pacific. Four other nuclear weapon powers followed with their own deadly blasts.

Less than three years before the first H-atom explosion shook the world a United Nations-sponsored International Conference on the Peaceful Uses of Atomic Energy opened in Geneva, Switzerland. The most spectacular single announcement that came out of this meeting of atomic scientists was that of Professor Homi J. Bhabha, then head of India's Atomic Energy Commission and head of the Conference. Bhabha represented a country where the energy problem is one of the many keystones of its future. It is a land where 80 per cent of its energy came from one of the most primitive methods still in use, the burning of dung, a product which is better put to use to improve the productivity of her soil. Bhabha was looking even further ahead than the nuclear fission of uranium.

"When we learn how to liberate *fusion* energy in a controlled manner," he told his fellow scientists, "the energy problems of the world will truly have been solved forever, for the fuel will be as plentiful as the heavy water in the oceans." But this was only still the fantasy of another dreamer from India.

Work on fusion energy had unknowingly begun actually thirty years before this conference when Irving Langmuir, the Nobel laureate, had begun his studies on ionized gases by applying systematically the principles of atomic and statistical physics to ionized gases. He coined the word *plasmas*, which are hot gases which owe their conductivity to ionization. Plasma is a mixture of free electrons and positively charged nuclei (the interior matter of a star).

It was not very long after that attempts were made in the United States, the U.S.S.R., and Great Britian to try to obtain a controlled fusion reaction. To achieve this it is necessary to heat the fusion fuel (deuterium and tritium) to about 100 million degrees C., that is, till it reaches the plasma state and can then fuse to helium + neutron. Then this reacting mixture must be confined away from any material walls (which would vaporize) and free from impurities long enough for a substantial fraction to react. This containment may be attained within a magnetic field or "bottle."

Such a magnetic configuration was the basis of a new device called a *Stellarator* originally built at Princeton University by Lyman Spitzer in 1961. With this new 1975 *Large Torus* machine modified and refined several times some progress was achieved. A second device also of the magnetic configuration concept was the basis of a machine first constructed by Artsimovitch and his coworkers in the Soviet Union and called *Tokamak*.

Another concept lately introduced includes the use of laser beams. A laser (acronym formed from the words *Light Amplification by the Stimulated Emission of Radiation*) is a newly developed device for producing an intense beam of coherent light of the same wavelength and frequency. This is based on the fact that the gamma ray laser (graser) energy can be focussed to very small dimensions and produced in short optical pulses, thus achieving intense power with microscopic fission reactions.

Plasma physics which Artsimovitch called "the latest infant of the classics" is being researched with vigor. But most scientists involved in this new field do not see a final solution of controlled fusion power for many, many years to come. The problems facing them are still almost overwhelming.

Even before the quest for a practical fusion energy machine was begun, theoretical scientists were continuing their battles to unravel the complicated structure and meaning of the atom. What they needed were more powerful cyclotrons and other devices to step up the pace of their subatomic particles and to trap the particles that resulted from the collision of these particles with carefully selected targets. New and more effective energy machines were devised and built in several laboratories around the world. These new machines were added to the arsenal of subatomic artillery already in use. Included were the betatron, synchrotron, bevatron, linear accelerator, and the *bubble chamber*. The *betatron* was developed in secrecy during World War II by Donald Kerst of the University of Illinois. It was an electron (beta ray) accelerator shaped like a huge doughnut. Electrons from a hot filament were injected into this vacuum glass doughnut and speeded around it by an alternating current until they reached a speed approaching that of light. At the end of their journey the electrons struck their target.

The *synchrocyclotron*, in general appearance and operation, resembled the betatron. It employed modulated frequency rather than power at a fixed frequency in order to overcome the limitation of the betatron when energies higher than 100,000,000 volts were involved. The strength of the magnetic field was changed while the electrons gathered energy.

The *linear accelerator* was another of these machines devised as early as 1931 and abandoned in favor of the cyclotron. Many years later, while still a young University of California physicist, Luis W. Alvarez (who later received the Nobel Prize) introduced several changes in this machine which attracted a new interest in it. The *linear accelerator* is used for speeding both electrons and protons; it employs no magnetic field at all. The moving bullets are exposed to accelerating electric fields which are applied at various points along the straight path just at the moment when they can act most effectively as the particle reaches that point. Radar techniques were applied in its operation—electrons, for example, rode through the accelerator on microwaves.

The *bubble chamber*, based on C. T. R. Wilson's cloud chamber principle of 1896, was a new and different piece of equipment. It was conceived by Donald Glaser of the University of Michigan in 1952, and eight years later it was considered such an essential addition to atomic research that he was awarded the Nobel Prize at the age of thirty-four. In this new machine liquid hydrogen is kept under pressure and when the pressure is released the hydrogen boils, and bubbles form especially along the paths of electrically charged particles or ionizing radiation. These tracks can be photographed.

Because this chamber is placed in a magnetic field, the tracks curve. From the curves the scientists can compute the mass and velocity of each of the particles. The largest of these devices is 15 feet in diameter and is located at the U.S. Atomic Energy's Fermi National Accelerator facility at Batavia, Illinois.

Larger and even more powerful accelerators were built in the years that followed these earliest inventions. Each transition from smaller to larger atom-smashers resulted in new and exciting discoveries. Among the newer accelerators were an electron accelerator at Cambridge, Massachusetts; one at the Brookhaven National Laboratory; a synchrocyclotron at Cornell University; the Lawrence bevatron (Billion electron-volt) at Berkeley, California; a super HILAC; a 12-Bev zero gradient synchrocyclotron at the Argonne National Laboratory; and an electron-positron storage ring accelerator at Stanford University in California.

The achievements of atomic scientists had been formidable. But many questions still remained unanswered. One of the most perplexing problems deals with the nature of the binding forces which keep the complex nucleus from flying apart. The nucleus, as it was pictured, consisted of free positively charged protons and electrically neutral neutrons. What held all the plus protons together since particles of similar electrical charge should repel each other? Then again the neutron was thought to be made up of an electron ($-$) and a proton ($+$), very close together. Why did not these plus and minus charged bodies present in the neutron annihilate one another since they possessed opposite charges? What kept them apart? What mysterious force kept all the protons and neutrons together? This nuclear force was apparently different from gravitational and electrical forces.

Leading nuclear physicists, including Lawrence and Rabi, believed that this cosmic cement that holds the different particles might be the meson particle mentioned earlier. This particle might offer the key to an understanding of these puzzling nuclear forces. We know something about mesons. Several kinds have already been identified. It is a particle of variable mass—of either positive or negative charge or it may even be neutral. The positively charged meson first reported in 1937 had a mass of 200 times that of an electron. This type of meson is formed when a billion-volt proton from outer space (*primary* cosmic ray) strikes the kernal of an atom in the lower atmosphere. Its lifetime is only about one-half of one-millionth of a second.

Men needed to know more about these mesons. Some climbed to the top of the Bolivian Andes in search of them. Cesare Mansueto Lattes, a twenty-three-year-old Sao Paulo scientist, worked with a group from Bristol, England, at this high altitude, and later joined them in Bristol for further investigations. While in Europe he worked with G. P. S. Occhialine and C. F. Powell in the development of a new and ingenious *emulsion* technique for trapping mesons photographically. In 1947 they found new types of mesons of masses 380 and 480 times that of an electron. Even before this announcement, a French scientist of the University of Paris photographed in the Alps the track of another new meson type 900 times the mass of the electron.

The night of Feb. 21, 1948, marked another even more exciting discovery. Lattes and an associate were at the controls of the new, redesigned 4000-ton, 184-inch synchrocyclotron of Lawrence's Radiation Laboratory at Berkeley. This was an even more powerful machine than the original 184-inch cyclotron which had been completed in 1942. The principle of frequency modulation used in radio transmission was applied to increase its power. The utilization of this principle was suggested by the Soviet scientist V. Veksler and independently by Edwin M. McMillan of Lawrence's laboratory. Its use is based on the fact that according to Einstein's theory of relativity a particle becomes heavier as its speed increases. A proton, for example, accelerated at 200,000,000 electron-volts moves at a speed of almost 80,000 miles a second, and its mass is increased about 10 per cent. This increase in weight causes the proton in the cyclotron to lag behind. To compensate for this lag the frequency of the oscillations is gradually reduced. This very significant change made available particles accelerated to energies equivalent to 400,000,000 electron volts.

The two men hurled helium nuclei with energies as high as 380 million electron volts in this machine at carbon and beryllium targets. The effects of this bombardment were studied on their emulsion plates attached to the new atomic colossus. What they saw turned out to be "the most significant event in fundamental nuclear studies since the discovery of uranium fission." The track of a negatively charged meson of mass 313 times that of the electron turned up in one of their photomicrographs.

Lawrence was at first skeptical of this strange newcomer, but two weeks later he announced this historic event at a press conference. For the first time in history men had *artificially created mesons*. A fast-moving helium nucleus containing only neutrons and protons had completely changed into only neutral neutrons and negatively charged mesons. Furthermore, the number of such mesons produced was 10 million times as plentiful as the occasional mesons laboriously trapped at high altitudes. A new milestone had been reached in nuclear physics. Scientists, at last, had a new tool—mesons in great numbers to help probe more deeply into that cosmic glue which holds the nucleus of the atom intact.

In an effort to explain many other paradoxes of the structure of the nucleus, more theoretical particles were postulated. For example, the *neutrino* was introduced to explain what happens when a meson disintegrated. The meson was supposed to release an electron and a neutrino, each carrying 50,000,000 volts. The neutrino was supposed to be a chargeless particle possessing energy, momentum and a mass less than one-tenth of the electron. It was predicted as early as 1927 by W. Pauli and finally actually discovered in 1956 by F. Reines and C. Cowan. But the observed changes still remained puzzling.

A comparable machine at CERN, the European Organization for Nuclear Research, had been operating a 30 billion electron volt proton synchrotron with intersecting storage rings since 1959.

With these new machines research in *particle physics* was stepped up. More than a score of new and exotic "elementary" particles were discovered. Among them were pi mesons (pions) of mass one-seventh that of a proton, either positively or negatively charged or electrically neutral. *Kaons* (K mesons) of mass one-half that of the proton, and positively, or negatively charged or neutral also appeared, as did the *positronium*, a system consisting of a positron and an electron which decays into two or three photons.

The lightweight and weightless particles such as electrons, neutrinos, positrons and muons are known collectively as *leptons*. The heavier particles (*hadrons*) are the proton, neutron and pions. These respond to the nuclear force, the strongest and most baffling force in nature.

Many attempts had been made before to clear up the mysteries of the atom. For example, in the early 1930's Werner Heisenberg, who led the German atomic-energy research during World War II, had suggested a theory to explain what prevented the positively charged protons and the negatively charged electron particles in neutrons from falling into each other. He postulated an *exchange force*, the electrical charge of the proton being tossed back and forth between the proton and the neutron in the nucleus. Each neutron when approached by a proton for an instantaneous moment became a proton, and the proton for an instant became a neutron. The effect of this toss was to prevent the positive charges in the nucleus from coming within repelling distance of each other.

In 1975, thirteen years after the death of Niels Bohr, his son Aage Bohr while working in the Niels Bohr Institute in Copenhagen was awarded the Nobel Prize for his researches on a new nuclear model. He shared the prize of $143,000 with Ben Mottelson of Nordita in Copenhagen and James Rainwater of Columbia University. Their new model combined the features of the two most prevalent pictures of the atomic nucleus—the shell model and the liquid-drop model. It explained many nuclear properties hitherto unaccounted for. Scores of other scientists are working in this field, but as the atom's nucleus is probed deeper and deeper its enigma still remains, and more mysteries continue to surface.

Just as ancient Egypt built huge pyramids and Europe during the Middle Ages erected magnificent cathedrals, so twentieth-century science continues to construct gigantic atom smashers. In 1975 the world's two largest were located in the United States and Switzerland. The huge complex of the Fermi National Accelerator Laboratory near the town of Batavia, Illinois, which took several years to build, at a cost of $250 million, operates at energies up to about 500 billion electron volts (Gev), a five-fold increase over earlier machines of the same type. Protons from this synchroton accelerator machine are directed against targets to produce a variety of secondary beams. It is, for example, able to produce a high energy neutrino and a *muon* (an unstable electron 207 times heavier than the normal electron) beam travelling at close to the speed of light. This machine is housed in a cavernous, concrete, vault-like auditorium from which three main beams (proton, neutrino, and meson) emerge. A tall building of unusual design rises at the center among fields of soya beans and corn in the Illinois prairie west of Chicago. Around its four-mile diameter ring subatomic particles rush at staggering speeds.

During these extensive investigations, a new puzzling atomic world had gradually come into focus. Dirac's equation which had accurately predicted the positron (positive electron) had foreshadowed this. For every naturally occurring elementary particle, he said, there seemed to be a similar particle of equal mass but of opposite charge. For example, the negative electron had its positive positron of equal mass, and the positive proton [+] had its counterpart negative antiproton [−] of the same mass whose existence was first demonstrated by E. Segrè and O. Chamberlain in 1959. These *antiparticles* are but two of the denizens of this new world of *antimatter*. Antimatter has never been found in nature since matter and its oppositely charged particle (antimatter) annihilate each other on contact. Thus far scientists have not stumbled upon any type of antimatter as part of the substance of the physical world. There are ways, however, of creating particles of antimatter in high-energy collisions produced in high-energy particle accelerators.

By the mid-1950's, too, another exotic group of subatomic particles turned up to further confuse the world of science. Murray Gell-Mann and George Zweig independently in the United States and K. Nishgima in Japan postulated the existence of these so-called "strange" particles. They conceived the *Quark* hypothesis to account for all of the many subatomic particles that had been reported. They thus tried to find some simple order among this multitude of new particles. *Quarks* is a name chosen by Gell-Mann, from a line in James Joyce's *Finnegans Wake*, "Three Quarks for Muster Mark." Quarks are supposed to be fragments of elementary particles such as the electron or the proton. They proposed that three quarks and antiquarks with fractional charges of $+2/3$, $+1/3$, and $-1/3$ plus a fourth one with a charge of $+2/3$ indirectly and seemingly observed in 1975 in the Fermilab. This fourth quark is called a "charmed" quark because it is an indestructible property that survives collisions. Quarks have a lifetime of $1/10,000$ of a billion of a second. These four subunits could be put together in several ways to give all the heavy, strongly interacting particles. Gell-Mann is a Nobel laureate, and Richard P. Feynman shared the Nobel Prize with Julian S. Schwinger of Harvard University and Shinichero Tomonaga of Japan for their efforts to find some order for these exotic particles and some underlying meaning and principles of their relationship. However, so far separated quarks have not been seen, and even this bold attempt proved to be an oversimplification. The particle hunters were still stalking deep in the woods.

While speculations and theories kept piling up, something new and totally unexpected suddenly appeared toward the close of 1974. Within twelve days of each other two new subnuclear particles were reported. The first one was named "J" or *psi-3105*. It decays at once into an electron and a positron. It was found during proton-proton collision experiments at the Brookhaven National Laboratory by a team headed by Samuel Ting, and simultaneously by another research group at the Stanford Linear Accelerator Center led by Burton Richter.

The second particle produced during electron-positron annihilation experiments at Stanford was named *psi-3700*. They are both very heavy with masses of 3.15G and 3.70G, more than three times that of the proton. Neither of them has an electric charge and they do not fit into any scheme or pattern yet proposed. There followed, of course, another burst of theorizing. "We were left," said one of the researchers, "with still more questions to answer and very little sleep for anyone." As with Zeno's paradox, every advance seems to carry the particle physics man closer to his goal but leaves him with still another step to take.

What are these particles? They do not seem to be either the already defined hadrons or leptons. Some believe they may be a newer kind of quark possessed of a new quality called "charm" predicted more than a decade before. Some think it is a property of all elementary particles, something like an electric charge. Right now it remains only a purely mathematical concept.

Another baffling problem was the internal motions or dynamics of the nucleus. About a quarter of a century ago Enrico Fermi had explored this problem. He and his collaborators had found a short-lived and highly excited state of both the proton and the neutron. The motion of the nucleons indicated to them that they, like the whole atom, possessed a shell structure and manifested a spectrum of their own. The term *baryon* was introduced for nucleons and their excited states. But again no acceptable model for this phenomenon has been found. This is still not clearly understood.

"It is," indeed, as V. F. Weiskopf of the Massachusetts Institute of Technology, a pioneer in the realm of the nucleus, wrote, "a strange world, full of new forms of matter, new transformations and reactions with an unexpected richness and variety. We are only at the beginning of the exploration of this unknown part of the universe."

Chemists and other men of science have already created the means to achieve a new world for mankind. But there is much to be done to expand and spread the fruits of the laboratory to more of the people of the lands and to bring to them a higher quality of life.

A new era in human health and comfort can be within our grasp. To hasten the coming of this day more opportunities should be made available to all of our potential scientific brains. Whatever barriers still remain against the utilization of capable minds must be removed. More education, jobs and opportunities are needed. Any society which through blind prejudice, prevents qualified members of its community from engaging in scientific research, will suffer the loss of valuable sources of progress. Further effort should be made to find those young minds which have special talent for scientific research. More gifted men and women should be channeled into scientific investigation by attractive scholarships and grants. The National Science Foundation was established in 1950 to implement these suggestions. It was authorized "to develop basic scientific research." This included, for example, money for building astronomical telescopes and particle accelerators.

There is a ferment in laboratories both here and abroad. Scientists are still picking the nucleus of the atom apart and trying to put together the more than a score of subatomic particles already discovered or predicted, to see how the atom really ticks. Creative chemistry is in the middle of this great adventure, too. And it will continue to be as fruitful in many other areas where chemists are searching for new products which nature in all her lavishness neglected to create.

More money needs to be poured into the streams that feed our scientific pools of peacetime research. Since science benefits everybody, it should be generously financed by our government. For a country as large and wealthy as our own there was, and still is, an unjustifiable niggardliness in this connection. During the years immediately preceding World War II, there was spent in the United States an estimated total of $250,000,000 for scientific research. Of this, only $25,000,000 (equivalent to about seven per cent of our annual cosmetics bill) was devoted to what might be designated as pure research. This money came primarily from such private sources as the Rockefeller Foundation, the General Education Board, and the Bamberger and Fuld gift of $5,000,000 for the establishment of the Institute for Advanced Study at Princeton, New Jersey. The work was done almost exclusively in university laboratories, private medical research centers such as the Rockefeller Institute for Medical Research, and in astronomical observatories such as the Mount Wilson Observatory.

The remainder of the quarter of a billion dollars was spent by industrial laboratories on researches which were pointed directly at immediate practical problems. Large as this figure might appear to some, it was still less than two per cent of the total volume of business of the combined industrial organizations which were farseeing enough to engage in scientific research, and less than two-tenths of one per cent of our national income. During World War II we increased this amount more than ten times. In 1954 the figure was about 5 billion. Of this amount private industry had spent 3.7 billion, about 40 per cent of which was contributed by the Federal Government. Government spending was exceptionally high in the fields of atomic energy, missiles, supersonic planes and other weapons of war. On the other hand expenditures in basic scientific research in the United States continued to be neglected.

The Russian launching of the first sputnik in 1957 was soon followed by a stepped-up scientific research funding in the United States. By 1974 Research and Development spending totalled $32 billion. The Federal Government's share was $17 billion and industries' $13 billion. The Government allocated $10 billion to cope with the nation's energy shortage, and more than half of this sum was earmarked for nuclear research.

In 1930 there were about 500,000 technical engineers and electricians in our country, another 100,000 draftsmen and inventors, and about 50,000 chemists and metallurgists. By 1974 the number had multiplied. These figures show more than a hundred-fold increase over those of the year 1870. But this phenomenal growth in numbers is not enough. To make a program of better living for the mass of our people more effective, it is necessary to give our scientists and engineers a different educational training. A democracy requires intelligent scientists and engineers in the broadest sense of that term. The education of these men and women must produce citizens with a clearer understanding of social forces and their own social responsibilities. Their schooling must ensure skilled scientists who have also been influenced by the social sciences and the humanities. Liberal arts training is extremely important and should be included. Said James B. Conant, then president of Harvard University and a distinguished worker in the field of organic chemistry:

> Through many advances gained by science we may hope that as never before man may be free—free from want. But science alone, untempered by other knowledge, can lead not to freedom but to slavery. At the root of the relation between science and society in the postwar world must lie a proper educational concept of the interconnecting of our new scientific knowledge and our older humanistic studies.

Karl T. Compton, another distinguished scientist and then president of the Massachusetts Institute of Technology, also agreed that we need engineers with a recognition of their social responsibilities. Robert A. Millikan while head of the California Institute of Technology had for many years insisted that the students at this scientific institution spend at least one-fourth of their time in such studies as economics, literature, history, political science, and philosophy. It is hoped that with this new trend in education our future engineers will have a clearer and more dynamic attitude toward their duty as citizen-scientists in a democracy.

Above and beyond the birth of a new source of power and the application of nucleonics in both industry and the field of human health and comfort is the challenge of something even bigger. "I wish I could produce a substance," said Alfred Nobel many years ago, "of such frightful efficacy for wholesale destruction that it would make wars impossible." Atomic fission and fusion have made this wish come true. "For this is what is new in the atomic age," declared J. R. Oppenheimer, "a world to be united in law, in common understanding, in common humanity, before a common peril."

The new atomic era was officially born on December 2, 1942, when the first successful chain-reacting pile was achieved. Five years later, a plaque was dedicated at the University of Chicago. On the same day, while Hiroshima halted work to observe the second anniversary of the fall of a new bomb, a large white wooden cross was raised on a wind-swept dune in the Alamogordo Desert of New Mexico where the first atomic bomb was exploded. A small party of men from various walks of life flew to this spot to raise a rainbow flag of all nations, and to call upon the people of the world to renounce atomic warfare even as the survivors of Hiroshima prayed that mankind would renounce all war.

Ancient alchemy failed to give men an elixir of life that would ward off old age and extend their life span. Modern alchemy through nuclear fission has brought this dream within our grasp if we act now to prevent mankind from muddling into a war of total destruction. If we do not act quickly, the long story of the advance of science will end, and the dire prophecy of Harlow Shapley, the eminent American astronomer, will come true. On our dismal planet depopulated of man, some termite or other insect, crawling out of the skull of the last man on earth, will be musing, "Alas, the creature did not understand the business of survival."

NOTES

[1] From Ben Jonson's *The Alchemist.*

[2] Quotation from *The Dying Alchemist* by N. P. Willis.

[3] From Browning's *Paracelsus.*

[4] From George Eliot's *Middlemarch.*

[5] From J. P. Cooke's *The New Chemistry.*

[6] Priestley, called the Father of Pneumatic Chemistry, also prepared and collected sulfur dioxide gas.

[7] Cavendish anticipated Coulomb in the discovery of the inverse square law of electrical attraction, but his work remained unpublished until long after his death.

[8] Quotation from Youmans' *New Chemistry.*

[9] Among his pupils was James Prescott Joule who, at ten, was sent by his father to study mathematics under Dalton. It was this boy who, twenty-five years later, gave the world its first complete explanation of the Law of the Conservation of Energy.

[10] Between 1821 and 1848, the year of his death, Berzelius published his famous *Jahresberichte,* twenty-seven volumes of complete yearly reports on the progress of chemistry and physics.

[11] Hans Oersted, a Danish scientist, probably made aluminum in 1825 by extracting it from aluminum chloride by means of potassium amalgam. Woehler repeated this experiment and obtained pure aluminum two years later.

[12] The basic principles of Arrhenius' theory are as firm as ever. They are easily explained by the new electron theory, for even solid crystals have ions which separate in water solution. However, the process of dissociation in concentrated solutions of strong electrolytes still presents an anomaly. Recently, P. Debye and Hückel advanced the theory of *complete* ionization even in strong electrolytes. By a thoroughly mathematical treatment, they showed that departures from the general law of dissociation were due to electrical forces between the ions.

[13] George Johnstone Stoney, Secretary of the Royal Dublin Society, suggested, in 1891, the name electron as the "natural unit of electricity which would liberate one atom of hydrogen."

[14] By R. A. S. Paget.

[15] From the "Ballad of Ryerson" by Edwin H. Lewis.

SOURCES

CHAPTER I

Chymische Schriften des Hn. Bernhardi (contains his autobiography), by Casper Horn. Nürnberg, 1747.

Histoire de la Chimie, by F. Hoefer. F. Didot Frères, Fils & Cie. Paris, 1866.

Lives of Alchemystical Philosophers, by A. E. Waite, 1888.

The Revival of Alchemy, by Henry C. Bolton. Smithsonian Institution Annual Report. Washington, 1897.

Francis Joseph, by Eugene Bagger. Putnam's. New York, 1927.

Alchemy: Ancient and Modern, by H. S. Redgrove. W. Rider and Son. London, 1922.

The Alchemical Essence, by M. P. Muir. Longmans, Green & Co. London, 1894.

The Power of the Charlatan, by Grete de Francesco. Yale University Press. New Haven, 1939.

Alchemy, Child of Greek Philosophy, by Arthur J. Hopkins. Columbia University Press. New York, 1934.

Three Famous Alchemists, by W. P. Swainson. Rider & Co. London, 1939.

CHAPTER II

The Life of Paracelsus, by Anna M. Stoddart. J Murray. London, 1911.

"Paracelse, sa vie et son œuvre," by Achille Ouy in *Revue internat. de sociologie*, Année 27. Paris, 1919.

Die Tragödie des Paracelsus, by Annie Francé. W. Seifert. Stuttgart, 1924.

"Paracelsus as a Reformer of Chemistry," by John M. Stillman in *The Monist*, vol. 29, Chicago, 1919.

The Hermetic and Alchemical Writings of Paracelsus. J. Elliott & Co. London, 1894.

Philosophie Magnae des Paracelsus, by A. E. Erber. Cöln, 1567.

Versuch einer Kritik der Echtheit der Paracelsischen Schriften, by Karl Sudhoff. G. Reimer. Berlin, 1894-99.

Four Treatises of Paracelsus, by Temkin, Rosen, Zilboorg, and Sigerist. The John Hopkins Press. Baltimore, 1942.

Theophrastus Paracelsus: eine Kritische Studie, by Friedrich Mook. Wurzburg, 1876.

T. Paracelsus, Idee und Problem seiner Weltanschauung, by Franz Strunz. Salzburg, A. Pustet, 1937.

Paracelse, le medecin maudit, by R. F. Allendy. Gallimard, Paris, 1937.

CHAPTER III

Narrische Weissheit und Weise Narrheit, by Johan J. Becher. Frankfort, 1682.

"Becher, Ein Beitrag zur Geschichte der Nationalökonomik," by Erdberg-Krczenciewski in *Sammlung Staatswissenschaftliches Studien,* Bd. 6. Jena, 1896.

"Becher, Ein Beitrag zur Geschichte des Merkantilismus," by Hans Rizzi in *Kultur,* Jahrg. 5. Vienna, 1904.

"Becher als Wirtschafts und Sozialpolitiker," by Emil Kauder in *Schmollers Jahrbuch,* Jahrg. 48, Heft 4. München, 1924.

Natur-Kündigung der Metallen, by J. J. Becher. Frankfort, 1661.

Physica Subterranea, by J. J. Becher. Leipzig, 1703.

Magnalia Naturae, by J. J. Becher. T. Dawks. London, 1680.

History of the Phlogiston Theory, by J. H. White. Arnold and Company. London, 1932.

Becher, der Erfinder der Gasbeleuchtung, by Konrad Ullrich. Speyer. 1936.

CHAPTER IV

Memoirs of Priestley, written by himself, with a continuation to the time of his decease by his son. (Reprint from ed. 1809.) H. R. Allenson. London, 1904.

Life and Correspondence of Joseph Priestley, by John T. Rutt. R. Hunter. London, 1831.

Priestley in America, by Edgar F. Smith. P. Blakiston's Son & Co. Philadelphia, 1920.

Experiments and Observations on Different Kinds of Airs, by Joseph Priestley. J. Johnson. London, 1776.

Considerations on Phlogiston and Decomposition of Water, by Joseph Priestley. J. Johnson. London, 1796.

Experiments Relating to Analysis of Atmospherical Air, by Joseph Priestley. Philadelphia, 1799.

The Doctrine of Phlogiston Established and that of the Composition of Water Refuted, by Joseph Priestley. Northumberland, 1800.

Writing on Philosophy, Science and Politics, by Joseph Priestley, edited by John Passmore. Collier, 1965.

A Scientific Autobiography of Joseph Priestley. Selecteted Scientific Correspondence, edited by Robert E. Schofield. Massachusetts Institute of Technology, Cambridge, Mass. 1966.

An Appeal to the Public on the Subject of the Riots in Birmingham, by Joseph Priestley. J. Johnson. London, 1792.

"The Lunar Society," by Henry Bolton in *Trans. N. Y. Acad. of Sciences,* vol. 7, No. 8. New York, 1888.

Chemistry in America, by E. F. Smith. Appleton, 1914.

Letters to Edmund Burke, by J. Priestley. J. Johnson. Birmingham, 1791.

Scientific Correspondence of Priestley, by Henry Bolton. (Privately printed.) New York, 1892.

History of Electricity, by Joseph Priestley. (3rd edition.) J. Johnson. London, 1775.

Benjamin Franklin, by Bernard Fay. Little, Brown and Co. New York, 1929.

A Life of Joseph Priestley, by Anne Holt. Oxford University Press, London, 1931.

The Priestley Family Collection, by Charles C. Sellers. Dickinson College, Carlisle, Pennsylvania 1965.

Joseph Priestley, Adventurer in Science, by F. W. Gibbs. 1965.

CHAPTER V

Life of Henry Cavendish, by George Wilson. Cavendish Society. London, 1851.

Lives of Men of Letters and Science in the time of George III, by Lord Brougham. Carey & Hart. Philadelphia, 1845

"Elogé de Cavendish," by Georges Cuvier. *Institute Acad. Sci. Mem.* Paris, 1811.

Experiments on Air, by H. Cavendish. London, 1785.

Experiments to Determine the Density of the Earth, by H. Cavendish. London, 1798.

Electrical Researches of Henry Cavendish, by J. C. Maxwell. Cambridge University Press. Cambridge, 1879.

Correspondence of Watt on the Discovery of the Theory and Composition of Water, by J. P. Muirhead. London, 1846.

Three Philosophers (Cavendish, Priestley, Lavoisier), by W. R. Aykroyd. W. Heinemann Ltd. London, 1935.

CHAPTER VI

Life of Lavoisier, by Mary L. Foster. Smith College monographs No. 1. Northampton, 1926.

Œuvres de Lavoisier, publiées par les sonins du Ministre de L'Instruction Publique. E. Grimaux, editor. Paris, 1862-93.

Traité Elémentaire de Chimie, by A. L. Lavoisier (translated by R. Kerr). Evert Duyckinck. New York, 1806.

La Révolution Chimique, by M. Berthelot. F. Alcan. Paris, 1890.

The Composition of Hydrogen and the Non-Decomposition of Water Incontrovertibly Established, by W. F. Stevenson. J. Ridgway. London, 1849.

Lavoisier d'après sa correspondance, ses papiers de famille et d'autres documents inédits, by E. Grimaux. F. Alcan. Paris, 1896.

Antoine Lavoisier, by Douglas McKie. Lippincott. Philadelphia, 1935.

Lavoisier, by J. A. Cochrane. Constable. London, 1931.

Combustion from Heraclitus to Lavoisier, by J. C. Gregory. Arnold and Company. London, 1935.

CHAPTER VII

Memoirs of the Life and Researches of Dalton, by Wm. C. Henry. Cavendish Society. London, 1854.

John Dalton and the Rise of Modern Chemistry, by Henry E. Roscoe. Cassell & Co. London, 1895.

John Dalton, (English Men of Science), by J. P. Millington. J. M. Dent & Co. London, 1906.

The Development of the Atomic Theory, by A. N. Meldrum. Oxford University Press. London, 1920.

Memoirs of John Dalton and History of Atomic Theory, by R. Angus Smith. 1856.

A New View of the Origin of Dalton's Atomic Theory, by Roscoe and Harden. Macmillan & Co. London, 1896.

Experiments and Observations on the Atomic Theory, by William Higgins. Dublin, 1814.

"William Higgins, Pioneer of Atomic Theory," by Reilly and MacSweeney in *Sc. Proc. Royal Dublin Soc.,* vol. 19. Dublin, 1929.

A Short History of Atomism, by J. C. Gregory. The Macmillan Company, 1931.

CHAPTER VIII

"Selbstbiographische Aufzeichnungen" (Berzelius), by Emilie Woehler and H. G. Söderbaum in *Monographieen aus der Geschichte der Chemie.* J. A. Barth. Leipzig, 1903.

Berzelius' Werden und Wachsen (1779-1821), by H. G. Söderbaum. Monographieen No. 3. Leipzig, 1899.

Encyclopaedia Londinensis (Chemistry, vol. 4), by John Wilkes. J. Adlard. London, 1810.

Lehrbuch der Chemie, by J. J. Berzelius (translated by F. Woehler). Dresden, 1831.

The Use of the Blowpipe, by J. J. Berzelius (translated by J. D. Whitney). Boston, 1845.

Elements of Chemistry vol. 1, by B. Silliman. Hezekiah Howe. New Haven, 1830.

Théorie des Proportions Chimiques et table Synoptique des Poids Atomiques des Corps Simples, by J. J. Berzelius. Didot Frères. Paris, 1835.

Sir Humphry Davy, by John Davy. Smith, Elder & Co. London, 1839.

"Berzelius," by H. Rose. *American Journal of Science,* vol. 16. 1853-4.

"Berzelius." *Proc. of Royal Society,* p. 822. London, 1849.

J. J. Berzelius, by Olof Larsell. Williams and Wilkins Co. Baltimore, 1934.

CHAPTER IX

Onoranze centenarie internazionali ad Amedeo Avogadro, by I. Guareschi. Unione Tipografico-Editrice Torinese. Turin, 1911.

Amedeo Avogadro und die Molekulartheorie, by I. Guareschi. (Monograph 7.) J. A. Barth. Leipzig, 1903.

Stanislao Cannizarro (nel centenario della nascite). Associazione Italiana di Chimica. Rome, 1926.

Molécules, atomes et notations chimiques. (Les Classiques de la Science). A. Colin. Paris, 1913.

Avogadro and Dalton, the Standing of Their Hypotheses, by A. N. Meldrum. Aberdeen University Studies No. 10. Aberdeen, 1904.

A History of Chemical Theories and Laws, by M. P. Muir. J. Wiley & Sons. New York, 1909.

Fisica de corpi ponderabili, by A. Avogadro. Turin, 1841.

The New Chemistry, by J. P. Cooke. D. Appleton & Co. New York, 1873.

Bernouilli ed Avogadro e la teoria cinetica dei gas, R. Accademia della Scienze di Torino. May 15, 1910.

"Short Biography of Avogadro," by Prof. Kuhnholtz in *Annales Chimiques de Montpelier.* September, 1856.

Atoms, by J. B. Perrin. Van Nostrand and Co. New York, 1917.

Charles Gerhardt, sa vie, son œuvre, by E. Grimaux. Masson et Cie. Paris, 1900.

"Amedeo Avogadro di Quaregna," by Alfonso Cossa in *Il Biellese,* pp. 181-8.

Determination des poids moleculaires, by R. Lespieau. Gauthier-Villars. Paris, 1938.

CHAPTER X

Friedrich Woehler, by H. Huebner. Abhandl. (vol. 29). Göttingen, 1882.

Zur Erinnerung an Vorangegangene Freunde (vol. 2), by A. W. von Hofmann. F. Vieweg und Sohn. Braunschweig, 1888.

Liebig, His Life and Work, by W. A. Shenstone. Macmillan & Co. New York, 1895.

Briefwechsel zwischen Berzelius und Woehler. W. Engelmann. Leipzig, 1901.

Grundriss der Chemie, by F. Woehler. Duncker und Humblot. Berlin, 1845.

"Reception of Woehler's Discovery of Synthesis of Urea," by W. H. Warren in *Journal of Chemical Education.* December, 1928.

Creative Chemistry, by E. E. Slosson. The Century Company. New York, 1920.

Chemistry in the Twentieth Century, by E. F. Armstrong. Longmans, Green & Co. New York, 1924.

Aus Meinem Leben, by E. Fischer. Springer. 1922.

Briefe von Liebig nach neuen Funden Gesellschaft Liebig-Museum und der Liebighaus-Stiftung in Darmstadt, by Ernst Berl. Darmstadt, 1928.

CHAPTER XI

"Dmitri Mendeléeff," by O. Lutz in *Ztschr. f. Angewandte Chemie,* Jahrg. 20. Leipzig, 1907.

"Dmitri Ivanovitch Mendeléeff," by W. A. Tilden in *Proceedings of Royal Society,* ser. A., vol. 84. London, 1910.

The Principles of Chemistry (two vols.), by D. I. Mendeléeff. Longmans, Green & Co. London, 1897.

An Attempt towards a Chemical Conception of the Ether, by D. I. Mendeléeff (translated by G. Kamensky). Longmans. London, 1904.

The Gases of the Atmosphere, by Wm. Ramsay. Macmillan & Co. New York, 1902.

"A Great Chemist, Sir William Ramsay," by Chas. Moureu. *Annual Report Smithsonian Institute.* Washington, 1921.

John Wm. Strutt (Lord Rayleigh), by R. J. Rayleigh. Longmans, Green & Co. New York, 1924.

Travaux du Congres Jubilaire Mendeleev. Moscow, 1936-37.

CHAPTER XII

"Arrhenius," by W. Ostwald in *Zukunft,* Jahrg. 18. Berlin, 1910.

J. H. van't Hoff, sein Leben und Wirken, by Ernst J. Cohen. Akademische Verlagsgesellschaft. Leipzig, 1912.

Major Prophets of To-day, by E. E. Slosson. Little, Brown & Co. Boston, 1914.

"Obituary of Arrhenius," by W. Ostwald in *Chem. Ztg.,* 51, 781 (1927). Cöthen, 1927.

"Obituary of Arrhenius," by Lorentz in *Zeitschrift für Angewandte Chemie,* vol. 40. Leipzig, 1927.

"Obituary of Arrhenius," by P. Walden in *Naturwissenschaften,* 16, 325. Berlin, 1928.

"Svante Arrhenius zur Feier des 25 jahrigen Bestandes seiner Theorie der Dissociation," by W. Ostwald in *Zeitschrift für Physikalische Ch.,* vols. 69-70. Leipzig, 1909-10.

Theories of Solutions, by S. Arrhenius. Yale University Press. New Haven, 1912.

"The Willard Gibbs Address," by S. Arrhenius. American Chemical Society (May 12, 1911). Chicago, 1911.

"Work of Arrhenius," by H. Aten in *Chem. Weekblad,* 25, 98 (1928).

"Obituary of Arrhenius" in *Proc. Royal Society,* 1928 (IX-XIX). London, 1928.

Die Entwicklung der Elektrochemie, by W. Ostwald. J. A. Barth. Leipzig, 1910.

Elements of Qualitative Chemical Analysis, by Julius Stieglitz. The Century Co. New York, 1921.

The Scientific Foundations of Analytical Chemistry, by W. Ostwald. Macmillan Co. New York, 1908.

Svante Arrhenius, by E. H. Risenfeld. Akademische Verlags. Leipzig, 1932.

CHAPTER XIII

Pierre Curie, by Marie Curie (translated by C. and V. Kellogg). The Macmillan Co. New York, 1923. (Dover reprint.)

"Pierre Curie Intime," by Marie Curie in *Rev. politique et littéraire.* Paris, 1923.

La Radiologie et la Guerre, by Marie Curie. F. Alcan. Paris, 1921.

"Radioactive Substances" (thesis) by Marie Curie in *Chemical News*. London, 1904.

"Commercial Production of Radium," by Moore in *Industrial & Engineering Chemistry*, vol. 18, No. 2, 1926.

"Commercial Production and Uses of Radium," by J. Viol in *Journal of Chemical Education*, vol. 3, No. 7, 1926.

Radioactive Substances and Their Radiations, by E. Rutherford. Cambridge University Press. Cambridge, 1913.

Madame Curie, by Eve Curie. Doubleday Doran. New York, 1937.

Radiations from Radioactive Substances, by Rutherford, Chadwick and Ellis. Macmillan. New York, 1932.

CHAPTER XIV

A Treatise on the Motion of Vortex Rings, by J. J. Thomson. Macmillan & Co. London, 1883.

Conduction of Electricity through Gases (3rd edition), by J. J. and G. P. Thomson. Macmillan & Co. New York, 1928.

The Atomic Theory, by J. J. Thomson. Clarendon Press. Oxford, 1914.

Electricity and Matter, by J. J. Thomson. C. Scribner's Sons. New York, 1924.

Rays of Positive Electricity and Their Application, by J. J. Thomson. Longmans, Green & Co. New York, 1921.

The Electron, by R. A. Millikan. University of Chicago Press. Chicago, 1924.

The A B C of Atoms, by B. Russell. Dutton. New York, 1923.

The Life of Sir William Crookes, by E. Fournier d'Albe. T. F. Unwin, Ltd. London, 1923.

The Interpretation of Radium and the Structure of the Atom, by F. Soddy. Putnam. New York, 1921.

Mass Spectra and Isotopes, by F. W. Aston. Longmans, Green and Co. New York, 1933.

The Nature of the Atom, by G. K. T. Conn. Chemical Publishing Co. New York, 1940.

Ernest Rutherford, by A. S. Eve. Macmillan. New York, 1939.

The Newer Alchemy, by Lord Rutherford. Macmillan. New York, 1937.

Recollections and Reflections, by Sir J. J. Thomson. Macmillan. New York, 1937.

The Life of Sir J. J. Thomson, by Lord Rayleigh. New York, 1943.

The Cavendish Laboratory, by Alexander Wood. Cambridge University Press, 1946.

J. J. Thomson, Discoverer of the Electron, by George Paget Thomson. Doubleday & Co. Garden City, N.Y., 1966.

Rutherford—Recollections of the Cambridge Days, by Mark Oliphant. American Elsevier Publishing Co., New York, 1972.

Rutherford and the Nature of the Atom, by E. N. da C. Andrade. Doubleday and Co. Garden City, N.Y., 1964.

CHAPTER XV

"Moseley, the Numbering of the Elements," by G. Sarton in *Isis*, vol. 9, No. 29. Bruxelles, 1927.

"Obituary Notice of H. G. J. Moseley," by E. Rutherford in *Proceedings of Royal Society*, A, vol. 93. London, 1916.

"Obituary of Henry Nottidge Moseley," by E. R. Lankester in *Nature*, vol. 45. London, 1891.

"Henry Moseley," by E. R. Lankester in *Philosophical Magazine*, vol. 31. London, 1916.

"Moseley's Work on X-Rays," by E. Rutherford *in Nature*, vol. 116. London, 1925.

"The Scattering of Alpha and Beta Particles," by E. Rutherford in *Philosophical Magazine*, vol. 21. London, 1911.

"Radioactive Products of Short Life," by H. Moseley in *Philosophical Magazine*. London, 1911.

"Number of Beta Particles Emitted by Radium," by H. Moseley in *Proceedings of Royal Society*, A, vol. 87. London, 1912.

"Attainment of High Potentials by Use of Radium," by H. Moseley in *Proceedings of Royal Society*, A, vol. 88. London, 1913.

"The High Frequency Spectra of the Elements," by H. Moseley in *Philosophical Magazine*, vols. 26 and 27. London, 1913-14.

"Atomic Models and X-Ray Spectra," by H. Moseley in *Nature*, vol. 92. London, 1914.

Isotopes, by F. W. Aston. Ed. Arnold & Co. London, 1924.

Chemistry of the Rarer Elements, by B. S. Hopkins. D. C. Heath & Co. New York, 1923.

Private Letters from Mrs. Amabel Nevill Sollas.

Moseley and the Numbering of the Elements, by Bernard Jaffe. Doubleday & Co., Garden City, N.Y., 1971.

H. G. J. Moseley, by John L Heilbron. University of California Press, Berkeley, Cal. 1974.

Niels Bohr, The Man, His Science, and the World They Changed, by Ruth Moore. Alfred A. Knopf. New York, 1966.

CHAPTER XVI

"Langmuir's Work," by W. R. Whitney in *Industrial & Engineering Chemistry*, vol. 20, No. 3. New York, 1928.

"Atomic Hydrogen as an Aid to Industrial Research," by I. Langmuir in *Ind. & Eng. Chemistry*, vol. 20, No. 3. March, 1928.

"Early Life of Irving Langmuir (Perkin Medalist)," by E. Hendrick. Meeting of N. Y. Sect. of Am. Ch. Soc. Jan. 13, 1928.

"Dr. I. Langmuir Had Own Laboratory at Eleven,"*Schenectady Gazette*, Apr. 8, 1926.

"Ueber Partielle Widervereinigung dissocierter Gase," by I. Langmuir. University of Göttingen. Göttingen, 1906.

"Arrangement of Electrons in Atoms and Molecules," by I. Langmuir in *Journal of Am. Chem. Soc.*, vol 41. 1919.

"The Atomic Theory," by I. Langmuir in *Ind. & Eng. Chem.*, vol. 12, No. 4. 1920.

"Structure of the Atom, and Types of Valence," by I. Langmuir in *Science*, vols. 53 and 54. 1921.

Valence and the Structure of Atoms and Molecules, by G. N. Lewis. Chemical Catalogue Co. New York, 1923.

The Theory of Spectra and Atomic Constitution, by N. Bohr. Cambridge University Press. Cambridge, 1924.

"Birth of the Elements," by R. A. Millikan in *Proceedings Nat. Acad. Sciences*, 14. Washington, 1928.

"The General Electric Research Lab," by Guy Bartlett in *J. of Chem. Education*, vol. 16, No. 10. October, 1929.

"Ancient and Modern Alchemy," by F. Paneth in *Science*, 66, 1661. 1924.

The Atom and the Bohr Theory, by Kramers and Holst. A. A. Knopf. New York, 1923.

Atomic Structure and Spectral Lines, by Sommerfeld. E. P. Dutton & Co. New York, 1923.

Matter and Energy, by F. Soddy. Henry Holt and Co. New York, 1912.

Forty Years with the General Electric, by J. T. Broderick. Fort Orange Press. Albany, 1929.

"Surface Chemistry," by I. Langmuir. *Chemical Reviews*, vol. 13: pp. 147-191. Baltimore, 1933.

Molecular Films, The Cyclotron and the New Biology, by Taylor, Lawrence, and Langmuir. Rutgers University Press. New Brunswick, N.J. 1945.

Willis R. Whitney, by J. T. Broderick. Fort Orange Press. Albany, N.Y., 1945.

Electronics Today and Tomorrow, by John Mills. Van Nostrand. New York, 1944.

"Optical Experiments with Electrons," by L. H. Germer in *Journal of the Franklin Institute*, 205. May, 1928.

Collected Papers on Wave Mechanics, by E. Schroedinger (translated by W. M. Deans). Blackie and Son, Ltd. London, 1928.

Selected Papers on Wave Mechanics, by Louis de Broglie (translated by W. M. Deans). Blackie and Son, Ltd. London, 1928.

"Ueber quantentheoretische Umdeuting Kinematischer und Mechanischer Beziehungen," by E. Heisenberg in *Zeitschrift für Physik*, vol. 33, page 879. Berlin, 1925.

The Nature of the Physical World, by A. S. Eddington. Macmillan and Co. New York, 1928.

The Universe Around Us, by J. H. Jeans. Macmillan and Co. New York, 1944.

Mr. Tomkins Explores the Atom, by George Gamow. Macmillan. New York, 1944.

Mr. Tomkins in Wonderland, by George Gamow. Macmillan. New York, 1940.

Atomic Theory and the Description of Nature, by Niels Bohr. Cambridge University Press. London, 1961.

Essays 1958 / 1962 on Atomic Physics and Human Knowledge, by Niels Bohr. John Wiley & Sons. New York, 1963.

CHAPTER XVII

"Science and Technology," by Ernest O. Lawrence. *Science,* October 1, 1937.

"Presentation of Nobel Prize to Lawrence," by Raymond T. Birge. *Science,* April 5, 1940.

Why Smash Atoms? by A. K. Solomon. Harvard University Press. Cambridge, 1945.

The Production and Properties of Positrons, by Carl D. Anderson. Les Prix Nobel en 1936. Stockholm, 1937.

The Cyclotron, by Wilfred B. Mann. Chemical Publishing Co. Easton, Pa., 1940.

Electrons (+ and −), Protons, Photons, Neutrons, Mesotrons and Cosmic Rays, by Robert A. Millikan. University of Chicago Press. Chicago, 1947.

Outposts of Science, by Bernard Jaffe. Simon and Schuster. New York, 1935.

The New Frontiers in the Atom, by Ernest O. Lawrence. Smithsonian Institution, Annual Report, 1941. Washington, D.C.

Modern Alchemy, by Noyes and Noyes. C. C. Thomas. Springfield, Illinois, 1932.

"Radioactivity Induced by Neutron Bombardment," by E. Fermi. *Nature,* London. May 19, 1934, and June 9, 1934.

CHAPTER XVIII

"Disintegration of Uranium by Neutrons," by L. Meitner and O. R. Frisch. *Nature,* London. March 18, 1939.

The Atomic Age Opens, edited by Donald P. Geddes. Pocket Books. New York, 1945.

Atomic Energy in the Coming Era, by David Dietz. Dodd, Mead and Co. New York, 1945.

Atomic Energy for Military Purposes, by Henry D. Smyth. Princeton University Press. Princeton. N.J., 1945.

Atomic Energy in War and Peace, by G. G. Hawley and S. W. Leifson. Reinhold Publishing Co. New York, 1945.

Applied Nuclear Physics, by E. Pollard and W. L. Davidson. John Wiley and Sons. New York, 1945.

Atomic Energy in Cosmic and Human Life, by George Gamow. Macmillan. New York, 1946.

Dawn Over Zero, by William L. Lawrence, Alfred A. Knopf. New York, 1946.

Elementary Nuclear Theory, by Hans A. Bethe. John Wiley and Sons. New York, 1947.

Energy Unlimited, by H. M. Davis. Murray Hill. New York, 1947.

Atomics for the Millions, by M. L. Eidinoff and H. Ruchlis. Whittlesey House. New York, 1947.

Alsos, by Samuel A. Goudsmit. Henry Schuman. New York, 1947.

Explaining the Atom, by Selig Hecht. Viking Press. New York, 1947.

Einstein, His Life and Times, by Philipp Frank. Knopf. New York, 1947.

CHAPTER XIX

A Report on the International Control of Atomic Energy, by David E. Lilienthal. U.S. Govt. Printing Office, Washington, D.C., 1946.

Science; the Endless Frontier, by Vannevar Bush. U.S. Govt. Printing Office. Washington, D.C., 1945.

Modern Man Is Obsolete, by N. Cousins. Viking. New York, 1945.

"Atomic Power in Peacetime Economy," by C. G. Suits, P. W. Swain, and J. A. Hutcheson. *Electrical Engineering,* April, 1946.

Bulletin of the Atomic Scientists. Chicago, Illinois, 1945-75.

"Future of Atomic Energy" by J. Robert Oppenheimer, W. E. Chamberlain, Enrico Fermi and Hugh S. Taylor. *Chemical and Engineering News,* May 25, 1946, Easton, Pa.

One World or None, edited by Dexter Masters and K. Way. McGraw-Hill Book Co. New York, 1946.

"Harnessing the Split-Atom," by Gerald Wendt, L. C. Beard, H. N. Rentschler. Radio Business Forum, Station WMCA. March 5, 1946, New York.

"Dr Einstein On the Atom Bomb." *Atlantic Monthly,* November, 1945. (as told to Raymond G. Swing).

The Social Impact of Science: A Select Bibliography, U.S. Govt. Printing Office, Washington, D.C., 1945.

Our Atomic World, by R. E. Marshak, E. C. Nelson, L. I. Schiff. University of New Mexico Press, 1946.

The Curve of Binding Energy, by John McPhee. Farrar Straus and Giroux. NewYork, 1974.

Must Destruction Be Our Destiny? by Harrison Brown. Simon and Schuster. New York, 1946.

The Absolute Weapon, by Bernard Brodie, Ed. Harcourt Brace and Co. New York, 1946.

Atomic Energy in International Politics, by Harold C. Urey. Foreign Policy Association (Pamphlet). New York, 1946.

The Atomic Revolution, by Robert D. Potter. McBride. New York, 1946.

The Social Relations of Science, by James G. Crowther. Macmillan. New York, 1941.

The Way of an Investigator, by Walter B. Cannon. Norton. New York, 1945.

The Social Function of Science, by John D. Bernal. Macmillan. New York, 1939.

Science and Social Needs, by Julian S. Huxley. Harper and Brothers. New York, 1935.

"Atomic Energy and Its Implications," Symposium. *Proceedings of the American Philosophical Society.* Vol. 90, No. 1. American Philosophical Society. 1947.

Hiroshima, by John Hersey. Knopf. New York, 1946.

Struggle for Atomic Control, by W. T. R. Fox. Public Affairs Committee. New York, 1946.

Atomic Challenge, by Wm. Higinbotham and E. K. Lindley. Foreign Policy Association. New York, 1946.

Science and Public Policy, by John R. Steelman. U.S. Govt. Printing Office, Washington, D.C. 1947.

Meson Theory of Nuclear Forces, by Wolfgang Pauli. Interscience Publishers. New York, 1947.

Control of Atomic Energy, by J. R. Newman and B. S. Miller. Whittlesey House. New York, 1948.

Nuclear Physics in Photographs, by C. F. Powell and G. P. S. Occhialini. Oxford University Press. New York, 1948.

David Lilienthal, by W. Whitman. Henry Holt. New York, 1948.

Physics—50 Years Later, edited by Sanborn C. Brown. National Academy of Sciences, Washington, D.C, 1973.

MISCELLANEOUS

Historical Introduction to Chemistry, by T. M. Lowry. Macmillan & Co. London, 1915.

Essays in Historical Chemistry, by T. E. Thorpe. Macmillan & Co. London, 1894.

Alembic Club Reprints. Alembic Club. Gurney & Jackson. London, 1923.

Famous Chemists, by W. A. Tilden. E. P. Dutton & Co. New York, 1921.

Popular History of Science, by R. Routledge. G. Routledge & Sons. London, 1881.

Geschichte der Chemie, (4 vol.), by H. Kopp. F. Vieweg und Sohn. Braunschweig, 1843-47.

Britain's Heritage of Science, by Schuster and Shipley. E. P. Dutton & Co. New York, 1917.

Essays Biographical and Chemical, by Wm. Ramsay. A. Constable & Co. London, 1908.

Story of Early Chemistry, by J. M. Stillman. D. Appleton & Co. New York, 1924.

Eminent Chemists of Our Time, by B. Harrow. D. Van Nostrand & Co. New York, 1920.

A History of Science, Technology and Philosophy in the 18th Century, by A. Wolf. Macmillan. New York, 1939.

A Short History of Chemistry, by J. R. Partington. New York, 1938.

Chemical Elements, by I. Nechaev. Coward McCann. New York, 1943.

A Short History of Science to the 19th Century, by Charles Singer. Oxford University Press. England, 1942.

The Contributions of Holland to the Sciences, by Barnouw and Landheer. Querido, 1944.

This Chemical Age, by Williams Haynes. Alfred Knopf. New York, 1942.

Magic in a Bottle, by Milton M. Silverman. Macmillan. New York, 1942.

Torch and Crucible (Life of Lavoisier), by S. J. French. Princeton University Press. Princeton, N.J., 1942.

Das Buch der Grosser Chemiker, by G. Bugge. Verlag Chemie. Berlin, 1929.

The Foundations of Chemical Theory, by R. M. Caven. Blackie & Son. London, 1921.

The Discovery of the Rare Gases, by M. W. Travers. Longmans, Green & Co. New York, 1928.

Introductory Theoretical Chemistry, by G. H. Cartledge. Ginn & Co. New York, 1929.

De Re Metallica, by G. Agricola (translated by Herbert and Lou Henry Hoover). Mining. London, 1912. (Dover reprint.)

A History of Chemistry, by F. J. Moore. McGraw-Hill Book Co. New York, 1939.

Three Centuries of Chemistry, by I. Masson. E. Benn, Ltd. London, 1925.

A History of Chemistry, by Ernst Meyer. Macmillan & Co: New York, 1906.

Concerning the Nature of Things, by W. H. Bragg. Harper & Bros. New York, 1925.

Les Prix Nobel. Nobelstiftelsen. Stockholm.

The Makers of Chemistry, by E. J. Holmyard. Oxford University Press. New York, 1931.

"Discovery of the Elements," by Mary Elvira Weeks. *Journal of Chemical Education.* Easton, Pa., 1946.

Klassiker der Exakten Wissenschaften, by Wilhelm Ostwald. Akad. Verlags. 1889.

The Origins and Growth of Chemical Science, by J. E. Marsh. John Murray. London, 1929.

Cambridge Readings in the Literature of Science, by Wm. C. Whetham. Macmillan & Co. New York, 1924.

A Short History of Chemistry, by F. P. Venable. D. C. Heath & Co. Boston, 1896.

INDEX

355

A CATALOGUE OF SELECTED DOVER BOOKS
IN ALL FIELDS OF INTEREST

A CATALOGUE OF SELECTED DOVER BOOKS
IN ALL FIELDS OF INTEREST

THE DEVIL'S DICTIONARY, Ambrose Bierce. Barbed, bitter, brilliant witticisms in the form of a dictionary. Best, most ferocious satire America has produced. 145pp. 20487-1 Pa. $1.50

ABSOLUTELY MAD INVENTIONS, A.E. Brown, H.A. Jeffcott. Hilarious, useless, or merely absurd inventions all granted patents by the U.S. Patent Office. Edible tie pin, mechanical hat tipper, etc. 57 illustrations. 125pp. 22596-8 Pa. $1.50

AMERICAN WILD FLOWERS COLORING BOOK, Paul Kennedy. Planned coverage of 48 most important wildflowers, from Rickett's collection; instructive as well as entertaining. Color versions on covers. 48pp. 8¼ x 11. 20095-7 Pa. $1.35

BIRDS OF AMERICA COLORING BOOK, John James Audubon. Rendered for coloring by Paul Kennedy. 46 of Audubon's noted illustrations: red-winged blackbird, cardinal, purple finch, towhee, etc. Original plates reproduced in full color on the covers. 48pp. 8¼ x 11. 23049-X Pa. $1.35

NORTH AMERICAN INDIAN DESIGN COLORING BOOK, Paul Kennedy. The finest examples from Indian masks, beadwork, pottery, etc. — selected and redrawn for coloring (with identifications) by well-known illustrator Paul Kennedy. 48pp. 8¼ x 11. 21125-8 Pa. $1.35

UNIFORMS OF THE AMERICAN REVOLUTION COLORING BOOK, Peter Copeland. 31 lively drawings reproduce whole panorama of military attire; each uniform has complete instructions for accurate coloring. (Not in the Pictorial Archives Series). 64pp. 8¼ x 11. 21850-3 Pa. $1.50

THE WONDERFUL WIZARD OF OZ COLORING BOOK, L. Frank Baum. Color the Yellow Brick Road and much more in 61 drawings adapted from W.W. Denslow's originals, accompanied by abridged version of text. Dorothy, Toto, Oz and the Emerald City. 61 illustrations. 64pp. 8¼ x 11. 20452-9 Pa. $1.50

CUT AND COLOR PAPER MASKS, Michael Grater. Clowns, animals, funny faces... simply color them in, cut them out, and put them together and you have 9 paper masks to play with and enjoy. Complete instructions. Assembled masks shown in full color on the covers. 32pp. 8¼ x 11. 23171-2 Pa. $1.50

STAINED GLASS CHRISTMAS ORNAMENT COLORING BOOK, Carol Belanger Grafton. Brighten your Christmas season with over 100 Christmas ornaments done in a stained glass effect on translucent paper. Color them in and then hang at windows, from lights, anywhere. 32pp. 8¼ x 11. 20707-2 Pa. $1.75

How to Solve Chess Problems, Kenneth S. Howard. Practical suggestions on problem solving for very beginners. 58 two-move problems, 46 3-movers, 8 4-movers for practice, plus hints. 171pp. 20748-X Pa. $2.00

A Guide to Fairy Chess, Anthony Dickins. 3-D chess, 4-D chess, chess on a cylindrical board, reflecting pieces that bounce off edges, cooperative chess, retrograde chess, maximummers, much more. Most based on work of great Dawson. Full handbook, 100 problems. 66pp. 7⅞ x 10¾. 22687-5 Pa. $2.00

Win at Backgammon, Millard Hopper. Best opening moves, running game, blocking game, back game, tables of odds, etc. Hopper makes the game clear enough for anyone to play, and win. 43 diagrams. 111pp. 22894-0 Pa. $1.50

Bidding a Bridge Hand, Terence Reese. Master player "thinks out loud" the binding of 75 hands that defy point count systems. Organized by bidding problem—no-fit situations, overbidding, underbidding, cueing your defense, etc. 254pp. EBE 22830-4 Pa. $2.50

The Precision Bidding System in Bridge, C.C. Wei, edited by Alan Truscott. Inventor of precision bidding presents average hands and hands from actual play, including games from 1969 Bermuda Bowl where system emerged. 114 exercises. 116pp. 21171-1 Pa. $1.75

Learn Magic, Henry Hay. 20 simple, easy-to-follow lessons on magic for the new magician: illusions, card tricks, silks, sleights of hand, coin manipulations, escapes, and more —all with a minimum amount of equipment. Final chapter explains the great stage illusions. 92 illustrations. 285pp. 21238-6 Pa. $2.95

The New Magician's Manual, Walter B. Gibson. Step-by-step instructions and clear illustrations guide the novice in mastering 36 tricks; much equipment supplied on 16 pages of cut-out materials. 36 additional tricks. 64 illustrations. 159pp. 6⅝ x 10. 23113-5 Pa. $3.00

Professional Magic for Amateurs, Walter B. Gibson. 50 easy, effective tricks used by professionals —cards, string, tumblers, handkerchiefs, mental magic, etc. 63 illustrations. 223pp. 23012-0 Pa. $2.50

Card Manipulations, Jean Hugard. Very rich collection of manipulations; has taught thousands of fine magicians tricks that are really workable, eye-catching. Easily followed, serious work. Over 200 illustrations. 163pp. 20539-8 Pa. $2.00

Abbott's Encyclopedia of Rope Tricks for Magicians, Stewart James. Complete reference book for amateur and professional magicians containing more than 150 tricks involving knots, penetrations, cut and restored rope, etc. 510 illustrations. Reprint of 3rd edition. 400pp. 23206-9 Pa. $3.50

The Secrets of Houdini, J.C. Cannell. Classic study of Houdini's incredible magic, exposing closely-kept professional secrets and revealing, in general terms, the whole art of stage magic. 67 illustrations. 279pp. 22913-0 Pa. $2.50

MOTHER GOOSE'S MELODIES. Facsimile of fabulously rare Munroe and Francis "copyright 1833" Boston edition. Familiar and unusual rhymes, wonderful old woodcut illustrations. Edited by E.F. Bleiler. 128pp. 4½ x 6⅜. 22577-1 Pa. $1.00

MOTHER GOOSE IN HIEROGLYPHICS. Favorite nursery rhymes presented in rebus form for children. Fascinating 1849 edition reproduced in toto, with key. Introduction by E.F. Bleiler. About 400 woodcuts. 64pp. 6⅞ x 5¼. 20745-5 Pa. $1.00

PETER PIPER'S PRACTICAL PRINCIPLES OF PLAIN & PERFECT PRONUNCIATION. Alliterative jingles and tongue-twisters. Reproduction in full of 1830 first American edition. 25 spirited woodcuts. 32pp. 4½ x 6⅜. 22560-7 Pa. $1.00

MARMADUKE MULTIPLY'S MERRY METHOD OF MAKING MINOR MATHEMATICIANS. Fellow to Peter Piper, it teaches multiplication table by catchy rhymes and woodcuts. 1841 Munroe & Francis edition. Edited by E.F. Bleiler. 103pp. 4⅝ x 6.
22773-1 Pa. $1.25
20171-6 Clothbd. $3.00

THE NIGHT BEFORE CHRISTMAS, Clement Moore. Full text, and woodcuts from original 1848 book. Also critical, historical material. 19 illustrations. 40pp. 4⅝ x 6. 22797-9 Pa. $1.00

THE KING OF THE GOLDEN RIVER, John Ruskin. Victorian children's classic of three brothers, their attempts to reach the Golden River, what becomes of them. Facsimile of original 1889 edition. 22 illustrations. 56pp. 4⅝ x 6⅜.
20066-3 Pa. $1.25

DREAMS OF THE RAREBIT FIEND, Winsor McCay. Pioneer cartoon strip, unexcelled for beauty, imagination, in 60 full sequences. Incredible technical virtuosity, wonderful visual wit. Historical introduction. 62pp. 8⅜ x 11¼. 21347-1 Pa. $2.00

THE KATZENJAMMER KIDS, Rudolf Dirks. In full color, 14 strips from 1906-7; full of imagination, characteristic humor. Classic of great historical importance. Introduction by August Derleth. 32pp. 9¼ x 12¼. 23005-8 Pa. $2.00

LITTLE ORPHAN ANNIE AND LITTLE ORPHAN ANNIE IN COSMIC CITY, Harold Gray. Two great sequences from the early strips: our curly-haired heroine defends the Warbucks' financial empire and, then, takes on meanie Phineas P. Pinchpenny. Leapin' lizards! 178pp. 6⅛ x 8⅜. 23107-0 Pa. $2.00

WHEN A FELLER NEEDS A FRIEND, Clare Briggs. 122 cartoons by one of the greatest newspaper cartoonists of the early 20th century — about growing up, making a living, family life, daily frustrations and occasional triumphs. 121pp. 8½ x 9½.
23148-8 Pa. $2.50

THE BEST OF GLUYAS WILLIAMS. 100 drawings by one of America's finest cartoonists: The Day a Cake of Ivory Soap Sank at Proctor & Gamble's, At the Life Insurance Agents' Banquet, and many other gems from the 20's and 30's. 118pp. 8⅜ x 11¼. 22737-5 Pa. $2.50

MATHEMATICAL PUZZLES FOR BEGINNERS AND ENTHUSIASTS, Geoffrey Mott-Smith. 189 puzzles from easy to difficult—involving arithmetic, logic, algebra, properties of digits, probability, etc.—for enjoyment and mental stimulus. Explanation of mathematical principles behind the puzzles. 135 illustrations. viii + 248pp.

20198-8 Paperbound $2.00

PAPER FOLDING FOR BEGINNERS, William D. Murray and Francis J. Rigney. Easiest book on the market, clearest instructions on making interesting, beautiful origami. Sail boats, cups, roosters, frogs that move legs, bonbon boxes, standing birds, etc. 40 projects; more than 275 diagrams and photographs. 94pp.

20713-7 Paperbound $1.00

TRICKS AND GAMES ON THE POOL TABLE, Fred Herrmann. 79 tricks and games— some solitaires, some for two or more players, some competitive games—to entertain you between formal games. Mystifying shots and throws, unusual caroms, tricks involving such props as cork, coins, a hat, etc. Formerly *Fun on the Pool Table*. 77 figures. 95pp. 21814-7 Paperbound $1.25

HAND SHADOWS TO BE THROWN UPON THE WALL: A SERIES OF NOVEL AND AMUSING FIGURES FORMED BY THE HAND, Henry Bursill. Delightful picturebook from great-grandfather's day shows how to make 18 different hand shadows: a bird that flies, duck that quacks, dog that wags his tail, camel, goose, deer, boy, turtle, etc. Only book of its sort. vi + 33pp. $6\frac{1}{2}$ x $9\frac{1}{4}$. 21779-5 Paperbound $1.00

WHITTLING AND WOODCARVING, E. J. Tangerman. 18th printing of best book on market. "If you can cut a potato you can carve" toys and puzzles, chains, chessmen, caricatures, masks, frames, woodcut blocks, surface patterns, much more. Information on tools, woods, techniques. Also goes into serious wood sculpture from Middle Ages to present, East and West. 464 photos, figures. x + 293pp.

20965-2 Paperbound $2.50

HISTORY OF PHILOSOPHY, Julián Marias. Possibly the clearest, most easily followed, best planned, most useful one-volume history of philosophy on the market; neither skimpy nor overfull. Full details on system of every major philosopher and dozens of less important thinkers from pre-Socratics up to Existentialism and later. Strong on many European figures usually omitted. Has gone through dozens of editions in Europe. 1966 edition, translated by Stanley Appelbaum and Clarence Strowbridge. xviii + 505pp. 21739-6 Paperbound $3.50

YOGA: A SCIENTIFIC EVALUATION, Kovoor T. Behanan. Scientific but non-technical study of physiological results of yoga exercises; done under auspices of Yale U. Relations to Indian thought, to psychoanalysis, etc. 16 photos. xxiii + 270pp.

20505-3 Paperbound $2.50

Prices subject to change without notice.
Available at your book dealer or write for free catalogue to Dept. GI, Dover Publications, Inc., 180 Varick St., N. Y., N. Y. 10014. Dover publishes more than 150 books each year on science, elementary and advanced mathematics, biology, music, art, literary history, social sciences and other areas.